I0053895

# Frontiers in Toxicity and Functionalization of Nanomaterials

Special Issue Editors

**Dong-Wook Han**
**Wojciech Chrzanowski**

MDPI • Basel • Beijing • Wuhan • Barcelona • Belgrade

**MDPI**

*Special Issue Editors*
Dong-Wook Han                      Wojciech Chrzanowski
Pusan National University (PNU)    The University of Sydney
Korea                              Australia

*Editorial Office*
MDPI AG
St. Alban-Anlage 66
Basel, Switzerland

This edition is a reprint of the Special Issue published online in the open access journal *Nanomaterials* (ISSN 2079-4991) in 2017 (available at: http://www.mdpi.com/journal/nanomaterials/special_issues/fron_nano_toxi).

For citation purposes, cite each article independently as indicated on the article page online and as indicated below:

Lastname, F.M.; Lastname, F.M. Article title. *Journal Name*. **Year**. *Article number, page range.*

**First Edition 2018**

**ISBN 978-3-03842-736-0 (Pbk)**
**ISBN 978-3-03842-735-3 (PDF)**

Cover photo courtesy of Dong-Woo Hank and Wojciech Chrzanowski

Articles in this volume are Open Access and distributed under the Creative Commons Attribution license (CC BY), which allows users to download, copy and build upon published articles even for commercial purposes, as long as the author and publisher are properly credited, which ensures maximum dissemination and a wider impact of our publications. The book taken as a whole is © 2018 MDPI, Basel, Switzerland, distributed under the terms and conditions of the Creative Commons license CC BY-NC-ND (http://creativecommons.org/licenses/by-nc-nd/4.0/).

# Table of Contents

# About the Special Issue Editors

**Dong-Wook Han**, PhD, DSc, is a professor at the Department of Optics and Mechatronics Engineering, College of Nanoscience & Nanotechnology, Pusan National University, Korea. He completed his Ph.D. degree in biomedical engineering from Yonsei University, Seoul in 2004 and did a post doc in the Institute for Frontier Medical Sciences, Kyoto University, Japan until 2007. Since he joined at PNU in 2008, he has been working on tissue engineering and regenerative medicine with smart nanobio-materials. His research interest concerns fabrication of tissue-mimetics via electrospinning and 3D bioprinting, development and biocompatibility evaluations of medical devices, (cryo)preservation of stem cells and tissues by using antifreeze protein, and assessment of nanomaterials toxicity (nanotox-icity). Recently, he has developed various hybrid scaffolds based on functional graphene nanomate-rials and cell-adhesive peptides and applied them for tissue regeneration.

**Wojciech Chrzanowski**, PhD, DSc, is an associate professor at the Faculty of Pharmacy, The University of Sydney, Australia. He received his Ph.D. degree from the Silesian University of Technology. In 2006 he was awarded with a prestigious Marie Curie Intra European Fellowship and moved to University College London, and then to the University of Glasgow. In 2014 he received DSc from the Polish Academy of Science. His research interest concerns nanomedicine, nanosafety, nanomaterials, and nano-bio-characterization. His work has made a significant contribution to the understanding interaction of nanomaterials with living cells and tissue-like models assembled using magnetic 3D bioprinting. Recently, he has developed stimuli responsive drug delivery systems and multifunctional surfaces that are capable of regulating cell responses.

# Preface to "Frontiers in Toxicity and Functionalization of Nanomaterials"

Over the last decade, various nanomaterials (NMs) have attracted tremendous attention with the incredible development in nanoscience and nanotechnology. Some NMs are explored increasingly for biomedical applications, including drug delivery carriers, imaging probes, antimicrobial agents, biosensors, and tissue engineering scaffolds. However, the in vitro and in vivo toxicities of NMs related to oxidative stress are the main obstacles to use them in biomedical fields. One of the most promising strategies to address these obstacles is functionalizing NMs with biocompatible molecules or materials.

In this Special Issue, we are especially interested in manuscripts that advance the understanding of the interaction of NMs with cells, such as cellular responses to NMs, intracellular behaviors of NMs, therapeutic and imaging potentials of NMs, as well as the functionalization of NMs through coating, patterning and hybridization with other biomolecules for multifaceted biomedical applications. In the 2 review papers and 10 original research papers compiled here, the reader is provided with new insights on valuable strategies for the successful design of NMs with desirable biosafety and biofunctionality in biomedical applications for targeted technologies. Moreover, the expanding interest in the toxicity and functionalization of NMs and their applications to biomedical engineering gives this Special Issue broad appeal to the audience.

A valuable review on multifaceted biomedical applications of graphene NMs presented by Shin et al. focuses on the bio-functionality of graphene NMs and summarize some of major literature concerning their biomedical potentiality. This inspiring review highlights three ways to apply graphene NMs in biomedical fields (graphene NM-coated substrates, -patterned arrays and -based hybrid scaffolds), and allows us to intuitively understand the promising potential of such functional graphene NMs. Several articles in this Special Issue focus on the antioxidant activity and antibacterial effect of nanoparticles (NPs). Choi et al. report the strong antibacterial effect of caffeic acid-functionalized ZnO NPs, while Lee et al. show the antioxidative and antibacterial activities of gallic acid-functionalized ZnO NPs. In the work of Yamaguchi et al., unique dual-function biointerfaces are prepared, and their antimicrobial and bone mineralization activities are studied. In this case, interfaces composed of gallium-containing calcium titanate or gallium titanate accompanied by rutile and anatase enhance the formation of bone-like apatite on the surface, while simultaneously exhibiting antibacterial activity against multidrug resistant Acinetobacter baumannii (MRAB12).

In addition, unique functionalization strategies for fabricating patterned gels and nanocrystalline this films are individually introduced herein. Bang et al. fabricate patterned gels based on hyaluronate (HA) with dimethyloxalylglycine (DMOG) and show that they lead to angiogenesis and osteogenesis for bone regeneration. The other work by Wang et al. report a sol-gel-derived nanocrystalline $Sr(Ti_{1-x}Fe_x)O_3$ (STF) thin films having room temperature tunable multiferroic properties, which gives an in-depth understanding of the effects of Fe doping and annealing environment on the room temperature multiferroic properties of STF thin films.

On the other hand, the promising potentials of mesoporous silica NPs as drug delivery vehicles in cancer therapies are reviewed by Anna et al. This review provides a basic understanding of mesoporous silica NP synthesis, characteristics and surface modifications, especially with regard to use of drug delivery vehicles for cancer therapies. In addition, the in vitro and in vivo biocompatibility of mesoporous silica NPs are also discussed. It will greatly help in the development of novel and promising tools for innovative cancer therapies, although clinical translation remains challenging. The evaluation of the toxicity and biocompatibility of NMs is the first priority to employ NMs in biomedical applications. The cellular responses to $TiO_2$ NPs in intestinal epithelial (Caco-2) cells are examined by Krüger et al. $TiO_2$ NPs enter the cell via epithelial growth factor receptor (EGFR)-associated endocytosis, followed by the activation of complete EGFR/ERK/ELK signaling pathway, which finally induces NF-κB activation. Kretowski et al. investigate the cellular and molecular effects of silica NPs on apoptosis and autophagy in glioblastoma cell lines (LBC3 and LN-18). Interestingly, they showed that the silica NPs can induce cytotoxicity in glioblastoma cell lines, but not in human skin fibroblasts, indicating that the silica NPs can act in a cell type-specific way, and can induce variable and complex cellular responses to their exposure.

Meanwhile, a variety of strategies have been suggested to employ NMs in biomedical applications. Choi et al. synthesize folic acid- and hematoporphyrin-conjugated magnetic NPs ($CoFe_2O_4$-HPs-FAs), and demonstrate their photodynamic anticancer activities in prostate cancer (PC-3) cells. In this work, LED irradiation after treatment of $CoFe_2O_4$-HPs-FAs can induce photo-killing effects on cancer cells through apoptosis-mediated cell death. Two articles in this Special Issue focus on the imaging potentials of NMs. Atabaev et al. prepare the lanthanide-doped $Y_2O_3$ NPs and polyethylene glycol (PEG)-functionalized $Ho_2O_3$ NPs for potential applications in optical and magnetic resonance imaging. They characterize the physicochemical and optical properties of both NPs, and reveal that the lanthanide-doped $Y_2O_3$ NPs ($Eu^{3+}$ and $Gd^{3+}$ codoped $Y_2O_3$ NPs) and PEG-functionalized $Ho_2O_3$ NPs have good biocompatibility with L-929 fibroblastic cells, and can be used as an efficient dual-imaging nanoprobe.

Overall, this Special Issue brings a compilation of articles that advance frontiers in toxicity and functionalization of NMs. The papers published in this Special Issue provide valuable insights into novel approaches and strategies to functionalize NMs with desired properties as well as the fundamental understanding of NMs.

**Dong-Wook Han and Wojciech Chrzanowski**

*Special Issue Editors*

*nanomaterials*

MDPI

*Review*

# Multifaceted Biomedical Applications of Functional Graphene Nanomaterials to Coated Substrates, Patterned Arrays and Hybrid Scaffolds

Yong Cheol Shin [1,†], Su-Jin Song [2,†], Suck Won Hong [2], Seung Jo Jeong [3], Wojciech Chrzanowski [4], Jae-Chang Lee [5,*] and Dong-Wook Han [2,*]

1  Research Center for Energy Convergence Technology, Pusan National University, Busan 46241, Korea; choel15@naver.com
2  Department of Cogno-Mechatronics Engineering, College of Nanoscience & Nanotechnology, Pusan National University, Busan 46241, Korea; songsj86@gmail.com (S.-J.S.); swhong@pusan.ac.kr (S.W.H.)
3  GS Medical Co., Ltd., Cheongju-si, Chungcheongbuk-do 28161, Korea; eric.jeong@gsmedi.com
4  Australian Institute for Nanoscale Science and Technology, Charles Perkins Centre, Faculty of Pharmacy, University of Sydney, Pharmacy and Bank Building A15, Sydney NSW 2006, Australia; wojciech.chrzanowski@sydney.edu.au
5  Research Center for Industrial Chemical Biotechnology, Korea Research Institute of Chemical Technology, Ulsan 44429, Korea
*  Correspondence: jclee@krict.re.kr (J.-C.L.); nanohan@pusan.ac.kr (D.-W.H.); Tel.: +82-52-241-6312 (J.-C.L.); +82-51-510-7725 (D.-W.H.)
†  The authors contributed equally.

Received: 28 September 2017; Accepted: 1 November 2017; Published: 4 November 2017

**Abstract:** Because of recent research advances in nanoscience and nanotechnology, there has been a growing interest in functional nanomaterials for biomedical applications, such as tissue engineering scaffolds, biosensors, bioimaging agents and drug delivery carriers. Among a great number of promising candidates, graphene and its derivatives—including graphene oxide and reduced graphene oxide—have particularly attracted plenty of attention from researchers as novel nanobiomaterials. Graphene and its derivatives, two-dimensional nanomaterials, have been found to have outstanding biocompatibility and biofunctionality as well as exceptional mechanical strength, electrical conductivity and thermal stability. Therefore, tremendous studies have been devoted to employ functional graphene nanomaterials in biomedical applications. Herein, we focus on the biological potentials of functional graphene nanomaterials and summarize some of major literature concerning the multifaceted biomedical applications of functional graphene nanomaterials to coated substrates, patterned arrays and hybrid scaffolds that have been reported in recent years.

**Keywords:** graphene nanomaterial; multifaceted biomedical application; coated substrate; patterned array; hybrid scaffold

## 1. Introduction

Over the last several decades, the development of nanoscience and nanotechnology has made dramatic progress in research towards the understanding of nanomaterials. In addition, numerous attempts have been devoted at that time to employ nanomaterials in various research and industrial fields. It is commonly acknowledged that nanomaterials are defined as materials having at least one dimension, such as length, thickness, width, or diameter, of smaller than 100 nm in size and the plenty of attentions has been paid to various nanomaterials, including carbon nanotube, nanoparticle, polymeric nanofiber, quantum dot, graphene and nanocomposite [1–4]. Nanomaterials exhibit unique properties, including extraordinary physicochemical, fluorescent, electrical and thermomechanical properties that do not appear in those bulk counterparts of the same composition [5–11]. Moreover, a very large surface area-to-volume ratio is probably the most distinctive property of nanomaterials, which makes them even more attractive for applications in diverse fields, such as physics, chemistry, biology, material science and electrical engineering. Therefore, many researchers have been interested in the unlimited potentials of nanomaterials and extensively studying to apply them to each research field.

Among a great number of nanomaterials, graphene is a relatively recently discovered and characterized nanomaterial that has a variety of advantages and has particularly attracted renewed interest in nanomaterial research. Graphene, a single atomic layer of graphite, is a two-dimensional (2D) nanomaterial, in which $sp^2$-bonded carbon atoms arranged in a honeycombed lattice structure. It was first described by Boehm and coworkers in 1986 [12]. After that, it was identified and isolated by Geim and Novoselov in 2004 and has been receiving an explosion in attention [13]. Graphene is a novel nanomaterial with remarkable properties, such as exceptional thermomechanical, excellent physicochemical, outstanding electrical and unique biological properties and its applicable fields are endless [14–19]. In addition to graphene, graphene derivatives—including graphene oxide (GO) and reduced graphene oxide (rGO)—also possess promising potentials as a graphene because they have many functional groups on their surface while maintaining the unique properties of graphene. GO can be obtained by oxidizing graphene and has many oxygen-containing functional groups, such as hydroxyl, carboxyl, epoxy and carbonyl groups, on its surface, which leads to facilitative interactions with biomolecules or cells. On the other hand, rGO can be prepared by reducing GO. rGO has been also actively studied because it has oxygen-containing functional groups on its surfaces as well as having relatively superior electrical properties as compared with GO [20–22]. Due to the obvious advantages of graphene and graphene derivatives, numerous studies on their toxic and biological effects have been reported and their applications to the biomedical fields have been constantly explored [17,23–27]. According to the recent findings, graphene and its derivatives have been revealed to have not only outstanding biocompatibility but also ability to enhance cellular behaviors, including cell growth, proliferation and differentiation [28–36]. However, unfortunately, there has been a lot of debate about the biosafety and biological effects of graphene and graphene derivatives, although much research has been conducted on these issues. In particular, the interactions of graphene and graphene derivatives with biological systems are quite varied depending on many parameters, including their size, shape, concentration, surface functional group, exposure time and preparation method; thereby, research on the biomedical applications of graphene nanomaterials is still in its infancy.

Herein, we focus on the key literature concerning the biomedical applications of graphene and its derivatives and are attempting to present valuable information that can provide guidance for future comprehensive research on their biomedical applications. Tremendous studies have been underway to employ graphene and graphene derivatives for biomedical applications by introducing functional groups on their surface, or by using them as surface coating, nanofiller and composite materials. Among these various types of applications, the aim of the present review is to summarize the recent studies concerning multifaceted biomedical applications of functional graphene nanomaterials to coated substrates, patterned arrays and hybrid scaffolds.

## 2. Graphene Nanomaterial-Coated Substrates

Until now, it has been widely believed that the toxicity of graphene nanomaterials is quite different depending on their shape, size, concentration, surface functional group and preparation method [37–39]. In particular, the biological effects of graphene nanomaterials have been reported to be strongly dependent on size, concentration, exposure time and cell type [37–43]. Therefore, many studies have been suggested to improve cell behaviors by coating graphene nanomaterials on substrates to minimize the influence of those various parameters (Figures 1 and 2) [28–30,32,44–52]. Ryoo et al. reported that the graphene nanomaterials, including GO and rGO, can be simply immobilized onto glass substrates treated with 3-aminopropyltriethoxysilane (3-APTES) via electrostatic interactions between GO and amine groups on the substrates (Figure 1a) [44]. In addition, they investigated the behaviors of NIH-3T3 fibroblasts on the graphene nanomaterial-coated substrates and revealed that the GO- and rGO-coated glass substrates can not only favorably support cell adhesion, spreading and proliferation, but can also improve the gene transfection efficiency of cells as compared to the glass substrates without graphene nanomaterial coatings (Figure 1b,c). These results indicated that the graphene nanomaterial-coated substrates are highly cell-friendliness and the graphene nanomaterials can be readily employed as surface coating materials in biomedical applications, such as implant, cell culture platform and cell-interfacing system.

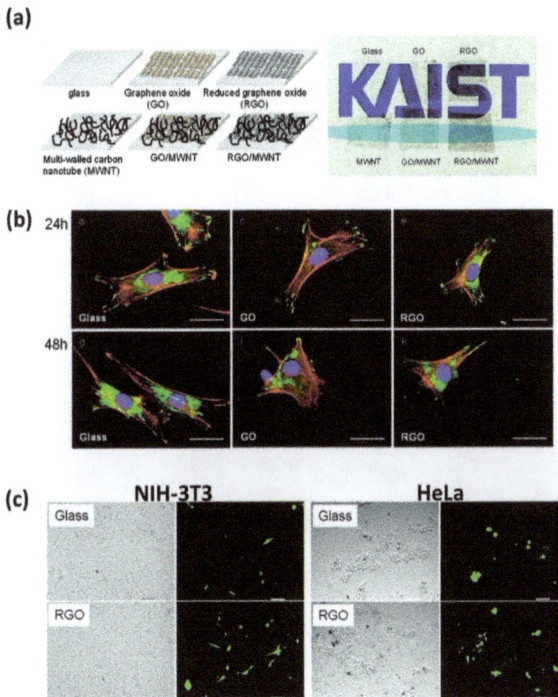

**Figure 1.** Stimulating effects of graphene nanomaterial-coated substrates on NIH-3T3 fibroblast behaviors. (**a**) Structural diagrams (left panel) and optical images (right panel) of graphene nanomaterial-coated substrate. (**b**) The fluorescence images of NIH-3T3 fibroblasts on each substrate for 24 and 48 h. Scale bars are 20 μm. (**c**) Improved gene transfection efficiency of NIH-3T3 fibroblasts and HeLa cells on rGO-coated substrates after 48 h incubation. Scale bars are 35 μm. Reproduced with permission from [44]. American Chemical Society, 2010.

Moreover, these stimulating effects of graphene nanomaterial-coated substrates on cell behaviors have also been examined in various cell types, including mesenchymal stem cells (MSCs), neural stem cells (NSCs), induced pluripotent stem cells (iPSCs) and myoblasts. Nayak et al. documented that the graphene coatings on substrates can effectively accelerate the differentiation of human MSCs without hampering cell proliferation. The human MSCs were successfully differentiated into osteogenic lineages only on graphene-coated regions. Moreover, it was shown that the graphene-coated polymeric substrates, including polydimethylsiloxane (PDMS) and polyethylene terephthalate (PET), could also increase osteogenic differentiation. In general, cellular behaviors are strongly dependent on the stiffness of substrates and the osteogenic differentiation is commonly promoted on stiff substrates rather than soft substrates, such as polymeric substrates [53–56]. However, the differentiation of human MSCs towards osteogenic lineage was increased on graphene-coated polymeric substrates (i.e., soft substrates) regardless of stiffness of underlying substrates, indicating that the graphene coatings can be a driving force of osteogenic differentiation.

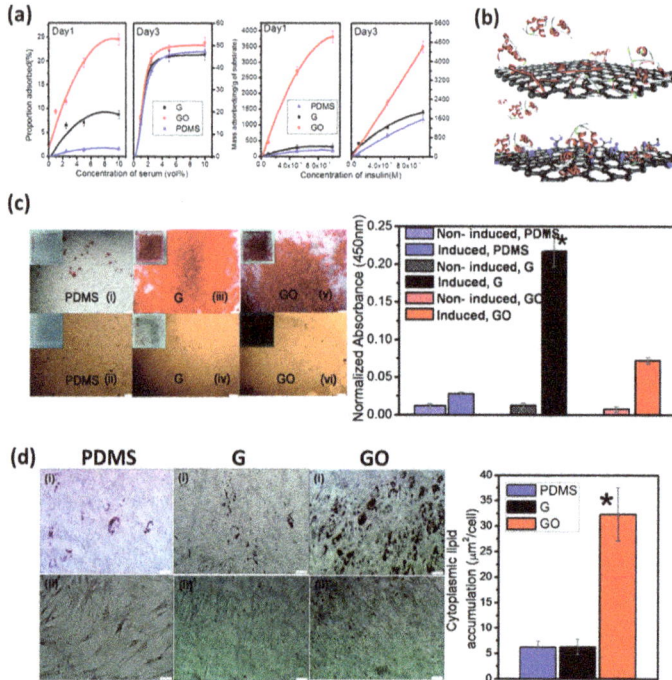

Figure 2. Molecular origin of enhanced cell behaviors on graphene nanomaterial-coated substrates. (a,b) Different binding interactions of graphene and GO with serum and insulin. (c) Osteogenic differentiation of MSCs after 12 days of incubation on PDMS (i) with induction and (ii) without induction, on graphene (iii) with induction and (iv) without induction and on GO (v) with induction (vi) and without induction. Scale bars are 200 μm. (d) Adipogenic differentiation of MSCs after 14 days of induction on PDMS (i) with induction and (ii) without induction, on graphene (i) with induction and (ii) without induction and on GO (i) with induction and (ii) without induction. Scale bars are 50 μm. Reproduced with permission from [29]. Copyright American Chemical Society, 2011.

Meanwhile, Lee et al. investigated the molecular origin of accelerated differentiation on graphene nanomaterial-coated substrates by comparing the binding abilities of graphene and GO to different growth factors [29]. They showed that the different binding interactions of graphene and GO with growth factor agents play a significant role in determining the stem cell growth and differentiation (Figure 2a,b). The osteogenic differentiation of human bone marrow-derived MSCs was enhanced on the graphene-coated PDMS substrates through π-π stacking interactions between graphene and osteogenic inducer, including dexamethasone and β-glycerophosphate, while GO-coated PDMS substrates could greatly enhance adipogenic differentiation via hydrogen bonding and electrostatic interactions with insulins (Figure 2c,d).

The specific binding affinity of GO for biomolecules can significantly promote the myoblast growth and myogenic differentiation. Ku et al. studied the myoblast behaviors on GO- and rGO-coated glass substrates and indicated that the GO- and rGO-coated substrates could enhance myogenic differentiation as well as supporting cell adhesion and proliferation (Figure 3) [32]. They suggested that the enhanced myogenic differentiation was attributed to both the unique physicochemical properties of graphene derivatives, such as ripples and wrinkles and the adsorption ability for serum proteins in culture media. Moreover, it was confirmed that the GO-coated substrates were more favorable for myogenic differentiation because GO has more oxygen-containing functional groups on its surface than rGO, which leads to further increase in serum protein adsorption.

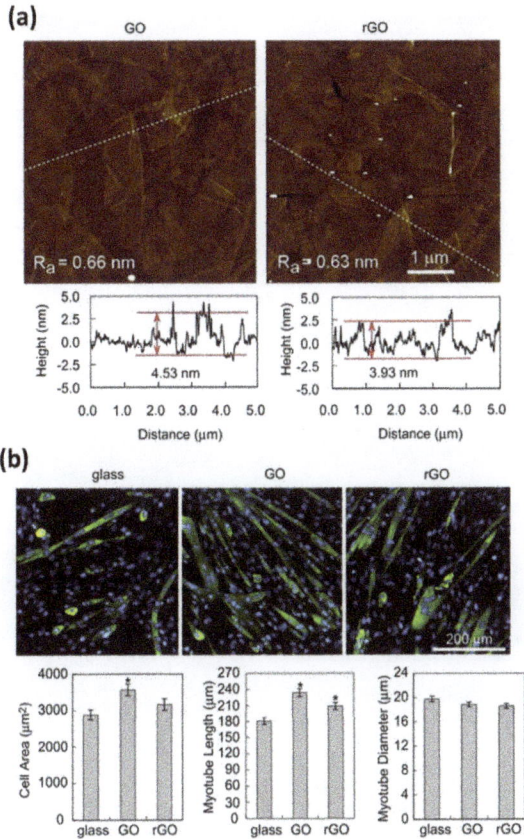

**Figure 3.** Enhanced growth and differentiation of myoblasts on graphene nanomaterial-coated substrates. (**a**) Characterizations of GO- and rGO-coated glass substrates by atomic force microscopy (AFM). (**b**) Myogenic differentiation on uncoated, GO- and rGO-coated glass substrates. C2C12 cells were grown in growth media (Dulbecco's modified Eagle's medium, DMEM, containing 10% fetal bovine serum and 1% antibiotic-antimyotic solution) for 1 day and then incubated in differentiation media (DMEM containing 2% horse serum and 1% antibiotic-antimyotic solution) for 5 days. Reproduced with permission from [32]. Copyright Elsevier Ltd, 2012.

In other studies, the graphene nanomaterial-coated substrates hold the potentials for neural cells [30,46,48,57]. In particular, it is worth noting that the graphene nanomaterials have superior electrical properties as compared with the standard graphite materials [58,59]. Qiu et al. demonstrated that the conductivity of graphene nanomaterials is quite a bit higher than that of graphite materials (highly oriented pyrolytic graphite crystal, HOPG), while the resistivity of graphene nanomaterials is much lower as compared with that of graphite materials [59]. They described that the conductivities were found to be 92, 407 and 2138 $(\Omega \cdot cm)^{-1}$ for the HOPG sample, the multi-layer graphene sample and the sample with a single-layer to few-layer graphene, respectively. On the other hand, there is also a significant increase in the carrier concentration, especially in the samples containing a mixture of single-layer to several layers of graphene: $3.58 \times 10^{18}$, $14.9 \times 10^{18}$ and $46.3 \times 10^{18}$ cm$^{-3}$ for HOPG sample, multi-layer graphene sample and samples containing a mixture of single-layer to several layers of graphene, respectively. In addition, it was revealed that the superior electrical properties of graphene nanomaterials result in highly sensitive detection of the micromolar concentration of dopamine—a neurotransmitter—on graphene-coated surfaces by Raman spectroscopy and microscopy.

Moreover, these superior electrical properties also allow graphene nanomaterials to be used in stimulating neural cells. Park et al. found that the differentiation of human NSCs into neurons was increased on graphene-coated substrates (Figure 4). Meanwhile, Tang et al. cultured NSCs on graphene-coated substrates and investigated the neural excitation by monitoring spontaneous Ca$^{2+}$ oscillations, which represents neural signal transmission [57]. The results indicated that the NSCs were able to form functionally active neural networks on graphene-coated substrates and the neural network activities, such as the intracellular spontaneous and synchronous Ca$^{2+}$ oscillations and spontaneous synaptic currents, were significantly improved on the graphene-coated substrates. Hence, it is indicated that graphene nanomaterial-coated substrates are a typical strategy for biomedical applications of graphene nanomaterials.

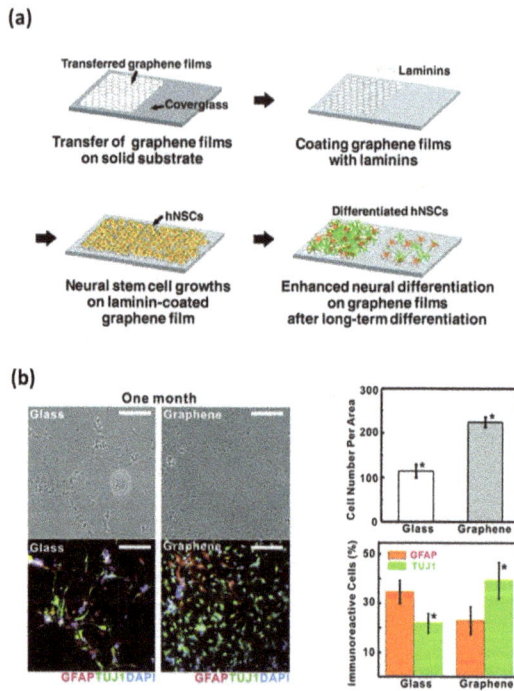

**Figure 4.** Enhanced growth and differentiation of human NSCs on graphene nanomaterial-coated substrates. (**a**) Schematic diagram depicting the growth and differentiation of human NSCs on graphene. (**b**) Enhanced neural differentiation of human NSCs on graphene-coated substrates. Scale bars are 200 μm. Reproduced with permission from [30]. Copyright John Wiley and Sons, 2011.

In addition to these, graphene-coated substrates or materials can be also applied for other biomedical applications, such as biosensor, implantable electrode, antibacterial system and composite graft material [50–52,60–65]. Several studies related to the other biomedical applications of graphene nanomaterial-coated substrates are summarized in Table 1.

**Table 1.** Summary of studies on the various biomedical applications of graphene nanomaterial-coated substrates to date.

| Applications | Target | Graphene Nanomaterial | Methods & Findings | Control | Ref. |
|---|---|---|---|---|---|
| Immunosensor | PSA-ACT | rGO on amine-SAM substrate | - Solution gated FET & anti-PSA<br>- 100 fg/mL of detection limit | CEA | [60] |
| Immunosensor | E. coli | TRMGO on SiO₂/Si | - Back gated FET & anti-E. coli<br>- 10 cfu/mL of detection limit | Non-pathogenic E. coli & plant-pathogenic bacterium | [61] |
| Immunosensor | DNA | PNA-rGO on SiO₂/Si | - Liquid-gated FET<br>- Label-free detection<br>- 100 fM of detection limit | Non-complementary DNA | [62] |
| Electrode array | Neural imaging | Four-layer graphene on Au or Pt | - Implantable on the brain surface in rodents<br>- Ability of in vivo 3D imaging & optogenetic stimulation | Pt micro-ECoG device | [64] |
| Antibacterial system | E. coli & S. aureus | GONWs | - Achieving GONWs by EPD of Mg²⁺-GONSs<br>- Toxic effects of GONWs on bacteria | RGNWs | [65] |
| GBR membrane | Rat calvarial defect | GO on Ti membrane | - Biocompatibility of GO at 10 μg/mL<br>- Bone regeneration effects of GO-Ti | Ti membrane | [50] |
| Bone graft material | BMSCs & Rat calvarial defect | rGO-HA & rGO-BCP | - Osteogenetic effects of rGO<br>- Bone regeneration ability of rGO | HA & BCP | [51,52] |

Abbreviations: ACT, α1-antichymotrypsin; BCP, biphasic calcium phosphate; BMSC, bone marrow-derived mesenchymal stem cell; CEA, carcinoembryonic antigen; DNA, deoxyribonucleic acid; ECoG, electrocorticography; *E. coli*, *Escherichia coli*; EPD, electrophoretic deposition; FET, field-effect transistor; GBR, guided bone regeneration; GO, graphene oxide; GONS, graphene oxide nanosheet; GONW, graphene oxide nanowall; HA, hydroxyapatite; Non-pathogenic *E. coli*, *E. coli* DH5α; Plant-pathogenic bacterium, *Dickeya dadantii* 3937; PNA, peptide nucleic acid; PSA, prostate specific antigen; RGNW, reduced graphene nanowall; rGO, reduced graphene oxide; SAM, self-assembly monolayer; *S. aureus*, *Staphylococcus aureus*; 3D, three-dimensional; TRMGO, thermally reduced monolayer graphene oxide.

## 3. Graphene Nanomaterial-Patterned Arrays

Up to now, much research has indicated that the beneficial effects of graphene nanomaterials in biomedical applications, such as promoting effects on cellular behaviors, including cell adhesion, proliferation, development, spreading and differentiation. Along with those findings previously reported, there have been considerable efforts to use the unique physicochemical and topographical properties of graphene nanomaterials in biomedical applications [64,66,67]. In particular, distinctive rippled or wrinkled features of graphene nanomaterials can provide specific topographical guidance cues for directing cell behaviors. The precisely controlled cell migration or orientation plays a crucial role in determining cell responses and fates [68–70]. Therefore, the studies concerning the regulation of cellular behaviors by graphene nanomaterials have been recently proposed and investigated for biomedical applications.

The graphene nanomaterial-patterned arrays have been especially spotlighted as a novel strategy for guiding and stimulating cellular behaviors, because the graphene nanomaterials can provide desirable topographical guidance cues as well as biochemical cues [71–79]. Bajaj et al. fabricated rectangular island-shaped graphene patterns on SiO₂/Si substrate using photolithography techniques and examined the myogenic differentiation of C2C12 skeletal muscle myoblasts (Figure 5) [71]. It was shown that most myotubes were formed on graphene patterns, while few cells were differentiated into myotubes on the SiO₂/Si substrate without graphene patterns (Figure 5a). In addition, the island-shaped graphene patterns were able to induce the spontaneous alignment of myotubes, which leads to a maximized the contractile power for muscle contractions (Figure 5b). Moreover, they evaluate the functionality of myotubes on graphene patterns and revealed that the myotubes on graphene patterns were mature and highly functional.

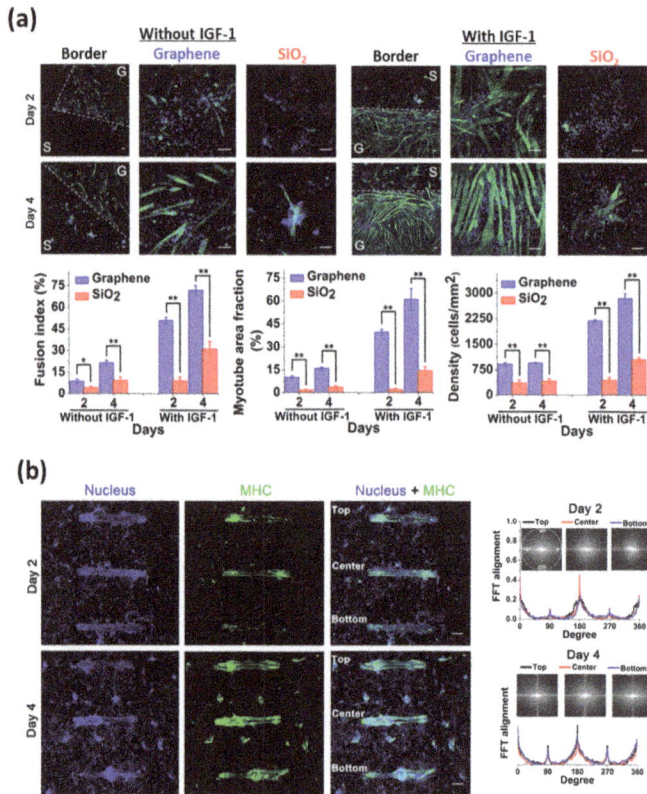

**Figure 5.** Graphene-based patterning and differentiation of myoblasts. (**a**) Myogenic differentiation of C2C12 skeletal muscle myoblasts on graphene-patterned substrates. The dashed white line in column one indicates the border of $SiO_2$ and graphene surfaces on the substrates. Scale bars are 100 μm. (**b**) Fluorescence images and alignment of the C2C12 myotubes on rectangular island-shaped graphene patterns. Scale bars are 250 μm. Reproduced with permission from [71]. Copyright John Wiley and Sons, 2013.

Akhavan et al. have also demonstrated that graphene patterns can be employed as selective 2D templates for accelerating the osteogenic differentiation of human MSCs (Figure 6) [72]. They fabricated aligned GO and rGO nanoribbon grid on $Si_3N_4$/Si(100) substrates by a paint-brushing method and investigated the osteogenic differentiation of human MSCs (Figure 6a). Their results indicated that both graphene nanogrids (GO and rGO nanoribbon grid) could enhance the actin cytoskeleton proliferations. Meanwhile, in the presence of chemical inducers, including dexamethasone, β-glycerophosphate and ascorbic acid, rGO nanoribbon grid especially accelerated the osteogenic differentiation of human MSCs (Figure 6b). They explained these findings by the fact that the rGO nanoribbon grids can highly adsorb chemical induces in culture media and can also provide physical stress induced by the surface topographical features of rGO nanogrids [29,80–82]. They also proved that those stimulating effects of rGO nanogrids are equally effective on human NSCs [73]. These results indicate that the graphene nanomaterial patterns can be readily applied in biomedical fields.

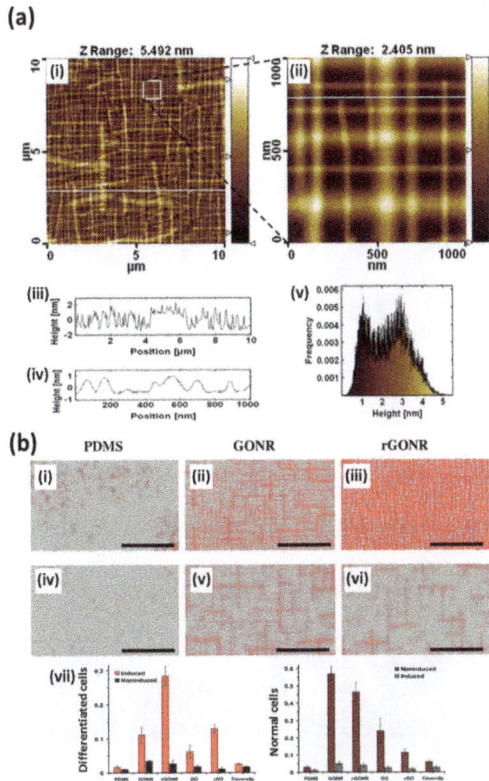

**Figure 6.** Graphene-based patterning and differentiation of human MSCs. (**a**) Characterizations of aligned GO nanoribbon grid on $Si_3N_4$/Si(100) substrates by AFM. AFM images of GO nanoribbon grid in a (i) wide and (ii) close window. (iii) and (iv) present height profiles of GO nanoribbon grid along the white lines marked in (i) and (ii), respectively. (v) exhibits the height profile histogram of GO nanoribbon grid. (**b**) Accelerated osteogenic differentiation of human MSCs determined by Alizarin Red staining after 1 week incubation (i–iii) with induction and (iv–vi) without induction. Scale bars are 10 µm. (vii) Normalized optical absorbance of the differentiated cells and normal cells. Reproduced with permission from [72]. Copyright Elsevier Ltd, 2013. Abbreviations: G, graphene; GONR, GO nanoribbon grid; MHC, myosin heavy chain; rGONR, rGO nanoribbon grid; S, $SiO_2$.

In several studies described above, the excellent biocompatibility and the applicability of graphene nanomaterial patterns in biomedical applications have been demonstrated. However, such patterning of graphene nanomaterials requires elaborate techniques, such as photolithography, dip-pen lithography and microcontact printing. These techniques, of course, are sufficiently efficient but simpler and more scalable methods have been reported by Wang et al. (Figure 7a) [77]. They simply fabricated wrinkled GO multilayer films by relaxation of GO sheets on pre-stretched elastomers. The wrinkled GO patterns can be easily removed by re-stretching the elastomer substrates and the fabrication process is reversible. In addition, the wavelength and height of GO wrinkled can be controlled by film thickness and pre-stretch. The cell alignment and morphology on the wrinkled GO patterns were also evaluated. It was observed that the fabricated GO wrinkles can effectively induce cell alignment and elongation by contact guidance provided from wrinkled GO patterns (Figure 7b). Hence, it was suggested that the wrinkled GO patterns are promising new approach for functional biomedical applications due to advantages, such as the simplicity and scalability of fabrication.

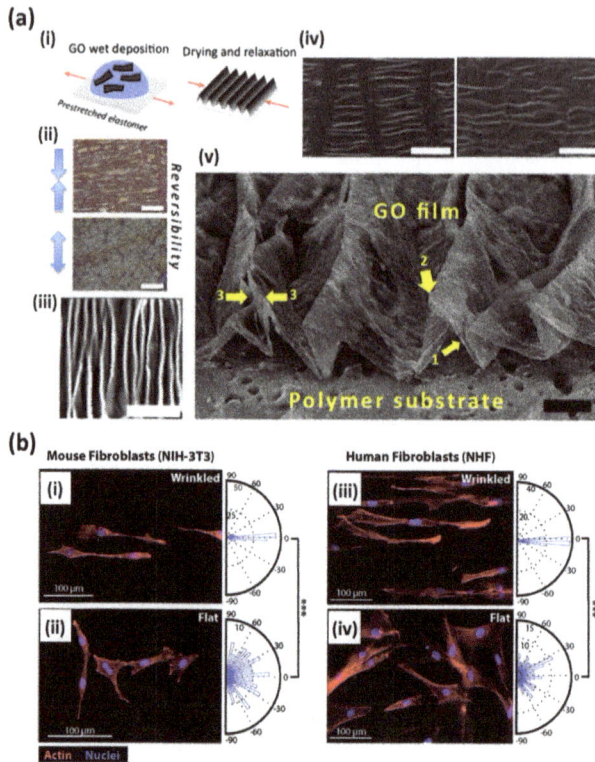

**Figure 7.** Controlling cellular behaviors using graphene nanomaterial-patterned arrays. (**a**) Fabrication process and morphology of wrinkled GO multilayer films. (i) Schematic diagram and (ii) reversibility of the fabrication process. Scale bars are 100 μm. (iii) AFM image, (iv) scanning electron microscopy (SEM) images and (v) high-resolution SEM image of wrinkled GO patterns. Scale bar are 20, 100 and 20 μm for (iii), (iv) and (v), respectively. (**b**) Fluorescence images of MIH-3T3 cells and human fibroblasts on (i,iii) 200 nm wrinkled GO substrates and (ii,iv) flat graphene substrates. Reproduced with permission from [77]. Copyright Elsevier Ltd, 2015.

On the other hand, intriguing results have been obtained by Kim et al. [76]. Kim et al. ascertained that the stem cell fate can be controlled by manipulating the sizes and geometries of patterned arrays (Figure 8a). They prepared GO-patterned arrays with different sizes and geometries on various types of substrates and examined the differentiation of human adipose-derived mesenchymal stem cells (ADMSCs) on those GO-patterned arrays. Interestingly, the differentiation of human ADMSCs was observed to strongly depend on the geometries of GO-patterned arrays. The GO line patterns promote the elongation and spreading of human ADMSCs following the geometry of GO line patterns, which results in the enhanced osteogenesis of human ADMSCs (Figure 8b). On the contrary, the GO grid patterns guide human ADMSCs to grow in a bipolar orientation and encourage the conversion of mesodermal stem cells to ectodermal neuronal cells (Figure 8c). These different cellular behaviors of human ADMSCs could be attributed to the physicochemical and geometric properties of GO-patterned arrays, indicating that the graphene nanomaterial-patterned arrays are particularly attractive for biomedical applications.

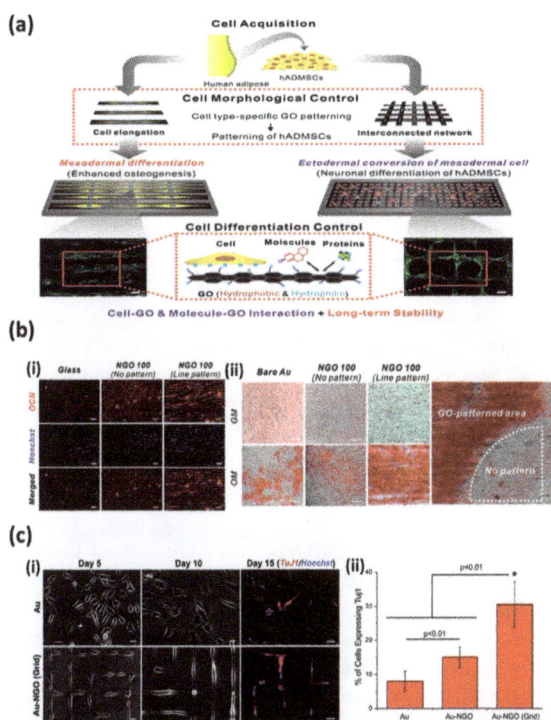

Figure 8. (a) Schematic diagram depicting control over the differentiation of human ADMSCs using the geometries of GO-patterned arrays. (b) Enhanced osteogenic differentiation of human ADMSCs using GO line-patterned arrays. (i) Fluorescence images of human ADMSCs on GO line-patterned arrays, showing elongated and well-spread morphology of human ADMSCs. Scale bars are 50 μm. (ii) Enhanced osteogenic differentiation of human ADMSCs confirmed by Alizarin Red staining at day 21. Scale bars are 50 μm. (c) Encouraged neuronal differentiation of human ADMSCs using GO grid-patterned arrays. (i) Fluorescence images of neural induced human ADMSCs on poly-L-lysine (PLL)-coated Au and GO grid-patterned arrays. Scale bars are 20 μm. (ii) Quantitative analysis of the percentage of cell expressing the neuronal marker (TuJ1). Reproduced with permission from [76]. Copyright American Chemical Society, 2015.

These guidance effects of graphene nanomaterial-patterned arrays are closely related to their size and shape. Zhang et al. confirmed that the width of GO-patterned arrays can directly affect the cell migration, alignment, morphology and cell adhesion [78]. They found that the cytoskeleton contractility, intracellular traction and actin filament elongation are significantly enhanced when the width of the GO-patterned arrays is similar to the cell dimension, which in turn cell migration is greatly increased. Kim et al. also revealed that the shape of GO-patterned arrays can determine cell morphology, migration distance, speed and directionality [79]. Therefore, graphene nanomaterial-patterned arrays fabricated with sophisticated control of structures and properties can provide unique opportunities for biomedical applications.

## 4. Graphene Nanomaterial-Based Hybrid Scaffolds

There have been astonishing advances in biomedical applications of graphene nanomaterials but at the same time, still there are many challenges remain to be solved. To employ graphene nanomaterials in biomedical applications, ensuring biosafety and biocompatibility of graphene nanomaterials is the first priority. To address this issue, there have been substantial efforts to develop graphene nanomaterials-based scaffolds through hybridization with biocompatible materials to maintain the unique properties of graphene nanomaterials and to improve biocompatibility [31,83–101].

Bai et al. reported that the GO and poly(vinyl alcohol) (PVA) composite hydrogels can be easily prepared and the GO/PVA composite hydrogels showed pH-sensitive gel-sol-gel transition behaviors (Figure 9) [83]. The GO/PVA composite hydrogels were decomposed with increasing pH value and gel-sol transition occurred. Meanwhile, when the pH value dropped again, the GO/PVA composite hydrogels underwent sol-gel transition. This could be attributed to the surface negative charge of GO originated from the carboxyl groups. The different pH values led to changes in surface charge densities of GO, which in turn, electrostatic repulsion forces between GO sheets were altered. These pH-sensitive properties make the GO/PVA composite hydrogels exceptionally useful as drug delivery carriers. On the other hand, graphene nanomaterial-hybridized hydrogels can be also utilized as cell scaffolds. Cha et al. showed that the mechanical properties of methacrylated gelatin (GelMA) hydrogels can be controlled by incorporation of methacrylate group-introduced GO (MeGO) and the GO-incorporated GelMA hydrogels showed good biocompatibility with fibroblasts [86]. The fracture strength of GelMA hydrogels could be enhanced by incorporating MeGO, while minimizing the changes in their rigidity. These enhanced mechanical properties were due to the interfacial bonding between GO and polymeric network [102,103]. In addition, the incorporated GO or MeGO did not detrimental effects on the viability and proliferation of encapsulated fibroblasts. Qiu et al. also proved the improving effects of graphene incorporation on the mechanical performance of polymeric hydrogels. They introduced the graphene aerogel into the poly(N-isopropylacrylamide) (PNIPAM) hydrogels. The results demonstrated that the good mechanical strength and electrical conductivity of graphene can significantly increase the mechanical performance of polymer hydrogels, suggesting that graphene nanomaterials can be used as reinforcing nanofillers for polymeric composites.

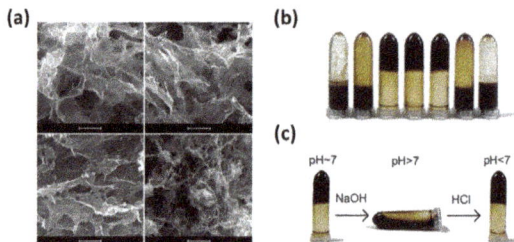

**Figure 9.** pH-sensitive GO/PVA composite hydrogel. (**a**) SEM images of GO/PVA composite hydrogel. Scale bars are 5 μm. (**b**) Digital photographs of GO/PVA composite hydrogel with different content ratio ($r_{P/G}$). From left to right, $r_{P/G}$ = 1:1, 1:1.5, 1:2, 1:5, 1:10, 1:20 and 1:40. (**c**) pH-sensitive gel-sol-gel transition behaviors according to the pH values. From left to right, pH ≈ 7, pH > 7 and pH < 7, respectively. Reproduced with permission from [83]. Copyright The Royal Society of Chemistry, 2010.

On the other hand, recently, the natural and synthetic polymers have been extensively used to fabricate biological scaffolds for tissue engineering applications. Polymeric biomaterials have superior biocompatibility and biodegradability but their intrinsic poor thermal and mechanical properties are often quoted as disadvantages. Therefore, much research has been suggested to compensate the poor thermal and mechanical properties of polymeric biomaterials by functionalization of graphene possessing exceptional thermomechanical properties. Shin et al. demonstrated that the impregnation of GO can not only reinforced the poor mechanical properties of polymer nanofiber scaffolds but can also promote myoblast growth and differentiation [93]. Despite excellent biocompatibility of collagen-based scaffold, it suffers from poor mechanical and rapidly degrading properties of collagen. However, the poor mechanical properties, including tensile strength and elastic modulus, of scaffolds could be remarkably improved by the incorporation of GO. These improved mechanical properties of scaffolds could be rationalized by the fact that the oxygen-containing functional groups on GO surface can strongly interact with hydroxyl or amine groups of polymeric substrates, which allows interfacial bonding between GO and polymeric substrates [100,102–105]. Moreover, the cellular behaviors of myoblasts, including proliferation and myogenic differentiation, were significantly promoted on the GO-hybridized scaffolds [94,96]. As mentioned above, graphene nanomaterials have great capability in adsorption of serum proteins from culture media, which leads to accelerated myogenic differentiation [29,32]. Furthermore, the promoting effects of graphene nanomaterial-based hybrid scaffolds were confirmed in various types of cells. Shah et al. documented the guided differentiation of NSCs towards oligodendrocytes using GO-coated polycaprolactone (PCL) nanofiber scaffolds (Figure 10a); meanwhile Serrano et al. described the promoted differentiation of embryonic neural progenitor cells into both neurons and glial cells on GO-based scaffolds (Figure 10b) [88,90].

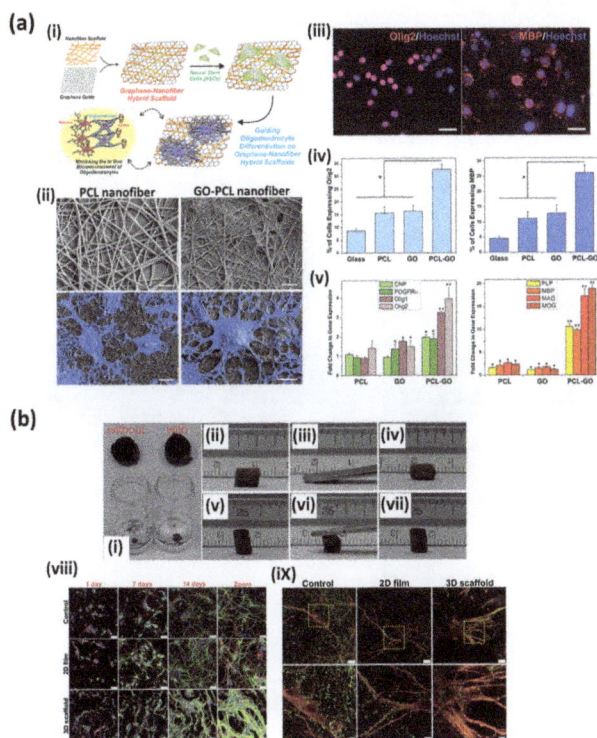

**Figure 10.** Multifaceted biomedical applications of functional graphene nanomaterials to hybrid scaffolds. (**a**) Guided differentiation of NSCs towards oligodendrocytes on GO-coated PCL nanofiber hybrid scaffolds. (i) Schematic illustration depicting the fabrication and application of GO-PCL nanofiber hybrid scaffolds. (ii) SEM images of NSCs on PCL and GO-PCL nanofiber scaffolds. Scale bars are 2 μm for upper row and 10 μm for lower row. (iii) Fluorescence images of NSCs on GO-PCL nanofiber scaffolds after 6 days of culture. Cells were stained for the early oligodendrocyte marker Olig2 and the mature oligodendrocyte marker myelin basic protein (MBP). (iv,v) Quantitative analysis of the expression of oligodendrocyte markers. Adapted with permission from Shah et al. [88]. Copyright (2014) John Wiley and Sons. (**b**) Promoted differentiation of embryonic neural progenitor cells into both neurons and glial cells on GO-based 3D porous scaffolds. (i) Digital photographs and (ii–vii) flexibility of GO-based 3D porous scaffolds. (viii) Neuronal differentiation and (ix) synapse formation of embryonic neural progenitor cells on GO-based 3D porous scaffolds. Reproduced with permission from [90]. Copyright The Royal Society of Chemistry, 2014.

More recently, studies concerning the development of three-dimensional (3D) scaffolds using graphene nanomaterials have been increasingly reported [31,85,95,97,99,101]. It has been well known that the cellular behaviors, including migration, growth, morphology, differentiation and protein expression, are definitely different in 2D and 3D environments [106–108]. Thus, developing 3D scaffolds that mimic the in vivo microenvironment of the natural extracellular matrix is critical to biomedical applications. Jakus et al. suggested the feasibility of 3D printing for the fabrication of graphene-based 3D scaffolds (Figure 11) [95,99]. They developed 3D printable graphene ink composed of poly(lactic-co-glycolic acid, PLGA) and graphene flakes and fabricated 3D-printed graphene scaffolds. It was verified that the mechanical integrity and electrical conductivity of graphene were maintained in the 3D-printed graphene scaffolds and the viability and proliferation of human MSCs were significantly increased on the 3D-printed graphene scaffolds as compared to the PLGA scaffolds. In addition, the expression of neurogenic relevant genes, such as glial fibrillary acidic protein, neuron-specific class III β-tubulin (Tuj1), nestin and microtubule-associated protein 2, was upregulated in human MSCs on 3D-printed graphene scaffolds. Further in vivo studies using a female BALB/c mouse model validated that the 3D-printed graphene scaffolds did not induce a severe immune response or fibrous capsule formation, indicating that the 3D-printed graphene scaffolds were highly biocompatible. These findings expand the versatility and applicability of graphene nanomaterials for emerging biomedical applications. Collectively, the graphene nanomaterial-based scaffolds can provide a great value for the potentials of graphene nanomaterials in biomedical applications.

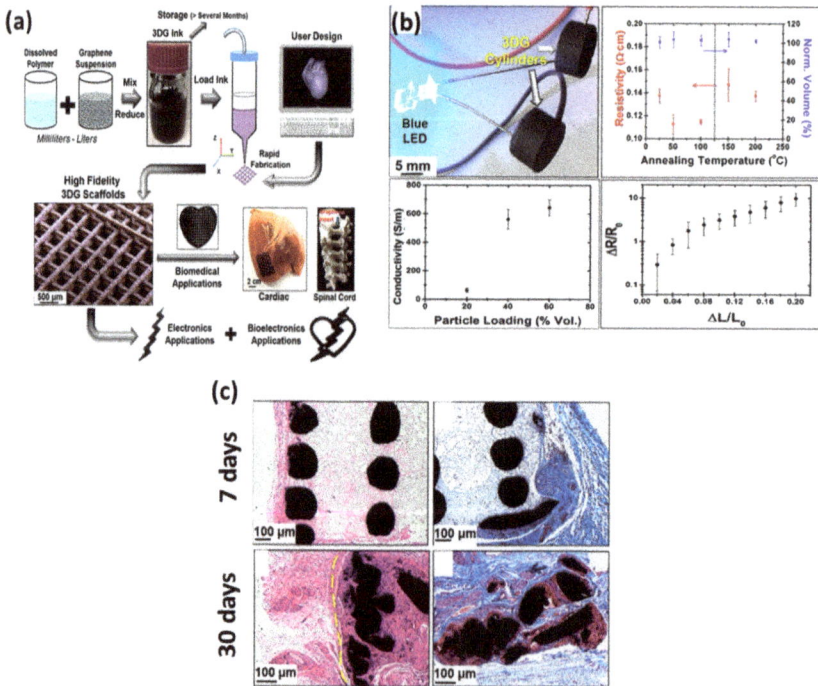

**Figure 11.** Feasibility of 3D printing for the fabrication of graphene-based 3D scaffolds. (**a**) Schematic diagram depicting the development of 3D printable graphene ink composed of PLGA and graphene flakes and the fabrication of 3D-printed graphene scaffolds. (**b**) Demonstration of the mechanical integrity and electrical conductivity in 3D-printed graphene scaffolds. (**c**) In vivo biocompatibility of 3D-printed graphene scaffolds. Reproduced with permission from [95]. Copyright American Chemical Society, 2015.

## 5. Conclusions

Splendid progress in nanoscience and nanotechnology has invigorated interest in application of nanomaterials to biomedical fields. Thanks to this interest, there has also been tremendous advances in research on graphene nanomaterials and their biocompatibility and biofunctionality having been gradually established. In this review, some of recent literature concerning the multifaceted biomedical applications of functional graphene nanomaterials was summarized and discussed. According to the recent studies, it is obvious that the functional graphene nanomaterials can be employed in a variety of ways to biomedical applications. In addition, the conventional approaches can be further developed by unique properties of graphene nanomaterials themselves, such as exceptional thermomechanical, excellent physicochemical, outstanding electrical and specific biological properties. In particular, many studies support the fact that graphene nanomaterial-coated substrates, -patterned arrays and hybrid scaffolds are typical approaches for biomedical applications of graphene nanomaterials, which allows us to more easily and intuitively understand the potential of graphene nanomaterials.

Although we have focused on three ways to apply graphene nanomaterials in biomedical applications, there are many other ways to employ them in biomedical fields and the potentials for application of graphene nanomaterials to biomedical fields will continue to evolve. Even if more research remains a significant challenge that should be addressed through comprehensive and systematic studies to fundamentally understand the functional graphene nanomaterials, we envision that the functional graphene nanomaterials will become promising novel candidates, which can open the way to handle unsolved problems in the current biomedical field.

**Acknowledgments:** This work was supported by the Financial Supporting Project of Long-term Overseas Dispatch of PNU's Tenure-track Faculty. Also, this research was partially supported by the Ministry of Trade, Industry & Energy (MOTIE), Korea Institute for Advancement of Technology (KIAT) through the Encouragement Program for The Industries of Economic Cooperation Region (No. R0004489), and by the Korea Research Institute of Chemical Technology (KRICT) and Ulsan government (Ulsan City-KRICT technical cooperation project).

**Author Contributions:** Y.C.S. and S.-J.S. developed the idea and structure of the review article. Y.C.S. wrote the paper using material supplied by S.W.H., S.J.J. and W.C. J.-C.L. and D.-W.H. revised and improved the manuscript. D.-W.H. supervised the manuscript. All authors have given approval to the final version of the manuscript.

**Conflicts of Interest:** The authors declare no conflict of interest.

## References

1. MacDiarmid, A.G. "Synthetic metals": A novel role for organic polymers (Nobel lecture). *Angew. Chem. Int. Ed.* **2001**, *40*, 2581–2590. [CrossRef]
2. Wang, J.; Qu, X. Recent progress in nanosensors for sensitive detection of biomolecules. *Nanoscale* **2013**, *5*, 3589–3600. [CrossRef] [PubMed]
3. Sun, J.; Xu, Z.; Li, W.; Shen, X. Effect of nano-$SiO_2$ on the early hydration of alite-sulphoaluminate cement. *Nanomaterials* **2017**, *7*, 102. [CrossRef] [PubMed]
4. Shin, Y.C.; Song, S.-J.; Shin, D.-M.; Oh, J.-W.; Hong, S.W.; Choi, Y.S.; Hyon, S.-H.; Han, D.-W. Nanocomposite scaffolds for myogenesis revisited: Functionalization with carbon nanomaterials and spectroscopic analysis. *Appl. Spectrosc. Rev.* **2017**, 1–28. [CrossRef]
5. Valiev, R. Materials science: Nanomaterial advantage. *Nature* **2002**, *419*, 887–889. [CrossRef] [PubMed]
6. Dobrovolskaia, M.A.; McNeil, S.E. Immunological properties of engineered nanomaterials. *Nat. Nanotechnol.* **2007**, *2*, 469–478. [CrossRef] [PubMed]
7. Liu, H.; Webster, T.J. Nanomedicine for implants: A review of studies and necessary experimental tools. *Biomaterials* **2007**, *28*, 354–369. [CrossRef] [PubMed]
8. Ray, P.C. Size and shape dependent second order nonlinear optical properties of nanomaterials and their application in biological and chemical sensing. *Chem. Rev.* **2010**, *110*, 5332–5365. [CrossRef] [PubMed]
9. Zhou, Z.-Y.; Tian, N.; Li, J.-T.; Broadwell, I.; Sun, S.-G. Nanomaterials of high surface energy with exceptional properties in catalysis and energy storage. *Chem. Soc. Rev.* **2011**, *40*, 4167–4185. [CrossRef] [PubMed]
10. Lin, N.; Huang, J.; Dufresne, A. Preparation, properties and applications of polysaccharide nanocrystals in advanced functional nanomaterials: A review. *Nanoscale* **2012**, *4*, 3274–3294. [CrossRef] [PubMed]

11. Cherukula, K.; Manickavasagam Lekshmi, K.; Uthaman, S.; Cho, K.; Cho, C.-S.; Park, I.-K. Multifunctional inorganic nanoparticles: Recent progress in thermal therapy and imaging. *Nanomaterials* **2016**, *6*, 76. [CrossRef] [PubMed]

12. Boehm, H.P.; Setton, R.; Stumpp, E. Nomenclature and terminology of graphite intercalation compounds. *Carbon* **1986**, *24*, 241–245. [CrossRef]

13. Novoselov, K.S.; Geim, A.K.; Morozov, S.V.; Jiang, D.; Zhang, Y.; Dubonos, S.V.; Grigorieva, I.V.; Firsov, A.A. Electric field effect in atomically thin carbon films. *Science* **2004**, *306*, 666–669. [CrossRef] [PubMed]

14. Gómez-Navarro, C.; Burghard, M.; Kern, K. Elastic properties of chemically derived single graphene sheets. *Nano Lett.* **2008**, *8*, 2045–2049. [CrossRef] [PubMed]

15. Tang, L.; Wang, Y.; Li, Y.; Feng, H.; Lu, J.; Li, J. Preparation, structure and electrochemical properties of reduced graphene sheet films. *Adv. Funct. Mater.* **2009**, *19*, 2782–2789. [CrossRef]

16. Wang, Y.; Li, Z.; Wang, J.; Li, J.; Lin, Y. Graphene and graphene oxide: Biofunctionalization and applications in biotechnology. *Trends Biotechnol.* **2011**, *29*, 205–212. [CrossRef] [PubMed]

17. Sanchez, V.C.; Jachak, A.; Hurt, R.H.; Kane, A.B. Biological interactions of graphene-family nanomaterials: An interdisciplinary review. *Chem. Res. Toxicol.* **2011**, *25*, 15–34. [CrossRef] [PubMed]

18. Wei, W.; Qu, X. Extraordinary physical properties of functionalized graphene. *Small* **2012**, *8*, 2138–2151. [CrossRef] [PubMed]

19. Kostarelos, K.; Novoselov, K.S. Exploring the interface of graphene and biology. *Science* **2014**, *344*, 261–263. [CrossRef] [PubMed]

20. Ambrosi, A.; Chua, C.K.; Bonanni, A.; Pumera, M. Electrochemistry of graphene and related materials. *Chem. Rev.* **2014**, *114*, 7150–7188. [CrossRef] [PubMed]

21. Storm, M.M.; Overgaard, M.; Younesi, R.; Reeler, N.E.A.; Vosch, T.; Nielsen, U.G.; Edström, K.; Norby, P. Reduced graphene oxide for Li-air batteries: The effect of oxidation time and reduction conditions for graphene oxide. *Carbon* **2015**, *85*, 233–244. [CrossRef]

22. Zhang, C.; Chen, S.; Alvarez, P.J.J.; Chen, W. Reduced graphene oxide enhances horseradish peroxidase stability by serving as radical scavenger and redox mediator. *Carbon* **2015**, *94*, 531–538. [CrossRef]

23. Aillon, K.L.; Xie, Y.; El-Gendy, N.; Berkland, C.J.; Forrest, M.L. Effects of nanomaterial physicochemical properties on in vivo toxicity. *Adv. Drug Deliv. Rev.* **2009**, *61*, 457–466. [CrossRef] [PubMed]

24. Yang, K.; Li, Y.; Tan, X.; Peng, R.; Liu, Z. Behavior and toxicity of graphene and its functionalized derivatives in biological systems. *Small* **2013**, *9*, 1492–1503. [CrossRef] [PubMed]

25. Cheng, L.-C.; Jiang, X.; Wang, J.; Chen, C.; Liu, R.-S. Nano-bio effects: Interaction of nanomaterials with cells. *Nanoscale* **2013**, *5*, 3547–3569. [CrossRef] [PubMed]

26. Chimene, D.; Alge, D.L.; Gaharwar, A.K. Two-dimensional nanomaterials for biomedical applications: Emerging trends and future prospects. *Adv. Mater.* **2015**, *27*, 7261–7284. [CrossRef] [PubMed]

27. Singh, Z. Applications and toxicity of graphene family nanomaterials and their composites. *Nanotechnol. Sci. Appl.* **2016**, *9*, 15. [CrossRef] [PubMed]

28. Nayak, T.R.; Andersen, H.; Makam, V.S.; Khaw, C.; Bae, S.; Xu, X.; Ee, P.-L.R.; Ahn, J.-H.; Hong, B.H.; Pastorin, G. Graphene for controlled and accelerated osteogenic differentiation of human mesenchymal stem cells. *ACS Nano* **2011**, *5*, 4670–4678. [CrossRef] [PubMed]

29. Lee, W.C.; Lim, C.H.Y.; Shi, H.; Tang, L.A.; Wang, Y.; Lim, C.T.; Loh, K.P. Origin of enhanced stem cell growth and differentiation on graphene and graphene oxide. *ACS Nano* **2011**, *5*, 7334–7341. [CrossRef] [PubMed]

30. Park, S.Y.; Park, J.; Sim, S.H.; Sung, M.G.; Kim, K.S.; Hong, B.H.; Hong, S. Enhanced differentiation of human neural stem cells into neurons on graphene. *Adv. Mater.* **2011**, *23*, H263–H267. [CrossRef] [PubMed]

31. Li, N.; Zhang, Q.; Gao, S.; Song, Q.; Huang, R.; Wang, L.; Liu, L.; Dai, J.; Tang, M.; Cheng, G. Three-dimensional graphene foam as a biocompatible and conductive scaffold for neural stem cells. *Sci. Rep.* **2013**, *3*, 1604. [CrossRef] [PubMed]

32. Ku, S.H.; Park, C.B. Myoblast differentiation on graphene oxide. *Biomaterials* **2013**, *34*, 2017–2023. [CrossRef] [PubMed]

33. Lee, J.H.; Shin, Y.C.; Jin, O.S.; Kang, S.H.; Hwang, Y.-S.; Park, J.-C.; Hong, S.W.; Han, D.-W. Reduced graphene oxide-coated hydroxyapatite composites stimulate spontaneous osteogenic differentiation of human mesenchymal stem cells. *Nanoscale* **2015**, *7*, 11642–11651. [CrossRef] [PubMed]

34. Lee, W.C.; Lim, C.H.; Su, C.; Loh, K.P.; Lim, C.T. Cell-assembled graphene biocomposite for enhanced chondrogenic differentiation. *Small* **2015**, *11*, 963–969. [CrossRef] [PubMed]

35.  Shin, Y.C.; Lee, J.H.; Jin, O.S.; Kang, S.H.; Hong, S.W.; Kim, B.; Park, J.-C.; Han, D.-W. Synergistic effects of reduced graphene oxide and hydroxyapatite on osteogenic differentiation of MC3T3-E1 preosteoblasts. *Carbon* **2015**, *95*, 1051–1060. [CrossRef]

36.  Lee, J.H.; Shin, Y.C.; Lee, S.-M.; Jin, O.S.; Kang, S.H.; Hong, S.W.; Jeong, C.-M.; Huh, J.B.; Han, D.-W. Enhanced osteogenesis by reduced graphene oxide/hydroxyapatite nanocomposites. *Sci. Rep.* **2015**, *5*, 18833. [CrossRef] [PubMed]

37.  Zhang, Y.; Ali, S.F.; Dervishi, E.; Xu, Y.; Li, Z.; Casciano, D.; Biris, A.S. Cytotoxicity effects of graphene and single-wall carbon nanotubes in neural phaeochromocytoma-derived PC12 cells. *ACS Nano* **2010**, *4*, 3181–3186. [CrossRef] [PubMed]

38.  Akhavan, O.; Ghaderi, E.; Akhavan, A. Size-dependent genotoxicity of graphene nanoplatelets in human stem cells. *Biomaterials* **2012**, *33*, 8017–8025. [CrossRef] [PubMed]

39.  Chng, E.L.K.; Pumera, M. The toxicity of graphene oxides: Dependence on the oxidative methods used. *Chem. Eur. J.* **2013**, *19*, 8227–8235. [CrossRef] [PubMed]

40.  Li, Y.; Liu, Y.; Fu, Y.; Wei, T.; Le Guyader, L.; Gao, G.; Liu, R.-S.; Chang, Y.-Z.; Chen, C. The triggering of apoptosis in macrophages by pristine graphene through the MAPK and TGF-beta signaling pathways. *Biomaterials* **2012**, *33*, 402–411. [CrossRef] [PubMed]

41.  Zhou, H.; Zhao, K.; Li, W.; Yang, N.; Liu, Y.; Chen, C.; Wei, T. The interactions between pristine graphene and macrophages and the production of cytokines/chemokines via TLR-and NF-κB-related signaling pathways. *Biomaterials* **2012**, *33*, 6933–6942. [CrossRef] [PubMed]

42.  Liao, K.-H.; Lin, Y.-S.; Macosko, C.W.; Haynes, C.L. Cytotoxicity of graphene oxide and graphene in human erythrocytes and skin fibroblasts. *ACS Appl. Mater. Interfaces* **2011**, *3*, 2607–2615. [CrossRef] [PubMed]

43.  Park, E.-J.; Lee, G.-H.; Han, B.S.; Lee, B.-S.; Lee, S.; Cho, M.-H.; Kim, J.-H.; Kim, D.-W. Toxic response of graphene nanoplatelets in vivo and in vitro. *Arch. Toxicol.* **2015**, *89*, 1557–1568. [CrossRef] [PubMed]

44.  Ryoo, S.-R.; Kim, Y.-K.; Kim, M.-H.; Min, D.-H. Behaviors of NIH-3T3 fibroblasts on graphene/carbon nanotubes: Proliferation, focal adhesion and gene transfection studies. *ACS Nano* **2010**, *4*, 6587–6598. [CrossRef] [PubMed]

45.  Chen, G.-Y.; Pang, D.-P.; Hwang, S.-M.; Tuan, H.-Y.; Hu, Y.-C. A graphene-based platform for induced pluripotent stem cells culture and differentiation. *Biomaterials* **2012**, *33*, 418–427. [CrossRef] [PubMed]

46.  Lee, J.H.; Shin, Y.C.; Jin, O.S.; Han, D.-W.; Kang, S.H.; Hong, S.W.; Kim, J.M. Enhanced neurite outgrowth of PC-12 cells on graphene-monolayer-coated substrates as biomimetic cues. *J. Korean Phys. Soc.* **2012**, *61*, 1696–1699. [CrossRef]

47.  Park, J.; Park, S.; Ryu, S.; Bhang, S.H.; Kim, J.; Yoon, J.K.; Park, Y.H.; Cho, S.P.; Lee, S.; Hong, B.H. Graphene—regulated cardiomyogenic differentiation process of mesenchymal stem cells by enhancing the expression of extracellular matrix proteins and cell signaling molecules. *Adv. Healthc. Mater.* **2014**, *3*, 176–181. [CrossRef] [PubMed]

48.  Akhavan, O.; Ghaderi, E.; Abouei, E.; Hatamie, S.; Ghasemi, E. Accelerated differentiation of neural stem cells into neurons on ginseng-reduced graphene oxide sheets. *Carbon* **2014**, *66*, 395–406. [CrossRef]

49.  Lee, T.-J.; Park, S.; Bhang, S.H.; Yoon, J.-K.; Jo, I.; Jeong, G.-J.; Hong, B.H.; Kim, B.-S. Graphene enhances the cardiomyogenic differentiation of human embryonic stem cells. *Biochem. Biophys. Res. Commun.* **2014**, *452*, 174–180. [CrossRef] [PubMed]

50.  Park, K.O.; Lee, J.H.; Park, J.H.; Shin, Y.C.; Huh, J.B.; Bae, J.-H.; Kang, S.H.; Hong, S.W.; Kim, B.; Yang, D.J. Graphene oxide-coated guided bone regeneration membranes with enhanced osteogenesis: Spectroscopic analysis and animal study. *Appl. Spectrosc. Rev.* **2016**, *51*, 540–551. [CrossRef]

51.  Lee, J.H.; Lee, S.-M.; Shin, Y.C.; Park, J.H.; Hong, S.W.; Kim, B.; Lee, J.J.; Lim, D.; Lim, Y.-J.; Huh, J.B. Spontaneous osteodifferentiation of bone marrow-derived mesenchymal stem cells by hydroxyapatite covered with graphene nanosheets. *J. Biomater. Tissue Eng.* **2016**, *6*, 818–825. [CrossRef]

52.  Kim, J.-W.; Shin, Y.C.; Lee, J.-J.; Bae, E.-B.; Jeon, Y.-C.; Jeong, C.-M.; Yun, M.-J.; Lee, S.-H.; Han, D.-W.; Huh, J.-B. The effect of reduced graphene oxide-coated biphasic calcium phosphate bone graft material on osteogenesis. *Int. J. Mol. Sci.* **2017**, *18*, 1725. [CrossRef] [PubMed]

53.  Discher, D.E.; Janmey, P.; Wang, Y.-l. Tissue cells feel and respond to the stiffness of their substrate. *Science* **2005**, *310*, 1139–1143. [CrossRef] [PubMed]

54. Rowlands, A.S.; George, P.A.; Cooper-White, J.J. Directing osteogenic and myogenic differentiation of MSCs: Interplay of stiffness and adhesive ligand presentation. *Am. J. Physiol. Cell Physiol.* **2008**, *295*, C1037–C1044. [CrossRef] [PubMed]

55. Shih, Y.R.V.; Tseng, K.F.; Lai, H.Y.; Lin, C.H.; Lee, O.K. Matrix stiffness regulation of integrin-mediated mechanotransduction during osteogenic differentiation of human mesenchymal stem cells. *J. Bone Miner. Res.* **2011**, *26*, 730–738. [CrossRef] [PubMed]

56. Hoon, J.L.; Tan, M.H.; Koh, C.-G. The regulation of cellular responses to mechanical cues by Rho GTPases. *Cells* **2016**, *5*, 17. [CrossRef] [PubMed]

57. Tang, M.; Song, Q.; Li, N.; Jiang, Z.; Huang, R.; Cheng, G. Enhancement of electrical signaling in neural networks on graphene films. *Biomaterials* **2013**, *34*, 6402–6411. [CrossRef] [PubMed]

58. Zhang, H.-B.; Zheng, W.-G.; Yan, Q.; Yang, Y.; Wang, J.-W.; Lu, Z.-H.; Ji, G.-Y.; Yu, Z.-Z. Electrically conductive polyethylene terephthalate/graphene nanocomposites prepared by melt compounding. *Polymer* **2010**, *51*, 1191–1196. [CrossRef]

59. Qiu, C.; Bennet, K.E.; Khan, T.; Ciubuc, J.D.; Manciu, F.S. Raman and conductivity analysis of graphene for biomedical applications. *Materials* **2016**, *9*, 897. [CrossRef] [PubMed]

60. Kim, D.-J.; Sohn, I.Y.; Jung, J.-H.; Yoon, O.J.; Lee, N.E.; Park, J.-S. Reduced graphene oxide field-effect transistor for label-free femtomolar protein detection. *Biosens. Bioelectron.* **2013**, *41*, 621–626. [CrossRef] [PubMed]

61. Chang, J.; Mao, S.; Zhang, Y.; Cui, S.; Zhou, G.; Wu, X.; Yang, C.-H.; Chen, J. Ultrasonic-assisted self-assembly of monolayer graphene oxide for rapid detection of *Escherichia coli* bacteria. *Nanoscale* **2013**, *5*, 3620–3626. [CrossRef] [PubMed]

62. Cai, B.; Wang, S.; Huang, L.; Ning, Y.; Zhang, Z.; Zhang, G.-J. Ultrasensitive label-free detection of PNA-DNA hybridization by reduced graphene oxide field-effect transistor biosensor. *ACS Nano* **2014**, *8*, 2632–2638. [CrossRef] [PubMed]

63. Zhan, B.; Li, C.; Yang, J.; Jenkins, G.; Huang, W.; Dong, X. Graphene field-effect transistor and its application for electronic sensing. *Small* **2014**, *10*, 4042–4065. [CrossRef] [PubMed]

64. Park, D.-W.; Schendel, A.A.; Mikael, S.; Brodnick, S.K.; Richner, T.J.; Ness, J.P.; Hayat, M.R.; Atry, F.; Frye, S.T.; Pashaie, R. Graphene-based carbon-layered electrode array technology for neural imaging and optogenetic applications. *Nat. Commun.* **2014**, *5*, 5258. [CrossRef] [PubMed]

65. Akhavan, O.; Ghaderi, E. Toxicity of graphene and graphene oxide nanowalls against bacteria. *ACS Nano* **2010**, *4*, 5731–5736. [CrossRef] [PubMed]

66. Yoon, H.J.; Kim, T.H.; Zhang, Z.; Azizi, E.; Pham, T.M.; Paoletti, C.; Lin, J.; Ramnath, N.; Wicha, M.S.; Hayes, D.F. Sensitive capture of circulating tumour cells by functionalized graphene oxide nanosheets. *Nat. Nanotechnol.* **2013**, *8*, 735–741. [CrossRef] [PubMed]

67. Yang, T.; Wang, W.; Zhang, H.; Li, X.; Shi, J.; He, Y.; Zheng, Q.-S.; Li, Z.; Zhu, H. Tactile sensing system based on arrays of graphene woven microfabrics: Electromechanical behavior and electronic skin application. *ACS Nano* **2015**, *9*, 10867–10875. [CrossRef] [PubMed]

68. Lauffenburger, D.A.; Horwitz, A.F. Cell migration: A physically integrated molecular process. *Cell* **1996**, *84*, 359–369. [CrossRef]

69. Franz, C.M.; Jones, G.E.; Ridley, A.J. Cell migration in development and disease. *Dev. Cell* **2002**, *2*, 153–158. [CrossRef]

70. Nie, F.-Q.; Yamada, M.; Kobayashi, J.; Yamato, M.; Kikuchi, A.; Okano, T. On-chip cell migration assay using microfluidic channels. *Biomaterials* **2007**, *28*, 4017–4022. [CrossRef] [PubMed]

71. Bajaj, P.; Rivera, J.A.; Marchwiany, D.; Solovyeva, V.; Bashir, R. Graphene-based patterning and differentiation of C2C12 myoblasts. *Adv. Healthc. Mater.* **2014**, *3*, 995–1000. [CrossRef] [PubMed]

72. Akhavan, O.; Ghaderi, E.; Shahsavar, M. Graphene nanogrids for selective and fast osteogenic differentiation of human mesenchymal stem cells. *Carbon* **2013**, *59*, 200–211. [CrossRef]

73. Akhavan, O.; Ghaderi, E. Differentiation of human neural stem cells into neural networks on graphene nanogrids. *J. Mat. Chem. B* **2013**, *1*, 6291–6301. [CrossRef]

74. Hong, D.; Bae, K.; Yoo, S.; Kang, K.; Jang, B.; Kim, J.; Kim, S.; Jeon, S.; Nam, Y.; Kim, Y.G. Generation of cellular micropatterns on a single-layered graphene film. *Macromol. Biosci.* **2014**, *14*, 314–319. [CrossRef] [PubMed]

75.  Kim, S.J.; Cho, H.R.; Cho, K.W.; Qiao, S.; Rhim, J.S.; Soh, M.; Kim, T.; Choi, M.K.; Choi, C.; Park, I. Multifunctional cell-culture platform for aligned cell sheet monitoring, transfer printing and therapy. *ACS Nano* **2015**, *9*, 2677–2688. [CrossRef] [PubMed]

76.  Kim, T.-H.; Shah, S.; Yang, L.; Yin, P.T.; Hossain, M.K.; Conley, B.; Choi, J.-W.; Lee, K.-B. Controlling differentiation of adipose-derived stem cells using combinatorial graphene hybrid-pattern arrays. *Acs Nano* **2015**, *9*, 3780–3790. [CrossRef] [PubMed]

77.  Wang, Z.; Tonderys, D.; Leggett, S.E.; Williams, E.K.; Kiani, M.T.; Steinberg, R.S.; Qiu, Y.; Wong, I.Y.; Hurt, R.H. Wrinkled, wavelength-tunable graphene-based surface topographies for directing cell alignment and morphology. *Carbon* **2016**, *97*, 14–24. [CrossRef] [PubMed]

78.  Zhang, H.; Hou, R.; Xiao, P.; Xing, R.; Chen, T.; Han, Y.; Ren, P.; Fu, J. Single cell migration dynamics mediated by geometric confinement. *Colloids Surf. B-Biointerfaces* **2016**, *145*, 72–78. [CrossRef] [PubMed]

79.  Kim, S.E.; Kim, M.S.; Shin, Y.C.; Eom, S.U.; Lee, J.H.; Shin, D.-M.; Hong, S.W.; Kim, B.; Park, J.-C.; Shin, B.S. Cell migration according to shape of graphene oxide micropatterns. *Micromachines* **2016**, *7*, 186. [CrossRef]

80.  Altman, G.; Horan, R.; Martin, I.; Farhadi, J.; Stark, P.; Volloch, V.; Vunjak-Novakovic, G.; Richmond, J.; Kaplan, D.L. Cell differentiation by mechanical stress. *FASEB J.* **2002**, *16*, 270–272. [CrossRef] [PubMed]

81.  Dalby, M.J.; McCloy, D.; Robertson, M.; Agheli, H.; Sutherland, D.; Affrossman, S.; Oreffo, R.O.C. Osteoprogenitor response to semi-ordered and random nanotopographies. *Biomaterials* **2006**, *27*, 2980–2987. [CrossRef] [PubMed]

82.  Yamamoto, K.; Sokabe, T.; Watabe, T.; Miyazono, K.; Yamashita, J.K.; Obi, S.; Ohura, N.; Matsushita, A.; Kamiya, A.; Ando, J. Fluid shear stress induces differentiation of Flk-1-positive embryonic stem cells into vascular endothelial cells in vitro. *Am. J. Physiol. Heart Circ. Physiol.* **2005**, *288*, H1915–H1924. [CrossRef] [PubMed]

83.  Bai, H.; Li, C.; Wang, X.; Shi, G. A pH-sensitive graphene oxide composite hydrogel. *Chem. Commun.* **2010**, *46*, 2376–2378. [CrossRef] [PubMed]

84.  Luo, X.; Weaver, C.L.; Tan, S.; Cui, X.T. Pure graphene oxide doped conducting polymer nanocomposite for bio-interfacing. *J. Mat. Chem. B* **2013**, *1*, 1340–1348. [CrossRef] [PubMed]

85.  Crowder, S.W.; Prasai, D.; Rath, R.; Balikov, D.A.; Bae, H.; Bolotin, K.I.; Sung, H.-J. Three-dimensional graphene foams promote osteogenic differentiation of human mesenchymal stem cells. *Nanoscale* **2013**, *5*, 4171–4176. [CrossRef] [PubMed]

86.  Cha, C.; Shin, S.R.; Gao, X.; Annabi, N.; Dokmeci, M.R.; Tang, X.S.; Khademhosseini, A. Controlling mechanical properties of cell-laden hydrogels by covalent incorporation of graphene oxide. *Small* **2014**, *10*, 514–523. [CrossRef] [PubMed]

87.  Qiu, L.; Liu, D.; Wang, Y.; Cheng, C.; Zhou, K.; Ding, J.; Truong, V.T.; Li, D. Mechanically robust, electrically conductive and stimuli-responsive binary network hydrogels enabled by superelastic graphene aerogels. *Adv. Mater.* **2014**, *26*, 3333–3337. [CrossRef] [PubMed]

88.  Shah, S.; Yin, P.T.; Uehara, T.M.; Chueng, S.T.D.; Yang, L.; Lee, K.B. Guiding stem cell differentiation into oligodendrocytes using graphene-nanofiber hybrid scaffolds. *Adv. Mater.* **2014**, *26*, 3673–3680. [CrossRef] [PubMed]

89.  Lee, E.J.; Lee, J.H.; Shin, Y.C.; Hwang, D.-G.; Kim, J.S.; Jin, O.S.; Jin, L.; Hong, S.W.; Han, D.-W. Graphene oxide-decorated PLGA/collagen hybrid fiber sheets for application to tissue engineering scaffolds. *Biomater. Res.* **2014**, *18*, 18–24.

90.  Serrano, M.C.; Patiño, J.; García-Rama, C.; Ferrer, M.L.; Fierro, J.L.G.; Tamayo, A.; Collazos-Castro, J.E.; del Monte, F.; Gutierrez, M.C. 3D free-standing porous scaffolds made of graphene oxide as substrates for neural cell growth. *J. Mat. Chem. B* **2014**, *2*, 5698–5706. [CrossRef]

91.  Weaver, C.L.; LaRosa, J.M.; Luo, X.; Cui, X.T. Electrically controlled drug delivery from graphene oxide nanocomposite films. *ACS Nano* **2014**, *8*, 1834–1843. [CrossRef] [PubMed]

92.  Paul, A.; Hasan, A.; Kindi, H.A.; Gaharwar, A.K.; Rao, V.T.S.; Nikkhah, M.; Shin, S.R.; Krafft, D.; Dokmeci, M.R.; Shum-Tim, D. Injectable graphene oxide/hydrogel-based angiogenic gene delivery system for vasculogenesis and cardiac repair. *ACS Nano* **2014**, *8*, 8050–8062. [CrossRef] [PubMed]

93.  Shin, Y.C.; Lee, J.H.; Jin, L.; Kim, M.J.; Kim, Y.J.; Hyun, J.K.; Jung, T.G.; Hong, S.W.; Han, D.W. Stimulated myoblast differentiation on graphene oxide-impregnated PLGA-collagen hybrid fibre matrices. *J. Nanobiotechnol.* **2015**, *13*, 21. [CrossRef] [PubMed]

94. Chaudhuri, B.; Bhadra, D.; Moroni, L.; Pramanik, K. Myoblast differentiation of human mesenchymal stem cells on graphene oxide and electrospun graphene oxide-polymer composite fibrous meshes: Importance of graphene oxide conductivity and dielectric constant on their biocompatibility. *Biofabrication* **2015**, *7*, 015009. [CrossRef] [PubMed]
95. Jakus, A.E.; Secor, E.B.; Rutz, A.L.; Jordan, S.W.; Hersam, M.C.; Shah, R.N. Three-dimensional printing of high-content graphene scaffolds for electronic and biomedical applications. *ACS Nano* **2015**, *9*, 4636–4648. [CrossRef] [PubMed]
96. Lee, J.H.; Lee, Y.; Shin, Y.C.; Kim, M.J.; Park, J.H.; Hong, S.W.; Kim, B.; Oh, J.-W.; Park, K.D.; Han, D.-W. In situ forming gelatin/graphene oxide hydrogels for facilitated C2C12 myoblast differentiation. *Appl. Spectrosc. Rev.* **2016**, *51*, 527–539. [CrossRef]
97. Krueger, E.; Chang, A.N.; Brown, D.; Eixenberger, J.; Brown, R.; Rastegar, S.; Yocham, K.M.; Cantley, K.D.; Estrada, D. Graphene foam as a three-dimensional platform for myotube growth. *ACS Biomater. Sci. Eng.* **2016**, *2*, 1234–1241. [CrossRef] [PubMed]
98. Shin, Y.C.; Kim, J.; Kim, S.E.; Song, S.-J.; Hong, S.W.; Oh, J.-W.; Lee, J.; Park, J.-C.; Hyon, S.-H.; Han, D.-W. RGD peptide and graphene oxide co-functionalized PLGA nanofiber scaffolds for vascular tissue engineering. *Regen. Biomater.* **2017**, *4*, 159–166. [CrossRef] [PubMed]
99. Jakus, A.E.; Shah, R. Multi and mixed 3D-printing of graphene-hydroxyapatite hybrid materials for complex tissue engineering. *J. Biomed. Mater. Res. A* **2017**, *105*, 274–283. [CrossRef] [PubMed]
100. Shin, Y.C.; Jin, L.; Lee, J.H.; Jun, S.; Hong, S.W.; Kim, C.-S.; Kim, Y.-J.; Hyun, J.K.; Han, D.-W. Graphene oxide-incorporated PLGA-collagen fibrous matrices as biomimetic scaffolds for vascular smooth muscle cells. *Sci. Adv. Mater.* **2017**, *9*, 232–237. [CrossRef]
101. Shin, Y.C.; Kang, S.H.; Lee, J.H.; Kim, B.; Hong, S.W.; Han, D.-W. Three-dimensional graphene oxide-coated polyurethane foams beneficial to myogenesis. *J. Biomater. Sci. Polym. Ed.* **2017**, 1–13. [CrossRef] [PubMed]
102. Yoon, O.J.; Jung, C.Y.; Sohn, I.Y.; Kim, H.J.; Hong, B.; Jhon, M.S.; Lee, N.-E. Nanocomposite nanofibers of poly (D, L-lactic-*co*-glycolic acid) and graphene oxide nanosheets. *Compos. A Appl. Sci. Manuf.* **2011**, *42*, 1978–1984. [CrossRef]
103. Yoon, O.J.; Sohn, I.Y.; Kim, D.J.; Lee, N.-E. Enhancement of thermomechanical properties of poly (D, L-lactic-*co*-glycolic acid) and graphene oxide composite films for scaffolds. *Macromol. Res.* **2012**, *20*, 789–794. [CrossRef]
104. Zhang, N.; Qiu, H.; Si, Y.; Wang, W.; Gao, J. Fabrication of highly porous biodegradable monoliths strengthened by graphene oxide and their adsorption of metal ions. *Carbon* **2011**, *49*, 827–837. [CrossRef]
105. Grinou, A.; Yun, Y.S.; Jin, H.-J. Polyaniline nanofiber-coated polystyrene/graphene oxide core-shell microsphere composites. *Macromol. Res.* **2012**, *20*, 84–92. [CrossRef]
106. Cukierman, E.; Pankov, R.; Stevens, D.R.; Yamada, K.M. Taking cell-matrix adhesions to the third dimension. *Science* **2001**, *294*, 1708–1712. [CrossRef] [PubMed]
107. Yamada, K.M.; Cukierman, E. Modeling tissue morphogenesis and cancer in 3D. *Cell* **2007**, *130*, 601–610. [CrossRef] [PubMed]
108. Pampaloni, F.; Reynaud, E.G.; Stelzer, E.H.K. The third dimension bridges the gap between cell culture and live tissue. *Nat. Rev. Mol. Cell Biol.* **2007**, *8*, 839–845. [CrossRef] [PubMed]

© 2017 by the authors. Licensee MDPI, Basel, Switzerland. This article is an open access article distributed under the terms and conditions of the Creative Commons Attribution (CC BY) license (http://creativecommons.org/licenses/by/4.0/).

*nanomaterials*

MDPI

*Article*

# Antioxidant Potential and Antibacterial Efficiency of Caffeic Acid-Functionalized ZnO Nanoparticles

Kyong-Hoon Choi [1], Ki Chang Nam [2], Sang-Yoon Lee [3], Guangsup Cho [1], Jin-Seung Jung [3,*], Ho-Joong Kim [4,*] and Bong Joo Park [1,*]

[1] Department of Electrical & Biological Physics, Kwangwoon University, 20 Kwangwoongil, Nowon-gu, Seoul 01897, Korea; solidchem@hanmail.net (K.-H.C.); gscho@kw.ac.kr (G.C.)
[2] Department of Medical Engineering, Dongguk University College of Medicine, Gyeonggi-do 10326, Korea; kichang.nam@gmail.com
[3] Department of Chemistry, Gangneung-Wonju National University, Gangneung 25457, Korea; nanochemistry@naver.com
[4] Department of Chemistry, Chosun University, Gwangju 61452, Korea
* Correspondence: jjscm@kangnung.ac.kr (J.-S.J.); hjkim@chosun.ac.kr (H.-J.K.); parkbj@kw.ac.kr (B.J.P.); Tel.: +82-33-640-2073 (J.-S.J.); +82-62-230-6643 (H.-J.K.); +82-2-940-8629 (B.J.P.)

Received: 2 May 2017; Accepted: 14 June 2017; Published: 16 June 2017

**Abstract:** We report a novel zinc oxide (ZnO) nanoparticle with antioxidant properties, prepared by immobilizing the antioxidant 3-(3,4-dihydroxyphenyl)-2-propenoic acid (caffeic acid, CA) on the surfaces of micro-dielectric barrier discharge (DBD) plasma-treated ZnO nanoparticles. The microstructure and physical properties of ZnO@CA nanoparticles were characterized by field emission scanning electron microscopy (FESEM), transmission electron microscopy (TEM), infrared spectroscopy, and steady state spectroscopic methods. The antioxidant activity of ZnO@CA nanoparticles was evaluated using an ABTS (3-ethyl-benzothiazoline-6-sulfonic acid) radical cation decolorization assay. ZnO@CA nanoparticles exhibited robust antioxidant activity. Moreover, ZnO@CA nanoparticles showed strong antibacterial activity against Gram-positive bacteria (*Staphylococcus aureus*) including resistant bacteria such as methicillin-resistant *S. aureus* and against Gram-negative bacteria (*Escherichia coli*). Although Gram-negative bacteria appeared to be more resistant to ZnO@CA nanoparticles than Gram-positive bacteria, the antibacterial activity of ZnO@CA nanoparticles was dependent on particle concentration. The antioxidant and antibacterial activity of ZnO@CA may be useful for various biomedical and nanoindustrial applications.

**Keywords:** multifunctional nanoparticle; antioxidant activity; antibacterial activity; caffeic acid; ZnO

## 1. Introduction

Antioxidants inhibit the oxidation process in biological systems and the environment by acting as free radical scavengers, reactive oxygen scavengers, or reducing agents [1,2]. Recently, a number of studies have investigated the antioxidant activities of synthetic as well as natural antioxidants [3–5]. Caffeic acid (CA) has been increasingly studied in pharmacological research owing to its antioxidant, cardiac, immunomodulatory, and anti-inflammatory activities [6–8]. CA is a natural phenolic acid widely found in plants and at high levels in some herbs, spices, and sunflower seeds. It is also one of the major natural phenols in argan oil.

In the pharmaceutical, biological, and food industries, natural antioxidants have been introduced into substrate materials to prevent or reduce oxidation in situations where free antioxidants cannot be used, such as under ambient $O_2$, volatilization, and thermal instability [9–11]. Nanoparticles are now widely used throughout the pharmaceutical, catalyst, electronics, nano-patterning, and tissue engineering industries [12–15]. Nanoparticles often possess unique nanoscale size-dependent physical and chemical properties that can be controlled in a manner that is not possible in the bulk state. Among the various nanoparticles, ZnO nanoparticles have been extensively investigated over the past decade in the fields of materials science, biotechnology, and medical engineering for their potential applications as catalysts and chemical absorbents [16,17]. A previous in vitro study of ZnO nanoparticles showed that particle size, particle morphology, surface modifications, and reactivity in aqueous solutions determined their biocompatibility [18,19]. In particular, ZnO nanoparticles exhibit high antimicrobial activity even at low concentrations. They offer many advantages as an antibacterial agent because of their good stability at high temperatures and pressures and long shelf life when compared to organic antibacterial agents. By employing various strategies to modify and tailor ZnO nanoparticle surfaces, multifunctional nanoparticles with improved aqueous dispersion and biocompatibility have been developed that can be used for targeted drug delivery and bioimaging [20–22]. However, although considerable progress has been made in the preparation of multifunctional ZnO nanoparticles for biological applications, ZnO nanoparticles with antioxidant properties have not been reported.

In the present study, we describe a simple surface modification process that functionalizes 20 nm ZnO nanoparticles with the antioxidant CA. The antioxidant activity of the resulting ZnO@CA was evaluated using an ABTS (3-ethyl-benzothiazoline-6-sulfonic acid) radical cation decolorization assay and compared to that of free CA. The prepared ZnO@CA was tested for its antimicrobial activity against *Escherichia coli*, *Staphylococcus aureus*, and methicillin-resistant *S. aureus* (MRSA). The physical and structural properties of ZnO@CA nanoparticles were also investigated. To the best of our knowledge, this is the first demonstration of antioxidant and antibacterial activities by a functionalized ZnO nanoparticle. This novel multifunctional nanoparticle represents a promising material for therapeutic applications in biomedical engineering.

## 2. Results and Discussion

The morphology, structure, and size of ZnO@CA nanoparticles were investigated using FESEM, high resolution transmission electron microscopy (HRTEM), and X-ray diffraction (XRD). Figure 1 shows FESEM images of ZnO@CA nanoparticles synthesized at a molar ratio of 1:2 (Zn nitrate:KOH). FESEM images of ZnO@CA nanoparticles revealed a spherical shape with a narrow size distribution. They were assembled in aggregates, as shown in Figure 1a. The high-magnification SEM image shows very small ZnO@CA crystallites with various spherical shapes. The average size of the ZnO@CA nanoparticles was about 20 nm, which is consistent with the size of the nanoparticles determined by XRD analysis of Sherrer's formula.

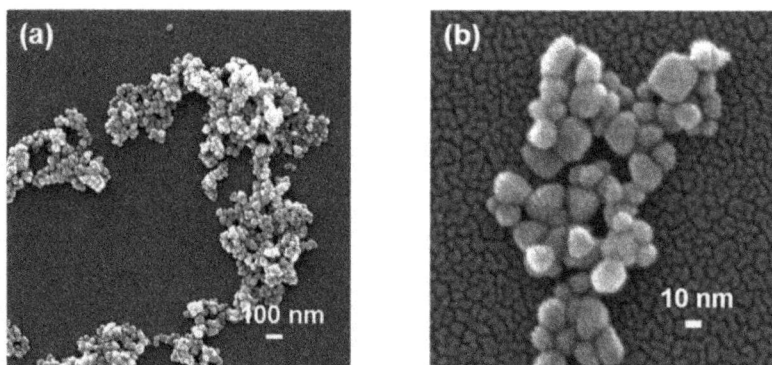

**Figure 1.** (**a**) Low and (**b**) high magnification FESEM images of ZnO@CA nanoparticles. The ZnO@CA nanoparticles have a spherical shape with a narrow size distribution. The mean size of the ZnO@CA nanoparticles is ~20 nm.

The microstructure of the ZnO@CA nanoparticles was characterized using HRTEM. As illustrated in Figure 2a, many ZnO@CA nanoparticles with a single crystalline nature were confirmed in the agglomerated particles. The mean size of a single nanoparticle was approximately 20 nm (Figure 2a). In Figure 2b, the ZnO@CA nanoparticles are single crystalline with an appropriate distance of 2.63 Å between two neighboring planes. This crystalline property is consistent with the neighboring crystal planes (002) of the wurtzite structure of ZnO.

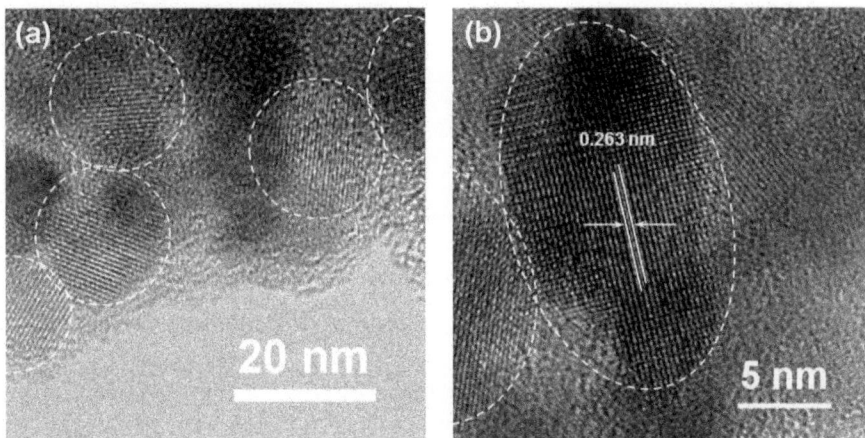

**Figure 2.** (a) Low magnification and (b) high magnification TEM micrographs of ZnO@CA nanoparticles. The interlayer distance of the lattice fringe is estimated to be ≈2.63 Å, which is comparable to those of the (002) planes in the typical wurtzite structure of ZnO.

Figure 3 shows a typical XRD pattern of ZnO@CA nanoparticles in the range of 10–80° at a scanning step of 0.01. A number of Bragg reflections with values of 31.7°, 34.3°, 36.2°, 47.5°, 56.5°, 62.8°, 66.3°, 67.9°, and 69.1° were observed corresponding to the (100), (002), (101), (102), (110), (103), (200), (112), and (201) planes, respectively [23]. All diffraction peaks in the XRD pattern can be indexed to the wurtzite ZnO structure (hexagonal phase, space group *P63mc*, and JCPDS No. 36-1451) [24]. The large amplitudes and small widths of the diffraction peaks indicate a high degree of crystallinity in ZnO@CA nanoparticles, as does the absence of other peaks in the XRD pattern, which might indicate the potential presence of impurities. To estimate the average crystallite size (D) of ZnO@CA nanoparticles, diffraction peak profiles were fit with a convolution of Lorentzian functions (inset of Figure 3), and the extent of line broadening was estimated using the Scherrer equation [25]

$$D = \frac{0.9\lambda}{\beta \times \cos\theta} \tag{1}$$

where D is the crystallite size (nm), $\lambda$ is the wavelength of incident X-ray (0.154 nm), $\beta$ is the full width at half maximum, and $\theta$ is the Bragg's diffraction angle. The mean crystallite size of ZnO@CA nanoparticles was 19.3 nm, which was calculated using the most intense peak (101) in the XRD pattern.

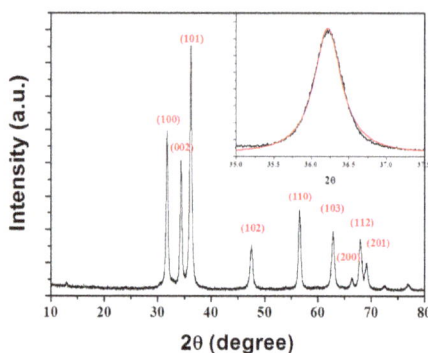

**Figure 3.** XRD spectrum of ZnO@CA nanoparticles. All diffraction peaks can be indexed to the wurtzite type lattice of ZnO, which matches well with the standard XRD pattern (JCPDS No. 36-1451).

To confirm that a covalent bond had formed between the carboxyl group of CA and the Zn ions on the surfaces of the ZnO nanoparticles, Fourier transform infrared (FT-IR) spectroscopy analyses of free ZnO, CA, and ZnO@CA nanoparticles were performed (Figure 4). First, the FT-IR spectrum of ZnO could not confirm any vibrational peaks of the CA molecules. Second, the IR spectrum of pure CA showed major absorption peaks at 1650, 1450, and 1278 cm$^{-1}$, which corresponded to the stretching modes of the free carboxyl double bond ($\upsilon_{C=O}$), the carbon–oxygen single bond ($\upsilon_{C-O}$), and the O-H deformation ($\upsilon_{C-OH}$), respectively (top panel of Figure 4) [26]. Absorption peaks at these positions are consistent with the IR spectrum of protonated carboxyl groups (COOH). In contrast, the IR spectrum of ZnO@CA nanoparticles showed new peaks at 1558 and 1378 cm$^{-1}$, which correspond to the asymmetric ($\upsilon_{as}$ = 1558 cm$^{-1}$) and symmetric ($\upsilon_s$ = 1378 cm$^{-1}$) stretching modes of the carboxylate group (bottom panel of Figure 4). These results indicate that the carboxyl group of CA is covalently bound to the Zn ions on the surfaces of ZnO nanoparticles.

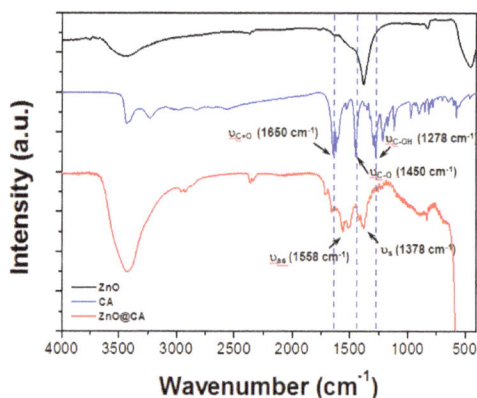

**Figure 4.** FT-IR spectra of free ZnO, CA, and ZnO@CA nanoparticles. The carboxylrate group of CA is bounded to the surface of the ZnO nanoparticle symmetrically through its two oxygen atoms.

ZnO@CA nanoparticles and free CA were evaluated for their ability to scavenge ABTS radicals. ZnO@CA nanoparticles were estimated to have on average four CA molecules per particle and exhibited favorable dispersion and stability in water. The abilities of ZnO@CA nanoparticles and free CA to scavenge ABTS radicals are shown in Table 1. Free CA scavenged ABTS radicals in a concentration-dependent manner, exhibiting its highest activity (93.25%) at the highest concentration tested (100 μM). The *o*-dihydroxybenzene moiety of CA exhibited high antioxidant activity because it is converted to stable *o*-quinone derivatives by reaction with ABTS radicals. At concentrations between 20 and 100 μM, ZnO@CA nanoparticles robustly scavenged ABTS radicals, with activities ranging from 44.99% to 73.68%, respectively. The reduced antioxidant activity of ZnO@CA nanoparticles compared to that of free CA may be attributed to steric repulsive forces between nanometer-sized ZnO and ABTS radicals. From these results, it was confirmed that the CA molecules were responsible for the antioxidant activity of the ZnO@CA nanoparticles. CA, which is a plant secondary metabolite, is well known as a natural antibiotic [27]. In addition, CA molecules are known to be biocompatible with a high affinity to cells. Therefore, the ZnO@CA nanoparticles will generate a significant synergy for the antimicrobial activity. In future, this functional coating technology could be an alternative for preventing microbism in clinical applications.

**Table 1.** ABTS radical scavenging activity.

| Sample | Concentration (μM) | % Inhibition |
| --- | --- | --- |
| CA:ZnO | 20 | 44.99 ± 0.48 |
| CA:ZnO | 40 | 73.68 ± 2.51 [a] |
| CA:ZnO | 100 | 51.88 ± 3.56 [a] |
| Caffeic acid | 20 | 47.98 ± 0.72 |
| Caffeic acid | 40 | 78.62 ± 0.73 [a] |
| Caffeic acid | 100 | 93.25 ± 0.43 [a,b] |

Results are shown as the means ± SD of three independent experiments. Significant differences between groups were determined using a one-way ANOVA with post-hoc Tukey's test, and $p < 0.05$ was considered significant. [a] Significant compared to activity at the lowest concentration of the same sample; [b] Significant difference between the two different samples at the same concentration.

For a quantitative antibacterial test of ZnO@CA nanoparticles, we used a static culture method after mixing the bacterial solution and the nanoparticles due to the properties, which tend to sink to the bottom depending on their weight. With this method, we assessed the antibacterial activity of the nanoparticles without any target molecules on the surface of them.

The antibacterial activities of ZnO@CA nanoparticles were evaluated by measuring the total number of viable bacterial cells against five bacterial strains, including both Gram-negative and Gram-positive strains and three MRSA strains. As shown in Figure 5, ZnO@CA nanoparticles showed more effective antibacterial activity than CA or ZnO nanoparticles. CA had no antibacterial activity, whereas ZnO nanoparticles showed antibacterial activity against Gram-negative and Gram-positive bacteria at high concentrations (over 100 μg/mL). However, the ZnO@CA nanoparticles exhibited strong bacterial killing activities against both Gram-negative and Gram-positive bacterial cells. In particular, Gram-positive bacteria such as *S. aureus* and MRSA, including two clinical MRSA isolates (MRSA-2 and MRSA-3), were completely inhibited by the ZnO@CA nanoparticles, which showed perfect killing efficiency. The antibacterial efficiencies of the ZnO@CA nanoparticles were dependent on the cell type, as they inhibited the growth of MRSA (including clinically isolated strains) more efficiently than that of Gram-negative bacteria (*E. coli*). The ZnO@CA nanoparticles showed greater antibacterial activity against Gram-positive strains than against Gram-negative bacteria with high selectivity. The highly selective antibacterial activity against Gram-positive bacteria, particularly against MRSA, was also confirmed by confocal fluorescence microscopy images (Figure 6). The images show live and dead bacterial cells stained with SYTO-9 (green) and propidium iodide (PI, red) fluorescent dyes, respectively. *S. aureus* and clinical isolates of MRSA were completely killed by the ZnO@CA nanoparticles, and many dead MRSA cells were observed after only 2 h of incubation. However, some *E. coli* cells survived after incubation with ZnO@CA nanoparticles. The fluorescent images of live and dead cells show the selectivity of the ZnO@CA nanoparticles for the Gram-positive bacterial cells of *S. aureus* and MRSA strains.

Overall, the results described above suggest that the ZnO@CA nanoparticles not only exhibit potent antioxidant activity but also completely inhibit the growth of Gram-positive bacteria, particularly MRSA, with high selectivity, although, the ZnO@CA nanoparitcles need to confirm their biocompatibility in vitro and in vivo and the functionality also have to be tested on the pathogenic bacterium in vivo in the future.

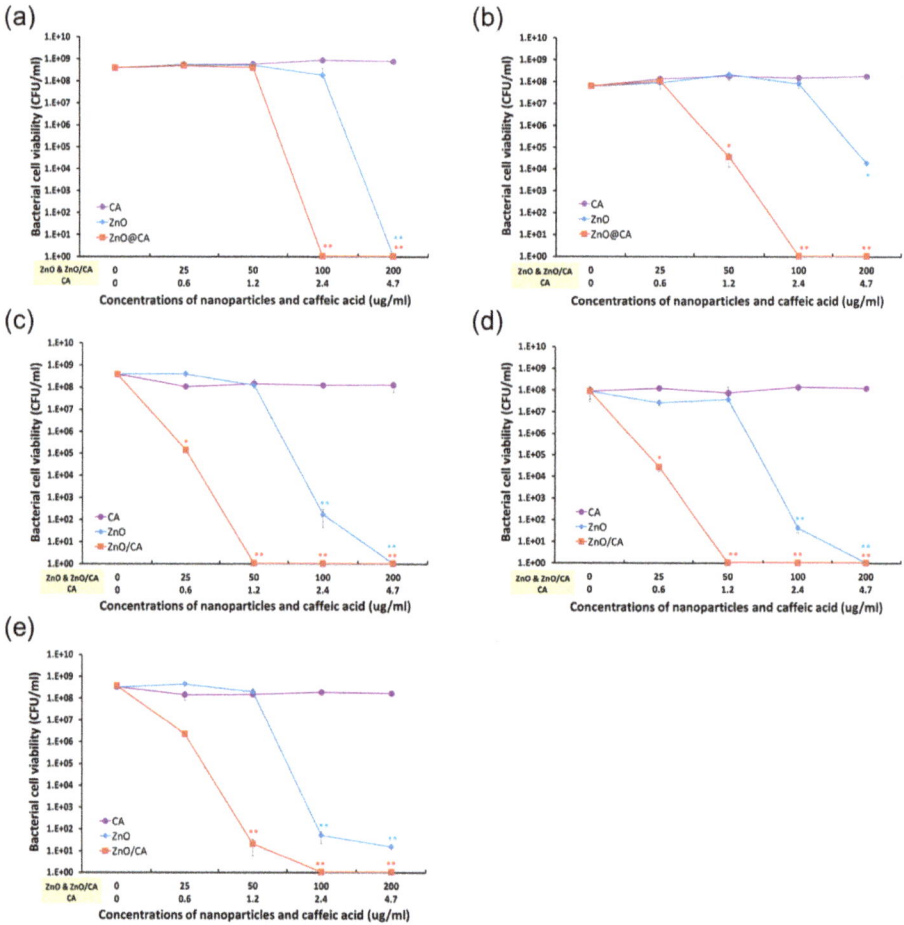

**Figure 5.** Antibacterial activities of ZnO@CA nanoparticles against Gram-negative and Gram-positive bacteria, including clinical isolates of methicillin-resistant *Staphylococcus aureus* (MRSA) strains. Antibacterial activities of the ZnO@CA nanoparticles against (**a**) *Escherichia coli*; (**b**) *Staphylococcus aureus*; (**c**) MRSA-1; (**d**) MRSA-2 (clinically isolated strain); and (**e**) MRSA-3 (clinically isolated strain). The killing efficiencies of the ZnO@CA nanoparticles against five bacterial strains, including clinical isolates of MRSA strains, are shown. The strains were incubated with the various concentrations of ZnO@CA nanoparticles and CA for 24 h. Data are expressed as means ± standard deviation (*n* = 6), as determined by the Student's *t*-test. $p < 0.05$ was considered to indicate statistical significance (\* $p < 0.05$, \*\* $p < 0.005$ vs. control).

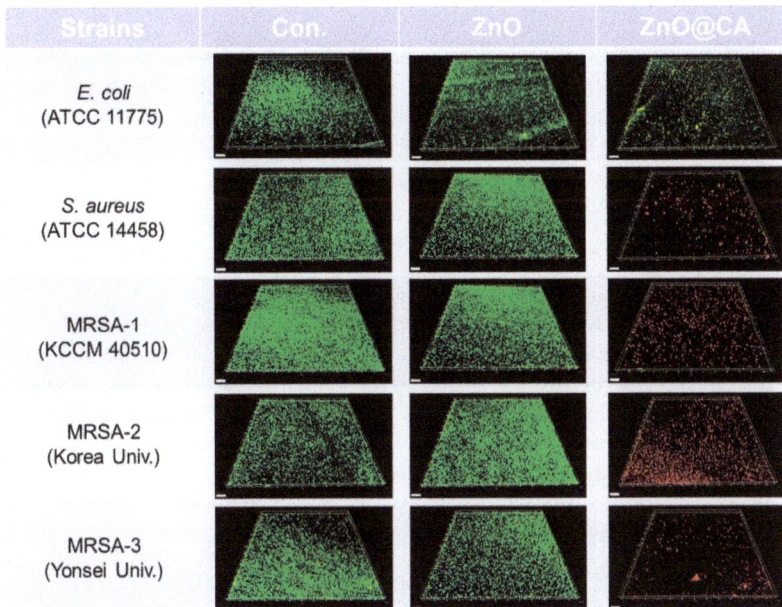

**Figure 6.** Live and dead cell images of bacteria (*E. coli*, *S. aureus*, MRSA-1, MRSA-2, and MRSA-3) using a confocal fluorescence microscope after incubation with ZnO and ZnO@CA nanoparticles. The bacterial cells of the five strains on the cover glass were incubated at 35 °C with ZnO and ZnO@CA nanoparticles for 24 h. The images show live and dead bacterial cells stained with SYTO-9 (green) and propidium iodide (PI, red) fluorescent dyes, respectively. All samples were tested in duplicate for each experiment, and each experiment was repeated three times (*n* = 6). There were no significant differences on live and dead cell imaging in each sample. Scale bars represent 50 µm.

## 3. Materials and Methods

### 3.1. Preparation of ZnO@CA Nanoparticles

ZnO nanoparticles with a diameter of 20 nm were prepared using previously described methods [28]. In a typical preparation, 0.03 M $Zn(NO_3)_2 \cdot 6H_2O$ was dissolved in 110 mL ethanol and agitated ultrasonically for 15 min. After 15 min, 40 mL KOH ethanol solution was added under vigorous stirring (the molar ratio of $Zn(NO_3)_2$ to KOH was fixed at 1:2). The solution was maintained at room temperature (RT) for 1 h until the solution temperature dropped. After reflux reaction at 80 °C for 10 h and returning the solution temperature to RT, the resulting amber-colored precipitate was separated by centrifugation, washed with deionized water and absolute alcohol several times, and then dried in a vacuum oven at 60 °C for 6 h.

In order to conjugate CA molecules, the surface of ZnO nanoparticles were treated with micro-dielectric barrier discharge (DBD) plasma for 30 min under an electrical discharge power of approximately <3 W (0.7 kV, 5 mA, and phase angle of ~1 radian), as previously reported [29]. Antioxidant functionality of the ZnO nanoparticles was provided by a wet chemical process with CA as follows. ZnO nanoparticles prepared in ethanol (EtOH) (20 mg/mL) were mixed with a $2.08 \times 10^{-5}$ M solution of CA/EtOH and agitated at RT for 24 h. After 24 h, the product was washed with EtOH several times. After the final wash, residual EtOH was further removed, and the product was dried at 60 °C for 12 h.

## 3.2. Physical Characterization of ZnO@CA Nanoparticles

The morphology and size of ZnO@CA nanoparticles were determined by field emission scanning electron microscopy (FESEM, SU–70, Hitachi, Tokyo, Japan) and transmission electron microscopy (TEM, JEM-2100F, JEOL, Tokyo, Japan). Crystallographic properties of ZnO@CA nanoparticles were investigated with an X-ray diffractometer (XRD, X' Pert Pro MPD, PANalytical, Almelo, The Netherlands) operated at 40 kV and 150 mA in a $2\theta$ range of 20–80°. IR spectra were obtained using a PerkinElmer Spectrum 100 FT-IR spectrometer.

## 3.3. Evaluation of Antioxidant Activity of ZnO@CA Nanoparticles

ABTS assays were performed using the following methods described by Arnao et al., but with modifications [30]. The ABTS radical cation (ABTS•+) was produced by mixing solutions of 7 mM ABTS and 2.4 mM potassium persulfate at equal ratios and incubating the solution at RT in the dark for 24 h. The solution was then diluted with methanol, and the absorbance was measured with a spectrophotometer until an absorbance of 0.7–1 units at a wavelength of 734 nm was achieved. Different concentrations of ZnO@CA nanoparticles were added to diluted ABTS•+ solutions for 30 min at 37 °C in the dark, after which the absorbance of the solution at a wavelength of 734 nm was measured using a spectrophotometer. The ability of ZnO nanoparticles to scavenge ABTS•+ radicals was calculated using the equation

$$\% \text{ inhibition} = [1 - (\text{Absorbance of sample} / \text{Absorbance of control})] \times 100 \qquad (2)$$

## 3.4. Assessment of Antibacterial Activity of ZnO@CA Nanoparticles against Bacterial Cells

To confirm the antibacterial activity of the ZnO@CA nanoparticles, we used five strains of bacteria including three strains of MRSA, as follows. Two strains each of the Gram-negative and Gram-positive bacteria *E. coli* ATCC 11775 and *S. aureus* ATCC 14458 were purchased from the American Type Culture Collection (ATCC, Rockville, MD, USA). MRSA-1 (KCCM 40510) was obtained from the Korean Culture Center of Microorganisms (KCCM, Sedaemun-Gu, Seoul, Korea). Two clinically isolated MRSA strains were acquired: MRSA-2, isolated from patients at the Korea University Anam Hospital, and MRSA-3, isolated at the Yonsei Medical Center in Seoul, Korea, were both kindly donated by the respective hospitals [31,32].

*E. coli* (ATCC 11775) and *S. aureus* (ATCC 14458) were grown on plate count agar (PCA, Becton, Dickinson and Company, Sparks, MD, USA), and the three MRSA strains were grown on Brain Heart Infusion agar (BHIA, Becton, Dickinson and Company, Sparks, MD, USA) at 35 °C for 24 h. All bacterial strains were passaged twice at 48 h intervals before use.

For assessing the antibacterial activity of ZnO@CA nanoparticles, we used two methods: a quantitative method and a qualitative method. For the quantitative method, bacterial cells from each bacterial colony were suspended at ~$10^{6-7}$ colony forming units (CFU)/mL in nutrient broth for *E. coli* and *S. aureus* and in BHI broth for MRSA strains. Bacterial cells in each broth solution were diluted 10-fold to ~$10^5$–$10^6$ CFU/mL and incubated with various concentrations of CA, ZnO, and ZnO@CA nanoparticles in 24-well plates at 35 °C for 24 h in static condition. After incubation with the samples, the bacterial cells in each well were inoculated onto BHI agar after serial 10-fold dilutions (10 to $10^7$) and incubated for 24 h. Bacterial cell viability was determined by counting the CFU, and the antibacterial activity of the ZnO@CA nanoparticles was determined by plotting the total number of viable bacterial cells as CFU/mL vs. the concentrations of the samples. Antibacterial activity of the ZnO@CA nanoparticles was defined as a >3 log decrease in CFU/mL.

The antibacterial activity of the ZnO@CA nanoparticles was confirmed with a confocal fluorescence microscope using live and dead bacterial cell images, after staining with two kinds of fluorescent dyes, green for live cells and red for dead cells. Each bacterial strain was suspended in normal saline solution. Bacterial cells at ~$10^6$–$10^7$ CFU/mL were inoculated onto sterilized cover glasses coated with poly-L-lysine in 24-well plates and incubated for 1 h to allow cells to attach to the cover glasses. Suspended bacterial cells were discarded after incubation, and the cover glasses in each well were gently rinsed three times with sterilized 0.9% saline solution to remove unattached bacterial cells. Bacterial cells on the cover glasses were incubated with ZnO and ZnO@CA nanoparticles for 24 h. After incubation, live and dead bacterial cells on the cover glass were stained with LIVE/DEAD BacLight Bacterial Viability Kits (Molecular Probes, Eugene, OR, USA) according to the manufacturer's instructions. Live and dead bacterial cells were analyzed with a laser scanning confocal fluorescence microscope (FV-1200, Olympus, Tokyo, Japan) with 20× objective lenses and fluorescence optics (excitation at 485 nm for SYTO 9 and PI and emission at 530 nm for SYTO 9 and 630 nm for PI). Confocal fluorescence images of live and dead bacterial cells were analyzed using imaging software (Imaris, Bitplane, Concord, MA, USA).

## 3.5. Statistical Analysis

All samples were tested in duplicate for each experiment, and each experiment was repeated three times ($n = 6$). Quantitative data are expressed as the means ± standard deviation (SD). Statistical comparisons were performed with a Student's $t$-test. A value of $p < 0.05$ was considered statistically significant.

## 4. Conclusions

In summary, we have for the first time, successfully fabricated ZnO nanoparticles with antioxidant and antibacterial activities. The ZnO nanoparticles were fabricated by covalently conjugating CA to ZnO nanoparticles using by a simple surface modification method. ZnO@CA nanoparticles with an average diameter of 20 nm were synthesized and characterized by FESEM, TEM, FT-IR, and XRD. Importantly, their antioxidant and antibacterial activities are robust and well suited for applications in biological technologies. In the future, nanoparticles of various sizes could be functionalized with widely used natural antioxidants and with specific targeting ligands to produce multifunctional nanomaterials for use in bioimaging and as antibacterial and anticancer therapeutic agents in the pharmaceutical industry.

**Acknowledgments:** This study was supported by Chosun University and the National Research Foundation of Korea (grant numbers NRF-2015M3A9E2066855 and NRF-2015M3A9E2066856) grant funded by the Korean government (MSIP). Professor Jin-Seung Jung's work has been financially supported by grants from the National Research Foundation of Korea (NRF-2014R1A1A2054928, 2014H1C1A10669352014).

**Author Contributions:** K.-H.C., K.C.N., and S.-Y.L. designed the study and wrote the manuscript. These authors contributed equally to this work. G. Cho performed the experiments and analyzed the data. J.-S.J., H.-J.K., and B.J.P. gave many suggestions during the project. All authors reviewed the manuscript.

**Conflicts of Interest:** The authors declare that there are no conflicts of interest regarding the publication of this paper.

## References

1. Skerget, M.; Kotnik, P.; Hadolin, M.; Hras, A.R.; Simonic, M.; Knez, Z. Phenols, proanthocyanidins, flavones and flavonols in some plant materials and their antioxidant activities. *Food Chem.* **2005**, *89*, 191–198. [CrossRef]
2. Gupta, V.K.; Sharma, S.K. Plants as natural antioxidants. *Nat. Prod. Radiance* **2006**, *5*, 326–334.
3. Ito, N.; Hirose, M.; Fukushima, S.; Tsuda, H.; Shirai, T.; Tatematsu, M. Studies on antioxidants: Their carcinogenic and modifying effects on chemical carcinogenesis. *Food Chem. Toxicol.* **1986**, *24*, 1071–1082. [CrossRef]
4. Mariod, A.A.; Ibrahim, R.M.; Ismail, M.; Ismail, N. Antioxidant activities of phenolic rich fractions (PRFs) obtained from black mahlab (*Monechma ciliatum*) and white mahlab (*Prunus mahaleb*) seedcakes. *Food Chem.* **2010**, *118*, 120–127. [CrossRef]
5. Erkan, N.; Ayranci, G.; Ayranci, E. Antioxidant activities of rosemary (*Rosmarinus officinalis* L.) extract, blackseed (*Nigella sativa* L.) essential oil, carnosic acid, rosmarinic acid and sesamol. *Food Chem.* **2008**, *110*, 76–82. [CrossRef]

6. Prasad, N.R.; Karthikeyan, A.; Karthikeyan, S.; Reddy, B.V. Inhibitory effect of caffeic acid on cancer cell proliferation by oxidative mechanism in human HT-1080 fibrosarcoma cell line. *Mol. Cell. Biochem.* **2011**, *349*, 11–19. [CrossRef]
7. Hirose, M.; Takesada, Y.; Tanaka, H.; Tamano, S.; Kato, T.; Shirai, T. Carcinogenicity of antioxidants BHA, caffeic acid, sesamol, 4-methoxyphenol and catechol at low doses, either alone or in combination, and modulation of their effects in a rat medium-term multi-organ carcinogenesis model. *Carcinogenesis* **1998**, *19*, 207–212. [CrossRef]
8. Chen, J.H.; Ho, C.T. Antioxidant activities of caffeic acid and its related hydroxycinnamic acid compounds. *J. Agric. Food Chem.* **1997**, *45*, 2374–2378. [CrossRef]
9. Scoponi, M.; Cimmino, S.; Kaci, M. Photo-stabilisation mechanism under natural weathering and accelerated photo-oxidative conditions of LDPE films for agricultural applications. *Polymer* **2000**, *41*, 7969–7980.
10. Giannakopoulos, E.; Christoforidis, K.C.; Tsipis, A.; Jerzykiewicz, M.; Deligiannakis, Y. Influence of Pb(II) on the radical properties of humic substances and model compounds. *J. Phys. Chem. A* **2005**, *109*, 2223–2232. [CrossRef]
11. Deligiannakis, Y.; Sotiriou, G.A.; Pratsinis, S.E. Antioxidant and antiradical SiO$_2$ nanoparticles covalently functionalized with gallic acid. *ACS Appl. Mater. Interfaces* **2012**, *4*, 6609–6617. [CrossRef]
12. Zhao, F.; Yao, D.; Guo, R.; Deng, L.; Dong, A.; Zhang, J. Composites of polymer hydrogels and nanoparticulate systems for bomedical and pharmaceutical applications. *Nanomaterials* **2015**, *5*, 2054–2130. [CrossRef]
13. Pardo-Yissar, V.; Katz, E.; Wasserman, J.; Willner, I. Acetylcholine esterase-labeled CdS nanoparticles on electrodes: Photoelectrochemical sensing of the enzyme inhibitors. *J. Am. Chem. Soc.* **2003**, *125*, 622–623. [CrossRef]
14. Park, S.J.; Taton, T.A.; Mirkin, C.A. Array-based electrical detection of DNA with nanoparticle probes. *Science* **2002**, *295*, 1503–1506.
15. Lin, J.; Raji, A.R.; Nan, K.; Peng, Z.; Yan, Z.; Samuel, E.L.; Natelson, D.; Tour, J.M. Iron oxide nanoparticle and graphene nanoribbon composite as an anode material for high-performance Li-ion batteries. *Adv. Func. Mater.* **2014**, *24*, 2044–2048. [CrossRef]
16. Zhou, Y.; Fang, X.; Gong, Y.; Xiao, A.; Xie, Y.; Liu, L.; Cao, Y. The interactions between ZnO nanoparticles (NPs) and α-linolenic acid (LNA) complexed to BSA did not influence the toxicity of ZnO NPs on HepG2 cells. *Nanomaterials* **2017**, *7*, 91. [CrossRef]
17. Chen, L.; Xu, J.; Holmes, J.D.; Morris, M.A. A facile route to ZnO nanoparticle superlattices: Synthesis, functionalization, and self-assembly. *J. Phys. Chem. C* **2010**, *114*, 2003–2011. [CrossRef]
18. Li, Z.; Yang, R.; Yu, M.; Bai, F.; Li, C.; Wang, Z.L. Cellular level biocompatibility and biosafety of ZnO nanowires. *J. Phys. Chem. C* **2008**, *112*, 20114–20117. [CrossRef]
19. Zheng, Y.; Li, R.; Wang, Y. In vitro and in vivo biocompatibility studies of ZnO nanoparticles. *Int. J. Mod. Phys. B* **2009**, *23*, 1566–1571. [CrossRef]
20. Wu, Y.L.; Lim, C.S.; Fu, S.; Tok, A.I.; Lau, H.M.; Boey, F.Y.; Zeng, X.T. Surface modifications of ZnO quantum dots for bio-imaging. *Nanotechnology* **2007**, *18*, 215604. [CrossRef]
21. Chakraborti, S.; Joshi, P.; Chakravarty, D.; Shanker, V.; Ansari, Z.A.; Singh, S.P.; Chakrabarti, P. Interaction of polyethyleneimine-functionalized ZnO nanoparticles with bovine serum albumin. *Langmuir* **2012**, *28*, 11142–11152. [CrossRef]
22. Xiong, H.M. ZnO nanoparticles applied to bioimaging and drug delivery. *Adv. Mater.* **2013**, *25*, 5329–5335. [CrossRef]
23. Nagarajan, S.; Kuppusamy, K.A. Extracellular synthesis of zinc oxide nanoparticle using seaweeds of gulf of Mannar, India. *J. Nanobiotechnol.* **2013**, *11*, 39. [CrossRef]
24. Zhang, R.; Kerr, L.L. A simple method for systematically controlling ZnO crystal size and growth orientation. *J. Solid State Chem.* **2007**, *180*, 988–994. [CrossRef]
25. Choi, K.H.; Wang, K.K.; Shin, E.P.; Oh, S.L.; Jung, J.S.; Kim, H.K.; Kim, Y.R. Water-soluble magnetic nanoparticles functionalized with photosensitizer for photocatalytic application. *J Phys. Chem. C* **2011**, *115*, 3212–3219. [CrossRef]
26. Park, B.J.; Choi, K.H.; Nam, K.C.; Min, J.E.; Lee, K.D.; Uhm, H.S.; Choi, E.H.; Kim, H.J.; Jung, J.S. Photodynamic anticancer activity of CoFe$_2$O$_4$ nanoparticles conjugated with hematoporphyrin. *J. Nanosci. Nanotechnol.* **2015**, *15*, 7900–7906. [CrossRef]

27. Maddox, C.E.; Laur, L.M.; Tian, L. Antibacterial activity of phenolic compounds against the phytopathogen *Xylella fastidiosa*. *Curr. Mocrobiol.* **2010**, *60*, 53–58. [CrossRef]

28. Shao, R.; Sun, L.; Tang, L.; Chen, Z. Preparation and characterization of magnetic core–shell $ZnFe_2O_4$@ZnO nanoparticles and their application for the photodegradation of methylene blue. *Chem. Eng. J.* **2013**, *217*, 185–191. [CrossRef]

29. Park, B.J.; Choi, K.H.; Nam, K.C.; Ali, A.; Min, J.E.; Son, H.; Uhm, H.S.; Kim, H.J.; Jung, J.S.; Choi, E.H. Photodynamic anticancer activities of multifunctional cobalt ferrite nanoparticles in various cancer cells. *J. Biomed. Nanotechnol.* **2015**, *11*, 226–235. [CrossRef]

30. Arnao, M.B.; Cano, A.; Acosta, M. The hydrophilic and lipophilic contribution to total antioxidant activity. *Food Chem.* **2001**, *73*, 239–244. [CrossRef]

31. Lee, K.Y.; Park, B.J.; Lee, D.H.; Lee, I.S.; Hyun, S.O.; Chung, K.H.; Park, J.C. Sterilization of *Escherichia coli* and MRSA using microwave-induced argon plasma at atmospheric pressure. *Surf. Coat. Technol.* **2005**, *193*, 35–38. [CrossRef]

32. Choi, K.H.; Lee, H.J.; Park, B.; Wang, K.K.; Shin, E.; Park, J.C.; Kim, Y.; Oh, M.K.; Kim, Y.R. Photosensitizer and vancomycin-conjugated novel multifunctional magnetic particles as photoinactivation agents for selective killing of pathogenic bacteria. *Chem. Commun.* **2012**, *48*, 4591–4593. [CrossRef]

© 2017 by the authors. Licensee MDPI, Basel, Switzerland. This article is an open access article distributed under the terms and conditions of the Creative Commons Attribution (CC BY) license (http://creativecommons.org/licenses/by/4.0/).

*nanomaterials*

MDPI

*Article*

# Functionalized ZnO Nanoparticles with Gallic Acid for Antioxidant and Antibacterial Activity against Methicillin-Resistant *S. aureus*

Joo Min Lee [1], Kyong-Hoon Choi [2], Jeeeun Min [3], Ho-Joong Kim [4,*], Jun-Pil Jee [5,*] and Bong Joo Park [2,6,*]

[1]   Department of Food and Nutrition, Chosun University, Gwangju 61452, Korea; joominlee@chosun.ac.kr
[2]   Institute of Biomaterials, Kwangwoon University, 20 Kwangwoongil, Nowon-gu, Seoul 01897, Korea; solidchem@hanmail.net
[3]   Halla Energy and Environment, Seoul 05769, Korea; iseeiget@naver.com
[4]   Department of Chemistry, Chosun University, Gwangju 61452, Korea
[5]   College of Pharmacy, Chosun University, Gwangju 61452, Korea
[6]   Department of Electrical & Biological Physics, Kwangwoon University, 20 Kwangwoongil, Nowon-gu, Seoul 01897, Korea
*   Correspondence: hjkim@chosun.ac.kr (H.-J.K.); Jee@chosun.ac.kr (J.-P.J.); parkbj@kw.ac.kr (B.J.P.); Tel.: +82-62-230-6643 (H.-J.K.); +82-2-940-8629 (B.J.P.)

Received: 30 September 2017; Accepted: 31 October 2017; Published: 2 November 2017

**Abstract:** In this study, we report a new multifunctional nanoparticle with antioxidative and antibacterial activities in vitro. ZnO@GA nanoparticles were fabricated by coordinated covalent bonding of the antioxidant gallic acid (GA) on the surface of ZnO nanoparticles. This addition imparts both antioxidant activity and high affinity for the bacterial cell membrane. Antioxidative activities at various concentrations were evaluated using a 2,2′-azino-bis(ethylbenzthiazoline-6-sulfonic acid) (ABTS) radical scavenging method. Antibacterial activities were evaluated against Gram-positive bacteria (*Staphylococcus aureus: S. aureus*), including several strains of methicillin-resistant *S. aureus* (MRSA), and Gram-negative bacteria (*Escherichia coli*). The functionalized ZnO@GA nanoparticles showed good antioxidative activity (69.71%), and the bactericidal activity of these nanoparticles was also increased compared to that of non-functionalized ZnO nanoparticles, with particularly effective inhibition and high selectivity for MRSA strains. The results indicate that multifunctional ZnO nanoparticles conjugated to GA molecules via a simple surface modification process displaying both antioxidant and antibacterial activity, suggesting a possibility to use it as an antibacterial agent for removing MRSA.

**Keywords:** ZnO@GA; antioxidative activity; antibacterial activity; gallic acid; antibiotic resistance; methicillin-resistant *Staphylococcus aureus*

## 1. Introduction

Infectious diseases caused by bacteria are a significant burden on public health and threaten the economic stability of societies worldwide [1]. In particular, the widespread incorrect use of conventional antibiotics has led to the adaptation of microorganisms to these therapies, and the appearance of antibiotic-resistant bacteria is a serious problem [2,3]. Currently, MRSA is the most commonly found antibiotic-resistant bacteria in many parts of the world [4]. Over the last decade, MRSA strains have become one of the main causes of mortality among hospital-acquired infectious diseases [5]. However, the development of novel antibiotics to solve this problem had limited progress. Therefore, the development of alternative methods to treat infections caused by antibiotic-resistant bacteria is an urgent challenge in medical biotechnology.

Material science researchers have focused on designing multifunctional nanoparticles, which combine various functionalities such as fluorescence, magnetism, and photoactivation to generate reactive oxygen species (ROS), and biotargetability [6–9]. In this respect, metal and metal oxide nanoparticles have been extensively researched for their photocatalytic, optoelectronic, bioengineering, magnetic, antimicrobial, wound-healing, and anti-inflammatory properties [10–13] arising from their unique physical and chemical characteristics, different from those of the bulk phase [14,15]. Among metal oxide nanoparticles, zinc oxide has extensive applications in various fields such as optics, piezoelectricity, bioimaging, and biosensing [16–18]. In particular, ZnO presents sufficient antimicrobial efficacy when the particle size is decreased to the nanometer range, as ZnO nanoparticles can interact with the bacterial core or surface as soon as they contact the cell and then exhibit distinct antibacterial mechanisms [19]. Therefore, ZnO nanoparticles are of great interest to biologists because of their excellent antibiotic properties and good biocompatibility. This distinct activity has opened new frontiers in the biological sciences. However, excessive ROS generated by ZnO nanoparticles may cause cell membrane disintegration, membrane protein damage, and genomic instability, which can initiate or enhance the development of many diseases [20,21]. Therefore, the development of novel multifunctional ZnO nanoparticles with antioxidant activity could minimize damage caused by excessive ROS.

In this study, we report a method for fabricating ZnO@GA nanoparticles by a simple surface modification process. The nanoparticles display strong antioxidant and antibiotic effects and were particularly effective against MRSA, suggesting that they can be a useful novel antibacterial agent for removing MRSA.

## 2. Results and Discussion

Antioxidant-functionalized ZnO@GA nanoparticles were synthesized via a simple coating of ZnO nanoparticles with gallic acid (GA). The ZnO nanoparticles were prepared by a sol-gel process. Then, the carboxyl group of GA was covalently bonded to surface $Zn^{2+}$ ions, forming multifunctional ZnO@GA nanoparticles.

Pure ZnO and ZnO@GA nanoparticles were analyzed using various analytical tools. Field emission scanning electron microscopy (FE-SEM) confirmed that the ZnO@GA nanoparticles were nearly spherical in shape, with good size uniformity (Figure 1a). High-magnification scanning electron microscopy (SEM) images of the ZnO@GA nanoparticles showed a variety of spherical shapes (inset of Figure 1a), indicating that the nanoparticles aggregated spontaneously. Figure 1b indicates the size distribution of pure ZnO nanoparticles, which was estimated by sampling 300 particles in the Transmission electron microscopy (TEM) image. The average size of the ZnO nanoparticles was $11.5 \pm 4.4$ nm. TEM confirmed that ZnO and ZnO@GA nanoparticles both had globular shapes and assembled in aggregates on the TEM grids (Figure 1c,d). It is notable that there was almost no change in particle size or morphology after the surface modification. However, comparing the images in Figure 1c,d, the ZnO@GA nanoparticles appeared blurrier than pure ZnO nanoparticles owing to the conjugated antioxidant molecules.

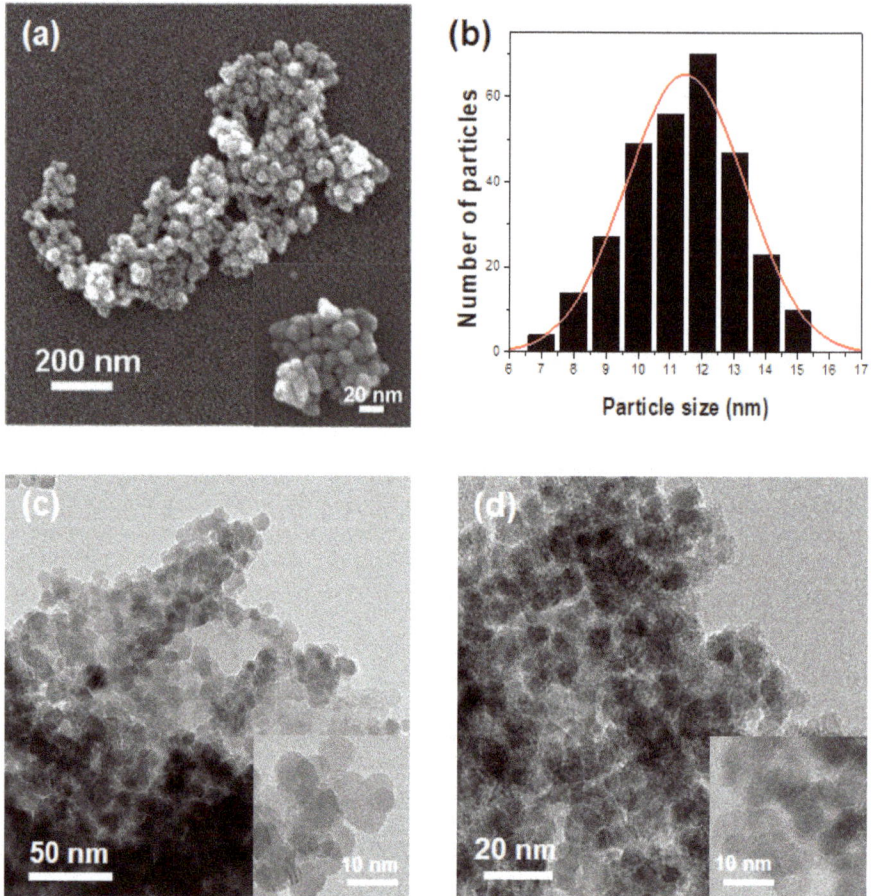

**Figure 1.** (a) Low-magnification FE-SEM image of the ZnO@gallic acid (GA) nanoparticles. A corresponding high-magnification FE-SEM image is shown in the inset; (b) Histogram of the ZnO@GA nanoparticle size distribution; TEM images of (c) ZnO nanoparticles and (d) ZnO@GA nanoparticles. Corresponding high-resolution TEM images are shown in the insets.

High-resolution TEM (HR-TEM) and X-ray diffraction (XRD) analysis were used to obtain a more detailed crystal structure of the ZnO@GA nanoparticles. The selected area electron diffraction (SAED) pattern is presented in Figure 2a. The interlayer distances of the ZnO@GA nanoparticles were calculated to be 0.282 and 0.259 nm, which are comparable to the (100) and (002) planes of hexagonal ZnO, respectively [22]. This result indicates that each of the ZnO@GA nanoparticles has a single crystalline nature; thus, the ZnO@GA nanoparticles displayed high crystallization. Figure 2b shows the powder XRD data of the ZnO@GA nanoparticles. The strong Bragg reflection peaks ($2\theta = 31.7°$, $34.4°$, $36.1°$, $47.6°$, $56.5°$, $62.8°$, $67.9°$ and $72.1°$), matched by their Miller indices ((100), (002), (101), (102), (110), (103), (112), and (004)), were obtained from a standard wurtzite ZnO structure (JCPDS Card No. 36-1451) [23]. Therefore, hexagonally structured ZnO was identified as a single crystalline phase in the ZnO@GA nanoparticles. The diffraction peak profile ($2\theta = 36.1°$) was fairly well fitted by a convolution of Lorentzian functions (inset of Figure 2b). The mean crystalline size of the ZnO@GA nanoparticles was 5.8 nm, calculated based on Scherrer's equation.

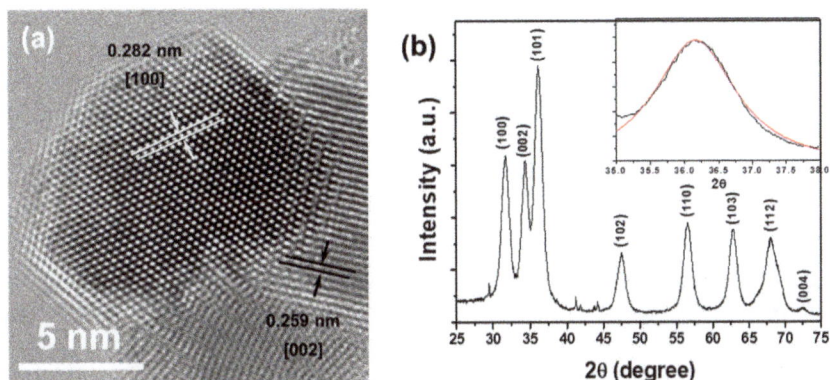

**Figure 2.** (a) High-resolution transmission electron micrograph of the ZnO@GA nanoparticles; (b) X-ray diffraction (XRD) pattern of ZnO@CA nanoparticles. a.u., arbitrary units.

To confirm the binding between the carboxyl group of the GA molecules and $Zn^{2+}$ cations on the surface of ZnO, Fourier-transform infrared (FT-IR) spectra of pure GA molecules and the ZnO@GA nanoparticles were compared (Figure 3a). The main peaks of the ZnO@GA nanoparticles and pure GA were very similar, resembling the characteristic peaks of GA. This indicates that GA molecules remain on the surface of ZnO nanoparticles even after washing with ethanol. In particular, both samples showed the presence of a carboxyl group (2700 to 3600 $cm^{-1}$), hydroxyl phenolic groups (3284, 3382 $cm^{-1}$), and an aromatic moiety (1541, 1618 $cm^{-1}$), as shown in Figure 3a [24]. However, the pure GA molecules had absorption peaks at 1613, 1427, and 1268 $cm^{-1}$, according to the stretching modes of the free carbonyl double bond ($\upsilon_{C=O}$), the C–O single bond ($\upsilon_{C-O}$), and the oxygen-hydrogen deformation ($\upsilon_{C-OH}$) (top panel of Figure 3a). This result indicates that pure GA molecules have protonated carboxyl groups (COOH), as expected. Conversely, the ZnO@GA nanoparticles displayed strong novel peaks at 1560 and 1376 $cm^{-1}$. These new bands can be attributed to the asymmetric ($\upsilon_{as}$ = 1560 $cm^{-1}$) and symmetric ($\upsilon_s$ = 1376 $cm^{-1}$) stretching modes of the carboxyl group, as shown in Figure 3a (bottom panel). These results indicate that the carboxyl group bound to the surface of the ZnO nanoparticle.

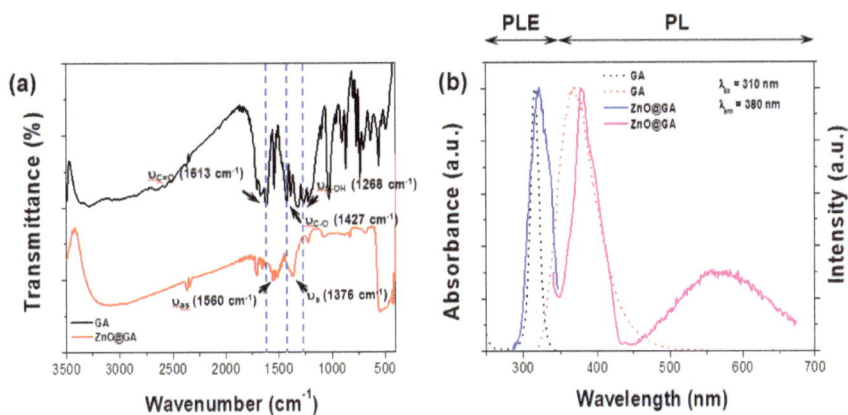

**Figure 3.** (a) Fourier-transform infrared spectroscopy spectra of pure GA molecules and ZnO@GA nanoparticles; (b) PL and PLE spectra of pure GA molecules and ZnO@GA nanoparticles in water. The excitation and detection wavelengths for both spectra were 310 and 380 nm. a.u., arbitrary units.

Figure 3b shows the photoluminescence and photoluminescence excitation (PL and PLE) spectra of pure GA molecules and ZnO@GA nanoparticles. The peak at 315 nm is an absorption band typical of pure GA molecules, which may be attributed to the aromatic ring. Also, pure GA molecules had one strong emission peak, located at 368 nm ($\lambda_{ex}$ = 310 nm). After surface modification, the PL and PLE spectra of the ZnO@GA nanoparticles exhibited characteristics very similar to those of pure GA molecules. However, the emission spectrum of the ZnO@GA nanoparticles also displayed emission at 450 to 650 nm. Broad, red-shifted emissions are typically observed with ZnO nanomaterials and are attributed to a recombination process through electronic states originating from oxygen vacancies or surface defects [25].

The antioxidant efficacies of pure GA molecules and ZnO@GA nanoparticles were evaluated using an ABTS radical scavenging method. The ZnO@GA nanoparticles contained an average of 2.89 GA molecules per particle and were suggested to have excellent diffusion and stability in water. The antioxidant activities of the ZnO@GA nanoparticles and pure GA are shown in Table 1. Pure GA molecules scavenged ABTS radicals proportionally to the concentration. The hydroxyl groups of GA are important for its free radical scavenging efficiency. In particular, the OH at the para-position to the carboxyl group appears to be essential for maintaining scavenging activity, as the scavenging activity is diminished by its methylation [26]. The ZnO@GA nanoparticles also robustly scavenged ABTS radicals. The decrease in the antioxidant activity of the ZnO@GA nanoparticles compared to that of pure GA molecules may be attributable to steric repulsion between the nanoparticles and ABTS radicals.

To confirm the antibacterial effects of GA, ZnO, and ZnO@GA nanoparticles, a static culture method was used after mixing the bacteria and nanoparticles, as previously described [27]. The antibacterial effect of each sample was tested in five bacterial strains (a Gram-negative strain, i.e., *E. coli*, and four Gram-positive strains, i.e., one *S. aureus* and three MRSA). The antibacterial activities of each sample were evaluated by counting the colony-forming units (CFUs) of each strain as a measure of the total number of viable bacteria (Figure 4). The ZnO@GA nanoparticles showed strong antibacterial activity, two to four fold higher than that of ZnO nanoparticles, and displayed higher antibacterial activity against *S. aureus* and MRSA than against *E. coli*. The ZnO@GA nanoparticles at 50 and 100 µg/mL completely inhibited MRSA-1 and MRSA-2 strains, and they more effectively killed the MRSA strains than the *S. aureus* strain at 50 µg/mL, suggesting that ZnO@GA nanoparticles are specifically effective against MRSA strains. The selective inhibition effects were also confirmed using confocal fluorescence microscopy, as shown in Figure 5. The confocal fluorescence microscopy showed that the ZnO@GA nanoparticles have strong killing effects against Gram-positive bacteria, with complete killing of the cells in the *S. aureus* and MRSA samples. The viability of Gram-negative *E. coli* decreased as well, although many cells remained viable. This result also confirmed that the ZnO@GA nanoparticles have selective inhibitory activity against Gram-positive bacteria and particularly against MRSA.

**Table 1.** ABTS radical scavenging activity.

| Sample | Concentration (µM) | % Inhibition |
|---|---|---|
| ZnO@GA | 20 | 33.29 ± 0.12 |
| ZnO@GA | 40 | 57.17 ± 0.96 [a] |
| ZnO@GA | 100 | 69.71 ± 5.26 [a] |
| Gallic acid | 20 | 43.38 ± 0.48 [b] |
| Gallic acid | 40 | 72.76 ± 0.12 [b] |
| Gallic acid | 100 | 93.25 ± 0.43 [b] |

Data are expressed as the mean ± standard deviation (SD) of independent experiments ($n = 3$). Statistical significance ($p < 0.05$) was analyzed using a one-way analysis of variance with a post-hoc Tukey's test. [a] Activity at the lowest concentration of the same sample. [b] Two different samples at the same concentration.

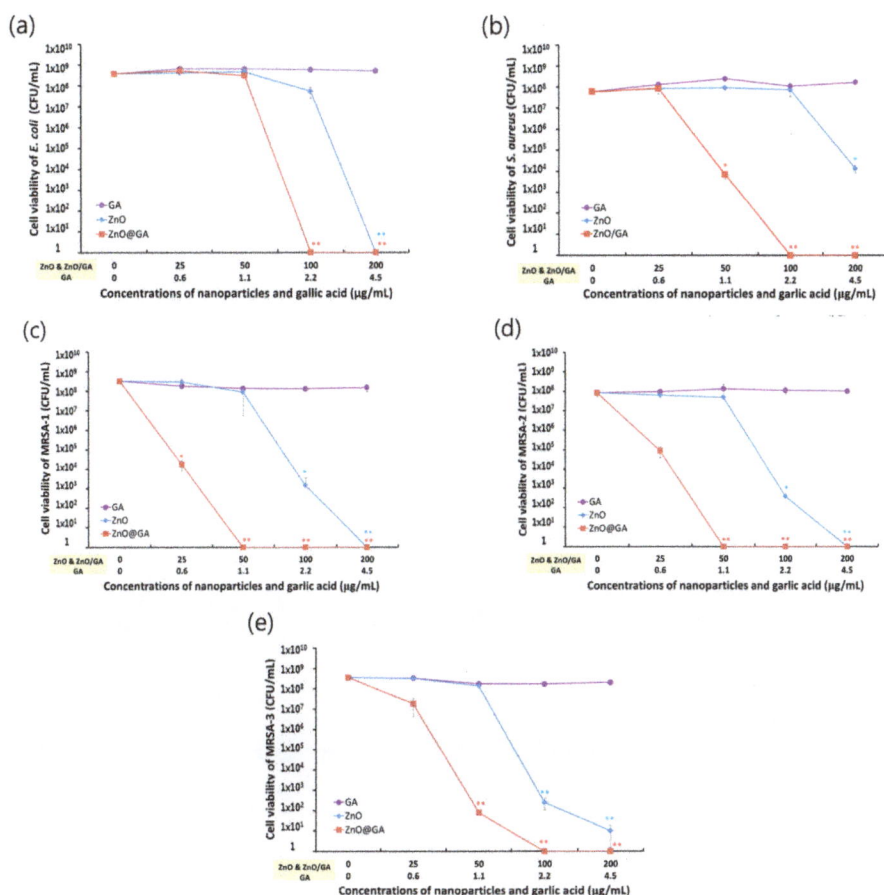

**Figure 4.** Antibacterial effects of ZnO@GA nanoparticles. (**a**) *E. coli*; (**b**) *S. aureus*; (**c**) MRSA-1; (**d**) MRSA-2; and (**e**) MRSA-3. Data are shown as the mean ± S.D ($n = 6$). Analysis of statistical significance (* $p < 0.05$, ** $p < 0.005$ versus control) was performed using Student's *t*-test.

As shown in Figure 4, although the bactericidal effects were relatively weak compared to those of the ZnO@GA nanoparticles, the ZnO nanoparticles also showed inhibitory effects against the five strains at 100 and 200 µg/mL. GA is known to have antibacterial activity, with a minimum inhibitory concentration of 8 mg/mL for *S. aureus* [28]. However, in this study, GA was used at 0.6 to 4.5 µg/mL. At these concentrations, GA alone did not display antibacterial activity against the bacterial strains (Figure 4).

Although GA treatment alone was not effective, the ZnO nanoparticles conjugated with low concentrations of GA, relatively lower concentrations than those in other published reports [28–31], had enhanced antibacterial properties compared to the ZnO nanoparticles, dramatically reducing the cell viability of Gram-positive bacteria, particularly MRSA. The strong antibacterial activity and selectivity may be attributed to the high affinity of GA for the bacterial cell membrane and the increased lipophilicity upon the addition of GA [31].

Overall, the results in this study suggest that ZnO nanoparticles functionalized with GA have antioxidant activity, as well as selective antibacterial activity, against MRSA. However, future studies will be required to evaluate their efficiency and bio-safety in vitro and in vivo.

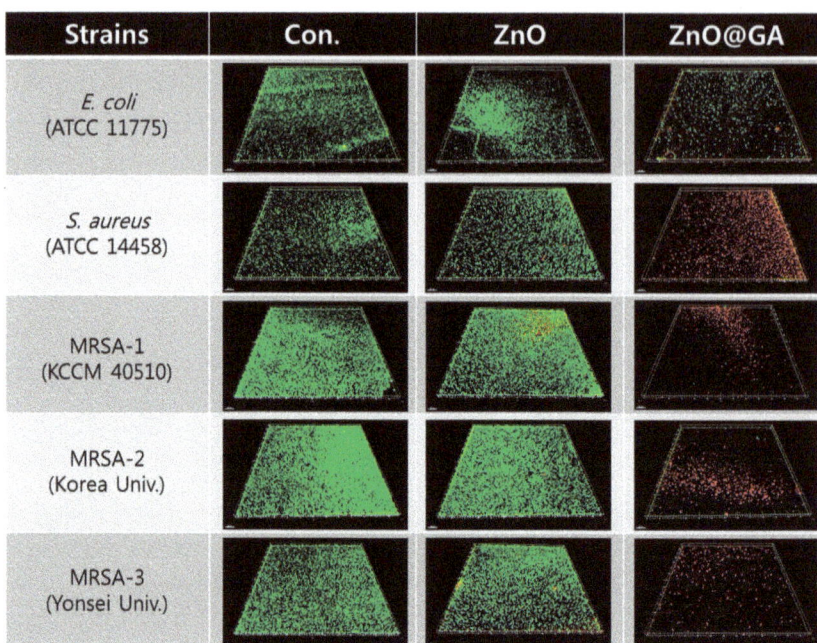

**Figure 5.** Qualitative assay of antibacterial activity using live and dead cell staining of Gram-negative and Gram-positive bacteria. Fluorescent images show live cells stained by SYTO-9 (green) and dead cells stained by PI (red) after 24 h of incubation. Scale bars represent 50 μm.

## 3. Materials and Methods

### 3.1. Preparation of ZnO@GA Nanoparticles

ZnO nanoparticles were prepared by a process similar to that in a previous report [23]. A stock solution of $Zn(CH_3COO)_2 \cdot 2H_2O$ (0.1 M) was dissolved in 50 mL of methanol under vigorous stirring. Then 25 mL of NaOH (0.2 M) in methanol was added to this mixture, and the pH was maintained at 8. This mixture was transferred to a Teflon-lined sealed stainless steel autoclave and maintained at 80 °C for 10 h. The resultant white solid products were washed with methanol several times, filtered, and then dried in a vacuum oven at 60 °C for 6 h.

A wet chemical process with GA was used to provide ZnO nanoparticles with antioxidant functionality, as follows. First, 20 mg of ZnO nanoparticles was diffused in EtOH (1 mL). This solution was added to a $2.08 \times 10^{-5}$ M solution of GA/EtOH, and vigorous stirring was applied for 24 h. The resulting product was washed several times in EtOH and dried at 60 °C.

### 3.2. Physical Characterization of ZnO@GA Nanoparticles

FE-SEM was performed on a SU-70 Analytical UltraHighResolution SEM (Hitachi, Tokyo, Japan). TEM SAED and high-resolution (HR) TEM were performed with a JEM-3100F TEM (JEOL, Tokyo, Japan) operating at 200 kV. TEM samples were prepared by placing a drop of sample suspension onto a standard carbon-coated copper grid. This grid was dried before the micrographs were recorded. The phase structures of the sample were identified by XRD using a X'Pert Pro MPD (PANalytical, Almelo, The Netherlands) with CuKα radiation (wavelength of the radiation, $k = 1.54$ Å). IR spectra were obtained using a FT-IR spectrometer (Perkin-Elmer PE 100, Waltham, MA, USA). For IR measurements, the samples were crushed on mortar and then prepared as pressed wafers (1% sample in KBr). The PL and PLE of the samples were measured on a F-4500 spectrofluorimeter (Hitachi, Tokyo, Japan), using a Xe arc lamp (150W, Abet Technologies, Milford, CT, USA) as the excitation source.

### 3.3. Evaluation of ZnO@GA Nanoparticle Antioxidant Activity

The radical cation decolorization assay determines the capacity of substances to scavenge ABTS [31]. The ABTS radical cation (ABTS $^+$) was prepared by mixing 2.45 mM potassium persulfate and 7 mM ABTS stock solution (1/1) and was placed in a dark place until the absorption peak was stabilized. After 20 h, this solution was diluted with MeOH, and the absorbance was maintained at 1 ($\lambda_{max}$ = 734 nm). For photo-detection, 0.9 mL of ABTS$\cdot^+$ solution was combined with 0.1 mL of different concentrations of GA and ZnO@GA nanoparticles, and mixed for 45 s. Measurements were taken at 734 nm after 15 min. The antioxidative activities of GA and ZnO@GA nanoparticles were estimated by detecting the decrease in absorbance at different concentrations using the following equation:

$$E = (A_c - A_s/A_c) \times 100 \tag{1}$$

where $A_s$ and $A_c$ are the respective absorbance of the samples and ABTS$\cdot^+$, expressed in μmol.

### 3.4. Evaluation of ZnO@GA Nanoparticle Antibacterial Activity

To evaluate the antibacterial effects of the ZnO@GA nanoparticles against a Gram-negative and four Gram-positive bacterial strains, we used previously described methods and bacterial strains [27]. Briefly, five bacterial strains were used: a Gram-negative bacteria, *E. coli* (ATCC 11775), and four Gram-positive bacterial strains, which were clinically isolated strains, namely, *S. aureus* (ATCC 14458), MRSA-1 (KCCM 40510), and MRSA-2 and MRSA-3, [32,33].

For the quantitative antibacterial test, *E. coli* and *S. aureus* were suspended at $10^6$ to $10^7$ CFU/mL in nutrient broth, and the MRSA strains were suspended in brain heart infusion (BHI) broth. After the 10-fold dilution of each bacterial suspension, the bacterial cells (approximately $10^5$ to $10^6$ CFU/mL) were inoculated in 24-well plates and incubated with various concentrations (0, 25, 50, 100, and 200 μg/mL) of GA, ZnO, and ZnO@GA samples at 35 °C. After 24 h of incubation, the bacterial cells in 24 well plates were diluted by a serial 10-fold dilution method and inoculated onto a plate count agar (PCA) plate for *E. coli* and *S. aureus* and a BHI agar plate for the MRSA strains. The PCA and BHI plates were incubated for a further 24 h, and the CFUs in each plate were counted. For the antibacterial activity of each sample, each bacterial cell viability was expressed as CFU/mL versus the concentrations of each sample, and an inhibition of >3 log in the cell viability of each strain was defined as a positive antibacterial effect.

For fluorescence microscopy to confirm the antibacterial effects of the samples, each bacterial cell type at $10^6$ to $10^7$ CFU/mL was plated onto a cover glass coated with poly-L-lysine and incubated for 1 h. Next, the cover glass was washed with 0.9% saline solution three times to remove the detached cells from the glass, before incubation with samples for 24 h. Finally, live and dead bacterial cells were stained with a LIVE/DEAD BacLight Bacterial Viability Kit (Molecular Probes, Eugene, OR, USA) according to the manufacturer's instructions. Each bacterial cell was analyzed using a confocal fluorescence microscope (FV-1200, Olympus, Tokyo, Japan) with a 20× objective lens, excitation filters at 485 nm for both SYTO 9 and propidium iodide (PI), and emission filters at 530 and 630 nm for SYTO 9 and PI, respectively. Each fluorescence image was analyzed using imaging software (Imaris, Bitplane, Concord, MA, USA).

### 3.5. Statistical Analysis

The experiments were repeated three times (*n* = 6), and the quantitative data are shown as the mean ± standard deviation (S.D.). The statistical significance ($p < 0.05$) was analyzed by Student's *t*-test.

## 4. Conclusions

In this study, we successfully fabricated multifunctional ZnO nanoparticles conjugated with GA via a simple surface modification process. These multifunctional ZnO@GA nanoparticles show high antioxidant and antibacterial activity, and the functionality and potentiality on antioxidant and antibacterial activity suggest that they can be useful as a novel antibacterial agent for MRSA.

**Acknowledgments:** This study was supported by research funds provided by Chosun University in 2014.

**Author Contributions:** J.L., K.-H.C. and J.M. designed the study and wrote the manuscript. These authors contributed equally to this work. H.-J.K., J.-P.J. and B.J.P. gave many suggestions during the project. All authors reviewed the manuscript.

**Conflicts of Interest:** The authors declare no conflicts of interest.

## References

1. Choi, K.-H.; Lee, H.-J.; Park, B.J.; Wang, K.-K.; Shin, E.P.; Park, J.-C.; Kim, Y.K.; Oh, M.-K.; Kim, Y.-R. Photosensitizer and vancomycin-conjugated novel multifunctional magnetic particles as photoinactivation agents for selective killing of pathogenic bacteria. *Chem. Commun.* **2012**, *48*, 4591–4593. [CrossRef] [PubMed]
2. Suay-García, B.; Pérez-Gracia, M.T. The antimicrobial therapy of the future: Combating resistances. *J. Infect. Dis. Ther.* **2014**, *2*, 1000146. [CrossRef]
3. Fonkwo, P.N. Pricing infectious disease; the economic and health implications of infectious diseases. *EMBO Rep.* **2008**, *9*, S13–S17. [CrossRef] [PubMed]
4. Grundmann, H.; Aires-de-Sousa, M.; Boyce, J.; Tiemersma, E. Emergence and resurgence of methicillin-resistant Staphylococcus aureus as a public-health threat. *Lancet* **2006**, *368*, 874–885. [CrossRef]
5. Ippolito, G.; Leone, S.; Lauria, F.N.; Nicastri, E.; Wenzel, R.P. Methicillin-resistant Staphylococcus aureus: The superbug. *Int. J. Infect. Dis.* **2010**, *14*, S7–S11. [CrossRef] [PubMed]
6. Kudr, J.; Haddad, Y.; Richtera, L.; Heger, Z.; Cernak, M.; Adam, V.; Zitka, O. Magnetic nanoparticles: From design and synthesis to real world applications. *Nanomaterials* **2017**, *7*, 243. [CrossRef] [PubMed]
7. Cherukula, K.; Manickavasagam Lekshmi, K.; Uthaman, S.; Cho, K.; Cho, C.-S.; Park, I.-K. Multifunctional inorganic nanoparticles: Recent progress in thermal therapy and imaging. *Nanomaterials* **2016**, *6*, 76. [CrossRef] [PubMed]
8. Lee, J.E.; Lee, N.; Kim, T.; Kim, J.; Hyeon, T. Multifunctional mesoporous silica nanocomposite nanoparticles for theranostic applications. *Acc. Chem. Res.* **2011**, *44*, 893. [CrossRef] [PubMed]
9. Kim, J.; Piao, Y.; Hyeon, T. Multifunctional nanostructured materials for multimodal imaging, and simultaneous imaging and therapy. *Chem. Soc. Rev.* **2009**, *38*, 372–390. [CrossRef] [PubMed]
10. Rosi, N.L.; Mirkin, C.A. Nanostructures in biodiagnostics. *Chem. Rev.* **2005**, *105*, 1547–1562. [CrossRef] [PubMed]
11. Taylor, P.L.; Ussher, A.L.; Burrell, R.E. Impact of heat on nanocrystalline silver dressings. Part I: Chemical and biological properties. *Biomaterials* **2005**, *26*, 7221–7229. [CrossRef] [PubMed]
12. Ingle, A.; Gade, A.; Pierrat, S.; Sönnichsen, C.; Rai, M. Mycosynthesis of silver nanoparticles using the fungus *Fusarium acuminatum* and its activity against some human pathogenic bacteria. *Curr. Nanosci.* **2008**, *4*, 141–144. [CrossRef]
13. Kolodziejczak-Radzimska, A.; Jesionowski, T. Zinc oxide-from synthesis to application: A review. *Materials* **2014**, *7*, 2833–2881. [CrossRef] [PubMed]
14. Sun, Y.G.; Mayers, B.; Herricks, T.; Xia, Y.N. Polyol synthesis of uniform silver nanowires: A plausible growth mechanism and the supporting evidence. *Nano Lett.* **2003**, *3*, 955–960. [CrossRef]
15. Navale, G.R.; Thripuranthaka, M.; Late, D.J.; Shinde, S.S. Antimicrobial activity of ZnO nanoparticles against pathogenic bacteria and fungi. *JSM Nanotechnol. Nanomed.* **2015**, *3*, 1033.
16. Butler, S.; Neogi, P.; Urban, B.; Fujita, Y.; Hu, Z.; Neogi, A. ZnO nanoparticles in hydrogel polymer network for bio-imaging. *Glob. J. Nanomed.* **2017**, *1*, 555572.
17. Sang, C.H.; Chou, S.J.; Pan, F.M.; Sheu, J.T. Fluorescence enhancement and multiple protein detection in ZnO nanostructure microfluidic devices. *Biosens. Bioelectron.* **2016**, *75*, 285–292. [CrossRef] [PubMed]
18. Chung, R.-J.; Wang, A.-N.; Liao, Q.-L.; Chuang, K.-Y. Non-enzymatic glucose sensor composed of carbon-coated nano-zinc oxide. *Nanomaterials* **2017**, *7*, 36. [CrossRef] [PubMed]

19. Sirelkhatim, A.; Mahmud, S.; Seeni, A.; Kaus, N.H.M.; Ann, L.C.; Bakhori, S.K.M.; Hasan, H.; Mohamad, D. Review on zinc oxide nanoparticles: Antibacterial activity and toxicity mechanism. *Nano-Micro Lett.* **2015**, *7*, 219–242. [CrossRef]

20. Valentão, P.; Fernandes, E.; Carvalho, F.; Andrade, P.B.; Seabra, R.M.; de Lourdes Bastos, M. Antioxidant activity of hypericum androsaemum infusion: Scavenging activity against superoxide radical, hydroxyl radical and hypochlorous acid. *Biol. Pharm. Bull.* **2002**, *25*, 1320–1323.

21. Borra, S.K.; Gurumurthy, P.; Mahendra, J.; Jayamathi, K.M.; Cherian, C.N.; Chand, R. Antioxidant and free radical scavenging activity of curcumin determined by using different in vitro and ex vivo models. *J. Med. Plant Res.* **2013**, *7*, 2680–2690. [CrossRef]

22. Du, X.-W.; Fu, Y.-S.; Sun, J.; Han, X.; Liu, J. Complete UV emission of ZnO nanoparticles in a PMMA matrix. *Semicond. Sci. Technol.* **2006**, *21*, 1202–1206. [CrossRef]

23. Aneesh, P.M.; Vanaja, K.A.; Jayaraj, M.K. Synthesis of ZnO nanoparticles by hydrothermal method. *Proc. SPIE* **2007**, *6639*, 66390J. [CrossRef]

24. Abri, A.; Maleki, M. Isolation and identification of Gallic acid from the Elaeagnus angustifolia leaves and determination of total phenolic, flavonoids contents and investigation of antioxidant activity. *Iran. Chem. Commun.* **2016**, *4*, 146–154.

25. Dutta, S.; Ganguly, B.N. Characterization of ZnO nanoparticles grown in presence of Folic acid template. *J. Nanobiotechnol.* **2012**, *10*, 29. [CrossRef] [PubMed]

26. Lu, Z.; Nie, G.; Belton, P.S.; Tang, H.; Zhao, B. Structure–activity relationship analysis of antioxidant ability and neuroprotective effect of gallic acid derivatives. *Neurochem. Int.* **2006**, *48*, 263–274. [CrossRef] [PubMed]

27. Choi, K.H.; Nam, K.C.; Lee, S.Y.; Cho, G.; Jung, J.S.; Kim, H.J.; Park, B.J. Antioxidant potential and antibacterial efficiency of caffeic acid-functionalized ZnO nanoparticles. *Nanomaterials* **2017**, *7*, 148. [CrossRef] [PubMed]

28. Akiyama, H.; Fujii, K.; Yamasaki, O.; Oono, T.; Iwatsuki, K. Antibacterial action of several tannins against *Staphylococcus aureus*. *J. Antimicrob. Chemother.* **2001**, *48*, 487–491. [CrossRef] [PubMed]

29. Li, A.; Chen, J.; Zhu, W.; Jiang, T.; Zhang, X.; Gu, Q. Antibacterial activity of gallic acid from the flowers of *Rosa chinensis* Jacq. against fish pathogens. *Aquacult. Res.* **2007**, *38*, 1110–1112. [CrossRef]

30. Nakamura, K.; Yamada, Y.; Ikai, H.; Kanno, T.; Sasaki, K.; Niwano, Y. Bactericidal action of photoirradiated gallic acid via reactive oxygen species formation. *J. Agric. Food Chem.* **2012**, *60*, 10048–10054. [CrossRef] [PubMed]

31. Shalaby, E.A.; Shanab, S.M.M. Comparison of DPPH and ABTS assays for determining antioxidant potential of water and methanol extracts of *Spirulina platensis*. *Indian J. Geo-Mar. Sci.* **2013**, *42*, 556–564.

32. Park, B.J.; Choi, K.H.; Nam, K.C.; Min, J.E.; Lee, K.D.; Uhm, H.S.; Choi, E.H.; Kim, H.J.; Jung, J.S. Photodynamic anticancer activity of CoFe$_2$O$_4$ nanoparticles conjugated with hematoporphyrin. *J. Nanosci. Nanotechnol.* **2015**, *15*, 7900–7906. [CrossRef] [PubMed]

33. Maddox, C.E.; Laur, L.M.; Tian, L. Antibacterial activity of phenolic compounds against the phytopathogen Xylella fastidiosa. *Curr. Microbiol.* **2010**, *60*, 53–58. [CrossRef] [PubMed]

© 2017 by the authors. Licensee MDPI, Basel, Switzerland. This article is an open access article distributed under the terms and conditions of the Creative Commons Attribution (CC BY) license (http://creativecommons.org/licenses/by/4.0/).

*nanomaterials*

MDPI

*Article*

# Two-in-One Biointerfaces—Antimicrobial and Bioactive Nanoporous Gallium Titanate Layers for Titanium Implants

Seiji Yamaguchi [1,*], Shekhar Nath [1], Yoko Sugawara [1], Kamini Divakarla [2], Theerthankar Das [3], Jim Manos [3], Wojciech Chrzanowski [2,*], Tomiharu Matsushita [1] and Tadashi Kokubo [1]

[1]   Department of Biomedical Sciences, College of Life and Health Sciences, Chubu University, Aichi Prefecture 487-8501, Japan; shekhar.nath@gmail.com (S.N.); cu33667@fsc.chubu.ac.jp (Y.S.); matsushi@isc.chubu.ac.jp (T.M.); kokubo@isc.chubu.ac.jp (T.K.)
[2]   Australian Institute for Nanoscale Science and Technology, Charles Perkins Centre, Faculty of Pharmacy, University of Sydney, Pharmacy and Bank Building A15, Sydney, NSW 2006, Australia; kdivakarla@gmail.com
[3]   Department of Infectious Diseases and Immunology, Sydney Medical School, University of Sydney, Sydney, NSW 2050, Australia; das.ashishkumar@sydney.edu.au (T.D.); jim.manos@sydney.edu.au (J.M.)
*    Correspondence: sy-esi@isc.chubu.ac.jp (S.Y.); wojciech.chrzanowski@sydney.edu.au (W.C.); Tel.: +81-568-51-6420 (S.Y.); +61-2-9351-5306 (W.C.)

Received: 10 July 2017; Accepted: 12 August 2017; Published: 20 August 2017

**Abstract:** The inhibitory effect of gallium (Ga) ions on bone resorption and their superior microbial activity are attractive and sought-after features for the vast majority of implantable devices, in particular for implants used for hard tissue. In our work, for the first time, Ga ions were successfully incorporated into the surface of titanium metal (Ti) by simple and cost-effective chemical and heat treatments. Ti samples were initially treated in NaOH solution to produce a nanostructured sodium hydrogen titanate layer approximately 1 μm thick. When the metal was subsequently soaked in a mixed solution of $CaCl_2$ and $GaCl_3$, its Na ions were replaced with Ca and Ga ions in a Ga/Ca ratio range of 0.09 to 2.33. 8.0% of the Ga ions were incorporated into the metal surface when the metal was soaked in a single solution of $GaCl_3$ after the NaOH treatment. The metal was then heat-treated at 600 °C to form Ga-containing calcium titanate (Ga–CT) or gallium titanate (GT), anatase and rutile on its surface. The metal with Ga–CT formed bone-like apatite in a simulated body fluid (SBF) within 3 days, but released only 0.23 ppm of the Ga ions in a phosphate-buffered saline (PBS) over a period of 14 days. In contrast, Ti with GT did not form apatite in SBF, but released 2.96 ppm of Ga ions in PBS. Subsequent soaking in hot water at 80 °C dramatically enhanced apatite formation of the metal by increasing the release of Ga ions up to 3.75 ppm. The treated metal exhibited very high antibacterial activity against multidrug resistant *Acinetobacter baumannii* (MRAB12). Unlike other antimicrobial coating on titanium implants, Ga–CT and GT interfaces were shown to have a unique combination of antimicrobial and bioactive properties. Such dual activity is essential for the next generation of orthopaedic and dental implants. The goal of combining both functions without inducing cytotoxicity is a major advance and has far reaching translational perspectives. This unique dual-function biointerfaces will inhibit bone resorption and show antimicrobial activity through the release of Ga ions, while tight bonding to the bone will be achieved through the apatite formed on the surface.

**Keywords:** gallium ion; apatite formation; gallium titanate; Ti metal; simulated body fluid; antibacterial

## 1. Introduction

Titanium metal (Ti) and its alloys are widely used as orthopaedic and dental implants because of their high degree of mechanical strength and good biocompatibility. However, a polished Ti surface cannot bond to living bone by forming the requisite layer of thin fibrous tissue at the interface of living bone and the metal [1]. Roughened Ti surface is able to come directly into contact with living bone, but still does not bond to it adequately. In order to achieve stable fixation, various kinds of surface modifications introducing a bone-bonding capacity into Ti and its alloys have been attempted. A plasma spray coating of calcium phosphate is often used to induce the bone-bonding capacity. However, this method does not produce a uniform bioactive surface layer, because only the surface exposed to the plasma is coated, so the calcium phosphate is liable to decomposition in the living body over time. A coating of calcium phosphate has also been achieved by sputtering, sol-gel or alternative soaking [2,3]. However, once again the resulting coat is not stable in the body environment over time. The incorporation of calcium ions into the surface of Ti has also been attempted using ion implantation, micro-arc oxidation and hydrothermal treatment [4–7] to provide the capacity of apatite formation to the metal, while titania nanotubes have been formed by anodic oxidation to stimulate osteoblast cells [8]. However, these techniques require a special apparatus and are not suitable for devices of complex structure and/or of large size.

The solution and subsequent thermal treatment does not require any special apparatus and allows the formation of a uniform bioactive surface layer, even on the inner surface of a porous body [9,10]. It has been demonstrated that a bioactive surface layer composed of sodium titanate and rutile is produced on Ti by simply soaking it in NaOH solution at 60 °C for 24 h and then heating it at 600 °C for 1 h [11]. The treated metal spontaneously forms a bone-like apatite on its surface in the body environment and directly bonds to living bone through this layer [1,11–14]. These treatments were applied to the porous Ti layer of an artificial hip joint that was commercialized in Japan in 2007 [15]. The long-term survivorship of the NaOH- and heat-treated total hip arthroplasty (THA) was recently reviewed for 70 primary THAs, of whom 67 were available for follow-up periods of 8–12 years [16]. It was histologically observed that direct bonding to bone took place within 2 weeks and was maintained for at least 8 years. No implant exhibited any radiographic signs of evident loosening, and the overall survival rate was 98% at 10 years.

However, for all the successful results of the NaOH- and heat-treated THAs, two joints were retrieved because of deep infection and periprosthetic femoral fracture [16]. In addition to bone bonding, the capacities of anti-bacterial and increasing bone density that may prevent bone infection and fracture are desirable for the next generation of THAs.

Recently, it was shown that various types of functional metal ions such as Sr, Mg and Ag can be incorporated into the Ti surface by modifying the NaOH and heat treatments [17,18]. In the case of Sr, the treated metal was expected to promote a growth of new bone surrounding Ti by releasing Sr ions and to form apatite on its surface by releasing Ca ions. Indeed, when the metal with Sr-containing calcium titanate was implanted into rabbit femur, it bonded to living bone in a shorter period of time compared to Ti with calcium titanate or sodium titanate, as expected [19,20]. Similar effects were observed for the Ti with Mg-containing calcium titanate that was produced by replacing $SrCl_2$ with $MgCl_2$ in the treatments [18–20]. While for Ag, Ti was soaked in a $CaCl_2$ solution after the NaOH treatment, then subjected to a heat and $AgNO_3$ treatment to form Ag-containing calcium titanate on its surface. This treated metal exhibited a strong anti-bacterial effect against *Staphylococcus aureus* as well as apatite formation; however, silver has long been recognised to be cytotoxic and recently recalled from many medical applications [21].

Gallium (Ga) is known for its inhibitory effect on calcium release from bone tissue, which is effective for preventing bone resorption. Gallium has also been shown to have clinical efficacy in suppressing osteolysis and bone pain and suggested as a treatment for osteoporosis [22]. In addition, recent studies have reported the anti-bacterial capability of Ga ions [23,24]. Valappil et al. reported the bactericidal activities of Ga-doped phosphate-based glass against both Gram-negative (*Escherichia coli* and

*Pseudomonas aeruginosa*) and Gram-positive (*Staphylococcus aureus*, methicillin-resistant *Staphylococcus aureus*, and *Clostridium difficile*) bacteria [23]. Cochis et al. reported that the Ga-doped Ti produced by anodic spark deposition attained better bacterial inhibition against *Acinetobacter baumannii* than Ag-doped Ti by the same method [24]. In contrast with Ag, Ga can active metabolically by substituting Fe in many biological systems, due to the chemical similarities of $Ga^{3+}$ with $Fe^{3+}$ in terms of charge, ionic radius, and electronic configuration [25]. As a result, Ga exhibits these beneficial effects without inducing cytotoxicity [26–29].

We have designed novel Ga-containing nanostructured interfaces that are capable of sustainably realising gallium ions. In this way, we will achieve highly desired antimicrobial activity without compromising the ability of the implant to bind to bone. In fact, gallium ions are likely to further improve bone bonding ability and ultimately lead to improved bone quality. This improved effectiveness in stimulation of the bone formation is of particular value to achieve stable integration in osteoporotic of Dorr bone type environment. For these patients, conventional approaches reach a very high level of failure, which continues to increase with aging society, and for which traditional surfaces are suboptimal. Statistical data suggest that implant applications will skyrocket over the next few decades. The high number of surgeries that need to be repeated as implants fail to integrate in the patient's body, becoming infected or ineffective (up to 17.5% of devices), imposes an additional and growing burden. Advances, such as that which we have developed, promoting rapid implant integration to improve function and reduce the risk of infection are therefore of great significance. The developed multifunctional interfaces will deliver rapid osseointegration of orthopaedic implants, and enable enhancement of the design of effective biomaterials in general.

## 2. Results

### 2.1. Surface Structures

The surface and cross-sectional FE-SEM (field emission scanning electron microscopy) observations showed that a fine network structure approximately 1 μm thick uniformly formed on the Ti surface with the first NaOH treatment, as reported in our previous work [30]. The nano-sized network morphology was retained even after the subsequent chemical and heat treatments, as shown in Figure 1a–h.

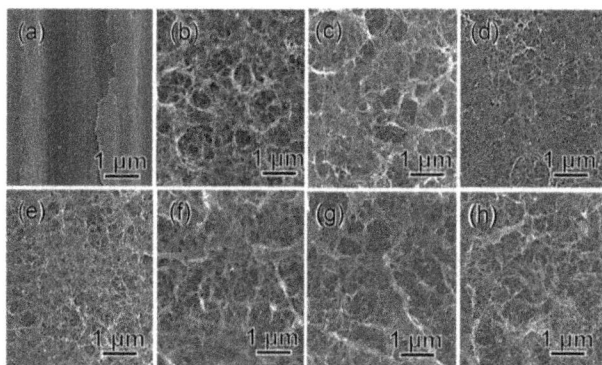

**Figure 1.** FE-SEM photographs of surfaces of Ti (**a**) untreated or subjected to (**b**) NaOH treatment, and subsequent (**c**) 100Ca + 0.05Ga and (**d**) heat treatment, and finally (**e**) water treatment, or (**f**) 100Ga after the NaOH treatment, and subsequent (**g**) heat and finally (**h**) water treatment.

Table 1 shows the chemical composition of the Ti surfaces taken by EDX (energy dispersive X-ray spectrometer) analysis after each chemical and heat treatment. It should be noted that the results show averaged values of the graded compositions from the surface to the Ti substrate (later shown in Figure 5). The first NaOH treatment incorporated approximately 5.5% of the Na into the surface of the metal. When the metal was subsequently soaked in a mixed solution of 100 mM $CaCl_2$ and 0.01 mM $GaCl_3$, the Na was replaced with 3.5% Ca and 0.3% Ga such that the Ga/Ca ratio was 0.09. The ratio of Ga/Ca in the surface region increased with an increasing concentration of $GaCl_3$ in the mixed solution. 8.0% of the Ga was incorporated into the surface layer when the metal sample was soaked in 100 mM $GaCl_3$ solution after the NaOH treatment.

**Table 1.** Chemical compositions of the surface layers of Ti metal subjected to NaOH, Ca + Ga, heat and water treatments, which were analysed by EDX.

| Treatment | Element (at. %) | | | | | Ga/Ca Ratio |
|---|---|---|---|---|---|---|
| | O | Ti | Na | Ca | Ga | |
| Untreated | 3.1 | 96.9 | 0 | 0 | 0 | - |
| NaOH | 66.8 | 27.7 | 5.5 | 0 | 0 | - |
| NaOH-100Ca + 0.01Ga | 68.6 | 27.6 | 0 | 3.5 | 0.3 | 0.09 |
| NaOH-100Ca + 0.05Ga | 67.9 | 29.2 | 0 | 2.3 | 0.6 | 0.26 |
| NaOH-100Ca + 0.10Ga | 68.3 | 27.7 | 0 | 1.2 | 2.8 | 2.33 |
| NaOH-100Ga | 69.3 | 22.8 | 0 | 0 | 8.0 | - |
| NaOH-100Ca + 0.05Ga-heat | 68.5 | 28.7 | 0 | 2.3 | 0.6 | 0.26 |
| NaOH-100Ca + 0.05Ga-heat-water | 68.6 | 29.5 | 0 | 1.2 | 0.7 | 0.58 |
| NaOH-100Ga-heat | 68.4 | 23.5 | 0 | 0 | 8.1 | - |
| NaOH-100Ga-heat-water | 69.8 | 22.4 | 0 | 0 | 7.8 | - |

The metal sample with 2.3% Ca and 0.6% Ga (Ga/Ca ratio 0.26) did not exhibit any change in Ca or Ga by subsequent heat treatment, but exhibited an appreciable decrease in Ca by subsequent water treatment. The metal sample with 8.0% Ga displayed little change in Ga by subsequent heat treatment, but a slight decrease in Ga resulted from the final water treatment.

Figure 2 shows the TF-XRD (thin-film X-ray diffraction) and FT-Raman (Fourier transform confocal laser Raman spectrometry) profiles of the sample surfaces subjected to NaOH, 100Ca + 0.05Ga or 100Ga, heat and final water treatments. The initial NaOH treatment resulted in broad XRD and Raman peaks attributed to sodium hydrogen titanate [31,32], $Na_xH_{2-x}Ti_3O_7$ (SHT). When the Ti was subsequently subjected to 100Ca + 0.05Ga treatment, the peak positions of SHT were essentially unchanged except for a slight shift in the Raman peak from approximately 920 to 900 cm$^{-1}$. This indicates that the sodium hydrogen titanate was isomorphously transformed into gallium-containing calcium hydrogen titanate, $Ga_xCa_yH_{2-(3x+2y)}Ti_3O_7$ (Ga–CHT), by substituting for the Na ions with Ca and Ga ions. The Raman peak around 920 cm$^{-1}$ was previously reported as Ti–O bonds involving nonbridging oxygen coordinated with Na ions [33], with its shift to a lower wave number potentially resulting from an exchange of Na ions with Ca and Ga ions. When the Ga–CHT was heat-treated, broad peaks at around 25° and 48°, as well as a peak of rutile around 27.5° in 2θ, appeared in the TF-XRD pattern. The broad peaks were well matched with those of the calcium titanate (CT), such as $CaTi_2O_4$ (JCPDS file 00-026-0333), $CaTi_2O_5$ (JCPDS file 01-072-1134), and $CaTi_4O_9$ (JCPDS file 00-025-1450), which were formed by the heat treatment of CHT [34–36]. This shows that Ga–CHT was transformed into gallium-containing calcium titanate (Ga–CT), such as $Ga_xCa_{1-x}Ti_2O_4$, $Ga_xCa_{1-x}Ti_2O_5$ and $Ga_xCa_{1-x}Ti_4O_9$. Certain amounts of anatase were also formed by the heat treatment, as seen in the peaks in Raman spectra at around 150 and 520 cm$^{-1}$. These crystalline phases were unchanged by the final water treatment. In contrast, gallium hydrogen titanate, $Ga_xH_{2-3x}Ti_3O_7$ (GHT), was formed on Ti surface when the metal was subjected to the 100Ga treatment after the NaOH treatment. The Raman spectrum in Figure 2f shows a slight shift of the peak at 290 cm$^{-1}$ toward a lower wave number, implying that a portion of the Ti$^{4+}$ in SHT was replaced by Ga ions. The GHT was transformed into gallium titanate, such as $Ga_2TiO_5$ (GT) (JCPDS file 00-020-0447), along with rutile accompanied by a small amount of anatase, which remained even after the final water treatment, as shown in Figure 2g,h.

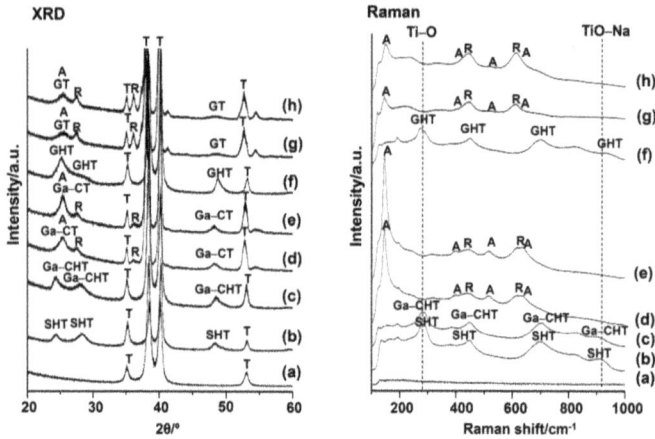

**Figure 2.** XRD and Raman spectra of surfaces of Ti (**a**) untreated or subjected to (**b**) NaOH treatment, and subsequent (**c**) 100Ca + 0.05Ga and (**d**) heat treatment, and finally (**e**) water treatment, or (**f**) 100Ga after the NaOH treatment, and subsequent (**g**) heat and finally (**h**) water treatment. T: α-Ti A: Anatase R: Rutile SHT: $Na_xH_{2-x}Ti_3O_7$ Ga–CHT: $Ga_xCa_yH_{2-(3x+2y)}Ti_3O_7$ GHT: $Ga_xH_{2-3x}Ti_3O_7$. Ga–CT: $Ga_xCa_{1-1.5x}Ti_4O_9$, $Ga_xCa_{1-1.5x}Ti_2O_4$, $Ga_xCa_{1-1.5x}Ti_4O_5$ GT: $Ga_2TiO_5$.

Figure 3 shows the high-resolution XPS (X-ray photoelectron spectroscopy) spectra of the Ca2p and Ga2p on the samples that were subjected to the NaOH and 100Ca + 0.05Ga or 100Ga treatments, subsequent heat treatment and final water treatment. The metal subjected to 100Ca + 0.05Ga following the NaOH treatment exhibited split peaks of approximately 347 and 350 eV in binding energy which were attributed to the $Ca2p_{3/2}$ and $Ca2p_{1/2}$ of CaO [37], and the peak range from 1115 to 1121 eV was deconvoluted into 1118.0, 1119.0 and 1116.5 eV of $Ga2p_{3/2}$. It is reported in the literature that the peak around 1117–1118 eV is attributable to the $Ga^{3+}$ of $Ga_2O_3$ [37,38], whereas the peak around 1119 eV is attributable to the $Ga^{3+}$ or $Ga^{4+}$ of the Ga–O located at the $Ti^{4+}$ site in $TiO_2$ [39]. The peak around 1116 eV was attributed to a metallic Ga bond [40]. Similar Ga 2p peaks were detected on the sample surfaces subjected to the 100Ga treatment after the NaOH treatment. These Ga peaks were not evidently changed by the subsequent heat and water treatments. The O 1s spectra on the same samples are shown in Figure 4. The spectra tailing to the higher binding energy side was deconvoluted to around 530 eV of Ti–O bond, 531 eV of physisorbed $H_2O$, and 532 eV of basic Ti–OH [41]. Abundant Ti–OH peaks were observed on the sample surfaces subjected to 100Ca + 0.05Ga or 100Ga treatment following the NaOH treatment. A quantitative analysis revealed that the ratio of Ti–OH to Ti–O was 0.26 for the former and 0.24 for the latter. These values were severely reduced by the subsequent heat treatment, but recovered to some degree with the final water treatment.

Figure 5 shows the depth profile of the XPS spectra of the Ti subjected to the 100Ca + 0.05Ga or 100Ga treatment after the NaOH treatment followed by heat and water treatments. Enrichment of Ca and Ga were observed near the surface in the former sample in addition to O, as shown in Figure 5a. They gradually decreased with increasing depth up to a thickness of approximately 1 μm. Substantially larger amounts of Ga were detected on the surface of the latter sample, as shown in Figure 5b.

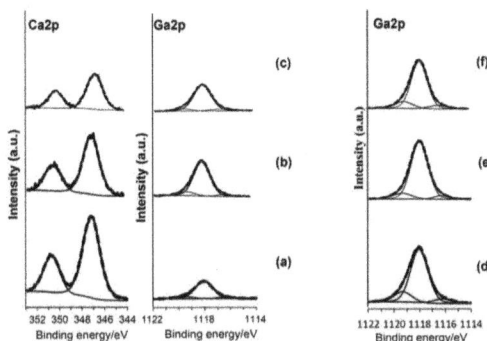

**Figure 3.** XPS profiles of Ca2p and Ga2p on Ti subjected to (**a**) 100Ca + 0.05Ga; (**b**) subsequent heat, and (**c**) final water treatment, or (**d**) 100Ga; (**e**) subsequent heat, and (**f**) final water treatment. Thick line represents spectrum. Thin lines represent deconvolution lines. Dot line represents composite line.

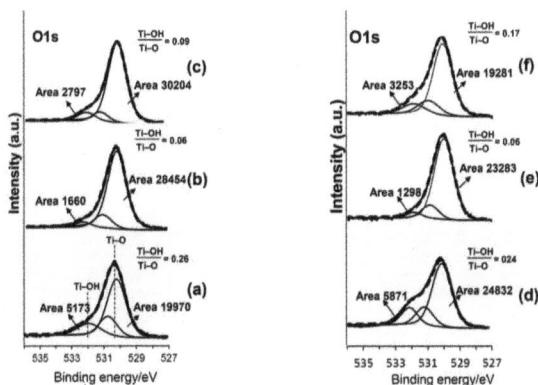

**Figure 4.** XPS profiles of O1s on Ti that was initially soaked in NaOH solution and then subjected to (**a**) 100Ca + 0.05Ga; (**b**) subsequent heat, and (**c**) final water treatment, or (**d**) 100Ga; (**e**) subsequent heat, and (**f**) final water treatment. Thick line represents spectrum. Thin lines represent deconvolution lines. Dot line represents composite line.

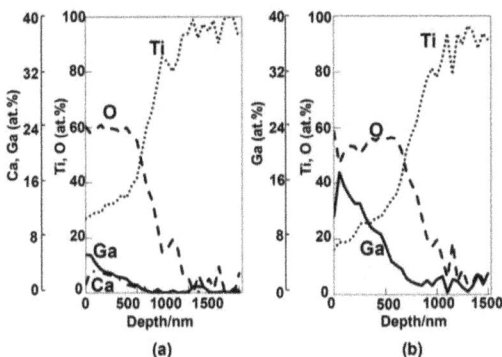

**Figure 5.** XPS depth profiles of Ti that was initially soaked in NaOH solution and then subjected to (**a**) 100Ca + 0.05Ga and (**b**) 100Ga treatment followed by heat and water treatments.

## 2.2. Ion Release

Figure 6 shows the concentration of the Ga and Ca ions released from the Ti subjected to 100Ca + 0.05Ga or 100Ga treatment after the NaOH treatment, followed by heat and water treatments, as a function of the square root of the soaking time in PBS. It can be seen in Figure 6c that the metal sample subjected to 100Ga and heat treatment following the NaOH treatment rapidly released 1.58 ppm of Ga ions within 1 h, and slowly released another 1.22 ppm up to 7 days in (a) proportion to the square root of the soaking time. Further release was not observed until 14 days. In total, 2.96 ppm of Ga ions was released. The Ga release increased up to 3.75 ppm when the metal was finally subjected to the water treatment, as shown in Figure 6d. In contrast, the metal subjected to 100Ca + 0.05Ga and heat treatment following the NaOH treatment released only a small amount of Ga ions, which was as low as 0.23 ppm, even after 14 days. The Ga release was slightly increased up to 0.35 ppm by the final water treatment, as shown in Figure 6b. A large amount of Ca ions, as high as 0.34 ppm, was released from the former sample, as seen in Figure 6a, but it decreased to 0.03 ppm with the final water treatment.

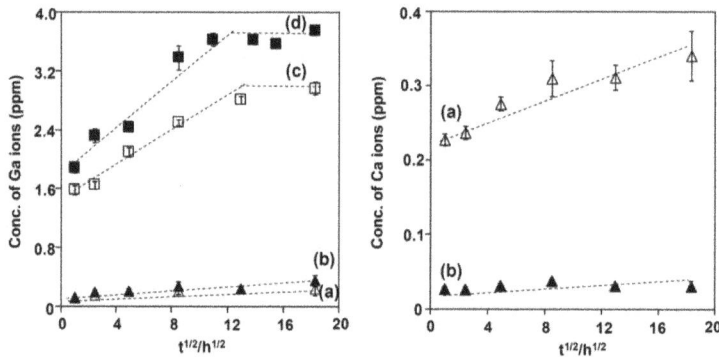

**Figure 6.** Concentrations of Ga and Ca ions measured by ICP (inductively coupled plasma emission spectroscopy), which was released from Ti subjected to (**a**) NaOH; 100Ca + 0.05Ga and heat treatment, and subsequent (**b**) water treatment; or (**c**) NaOH, 100Ga and heat treatment, and subsequent (**d**) water treatment as a function of square root of soaking time in PBS.

## 2.3. Apatite Formation

Figure 7 shows the FE-SEM photographs of the surfaces of the metal samples soaked in SBF after the chemical and heat treatments. When the metal was subjected to 100Ca + 0.05Ga and heat following the NaOH treatment, it formed many spherical precipitates on its surface within 3 days in SBF, as shown in Figure 7a. The amount of precipitate increased to fully cover the surface of the metal when additional water treatment was performed on the previously heat-treated samples, as shown in Figure 7b. Interestingly, when the metal was subjected to only 100Ga (without Ca) and heat following the NaOH treatment, no precipitates were observed on the metal surface. In contrast, considerable precipitate was observed on the surface of the heat-treated metal that was ultimately subjected to the water treatment, as shown in Figure 7d. The precipitate was also observed on the metal subjected to the same treatments followed by subsequent storage under 95% relative humidity at 80 °C for 7 days, as shown in Figure 7e. From high magnification images, it was quite clear that the precipitate that formed was composed of nano-sized crystals. No obvious difference was observed in terms of the crystal size of the apatite phase that formed on the various treated samples that were able to produce apatite in SBF.

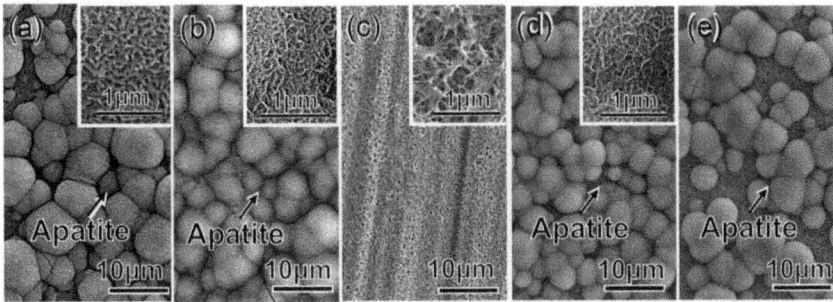

**Figure 7.** FE-SEM photographs of surfaces of Ti soaked in SBF for 3 days following (**a**) NaOH, 100Ca + 0.05Ga and heat treatments, and finally (**b**) water treatment, or (**c**) NaOH, 100Ga and heat treatments and finally (**d**) water treatment. Apatite formation of Ti subjected to the treatment of (**d**) followed by a storage under 95% relative humidity at 80 °C for 7 days is shown in (**e**). The insets in (**a–d**) show high-magnification images.

Figure 8 shows the TF-XRD profiles of the samples that were soaked in SBF for various numbers of days after the ultimate water treatment. Broad peaks around 26° and 32° in 2θ appeared on the surface of the samples that were subjected to 100Ca + 0.05Ga or 100Ga treatment in a second solution within 1 or 3 days, respectively. This indicates that the spherical precipitate observed under SEM was composed of poorly crystalline apatite. The intensity of the apatite peaks increased with the increase of soaking time in SBF in both cases. No obvious changes were observed in the apatite that formed on the sample surfaces for the same soaking periods of 3 or 7 days, respectively.

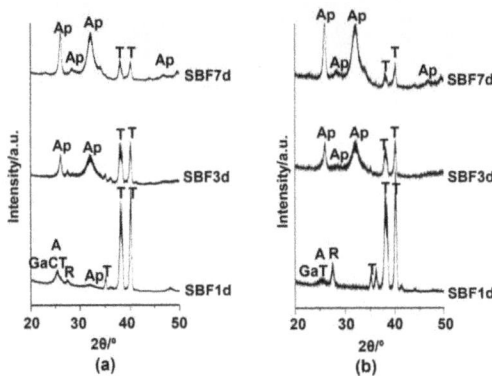

**Figure 8.** TF-XRD profiles of the Ti surfaces soaked in SBF for different days after (**a**) 100Ca + 0.05Ga, heat and water or (**b**) 100Ga, heat and water treatments following the NaOH treatment. T: α-Ti Ap: Apatite A: Anatase R: Rutile, Ga-CT: $Ga_xCa_{1-1.5x}Ti_4O_9$, $Ga_xCa_{1-1.5x}Ti_2O_4$, $Ga_xCa_{1-1.5x}Ti_4O_5$, GT: $Ga_2TiO_5$.

The sample formed with apatite by soaking in SBF for 7 days after the NaOH, 100Ga, heat and water treatments was subjected to cross-sectional FE-SEM observation and EDX line analysis to examine the Ga distribution on the surface of the metal (Figure 9). It was found from the FE-SEM photographs (Figure 9a) that apatite particles filled the spaces of the nano-structured surface layer and grew into a uniform layer approximately 4 μm in thickness. It was shown by EDX line analysis (Figure 9b) that a relatively large amount of Ga was detected in the region where Ti started to decrease, and it gradually decreased along with a decreasing Ti. A certain amount of Ga was detected even near the top surface, where Ca and P, which are components of apatite, were dominant. These results imply that Ga distribution was not limited in the surface GT layer, but also regions far from the GT layer. It was evident from the high-resolution XPS spectra of the Ga2p, Ca2p, P2p, and Ti2p taken on the top surface of the sample, shown in Figure 10, that a small amount of Ga was detected along with a large amount of Ca and P, even on the top surface of the apatite layer, while Ti was not detected.

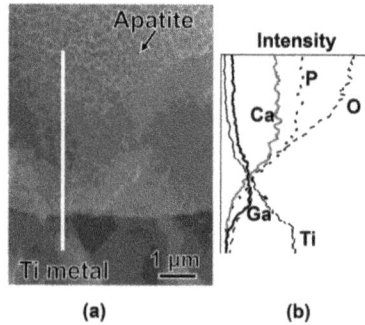

(a)                                        (b)

**Figure 9.** (a) FE-SEM photographs and (b) EDX line analysis of cross sections of Ti soaked in SBF for 7 days after 100Ga, heat and water treatments following the NaOH treatment. EDX line analysis was performed along with white line in the photographs.

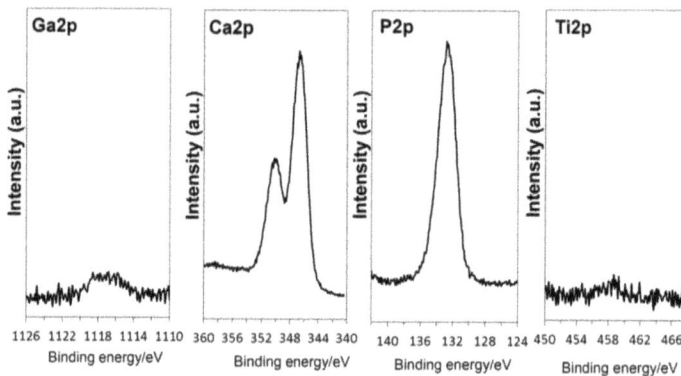

**Figure 10.** XPS profiles of Ga2p, Ca2p, P2p and Ti2p on Ti soaked in SBF for 7 days after NaOH-100Ga-heat-water treatment.

## 2.4. Antibacterial Activity

Figure 11 shows the confocal microscopy images of the Ti surfaces containing multi-resistant *Acinetobacter baumannii* (MRAB12). It was evident that the bacteria proliferated to form dense film on the surface of as-polished Ti, while they were almost entirely killed on the surfaces of Ti with Ga–CT or GT. Quantitative analysis revealed that the percentage of Live biofilm biomass on the control sample was significantly higher ($p < 0.05$) at 87.7% ± 4% in comparison to both the samples with Ga–CT (16.2% ± 5.5%) and those with GT (5.8% ± 2.9%), as shown in Figure 12. Concurrently, the percentage of Dead biofilm biomass was significantly higher ($p < 0.05$) on both the samples with Ga–CT (83.8% ± 5.5%) and those with GT (94.2% ± 2.4%).

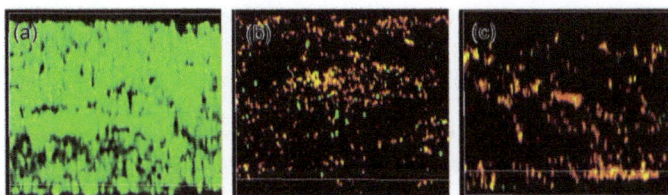

**Figure 11.** Confocal microscopy of Ti surfaces contacted with multi-resistant *Acinetobacter baumannii* (MRAB12) for 7 days (**a**) as-polished; (**b**) subjected to NaOH-100Ca + 0.05Ga-heat-water or (**c**) NaOH-100Ga-heat-water treatment. Green: live bacteria, Red: dead bacteria.

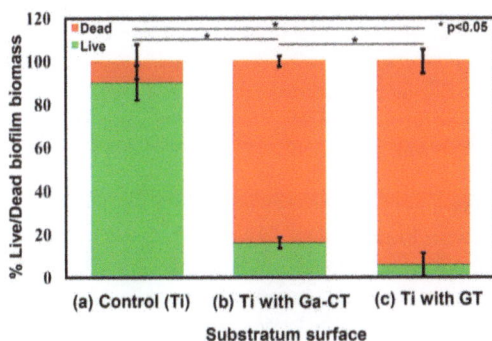

**Figure 12.** Live/Dead biofilm biomass on Ti surfaces contacted with multi-resistant *Acinetobacter baumannii* (MRAB12) for 7 days (**a**) as-polished; (**b**) subjected to NaOH-100Ca + 0.05Ga-heat-water or (**c**) NaOH-100Ga-heat-water treatment.

## 3. Discussion

The initial NaOH treatment formed a fine network layer that consisted of nano-sized sodium hydrogen titanate, $Na_xH_{2-x}Ti_3O_7$, on the surface of the metal. Since sodium hydrogen titanate (SHT) assumes a layered structure [34], its Na ions were easily substituted by Ca and Ga ions so as to form a gallium-containing calcium hydrogen titanate (Ga–CHT) when soaked in a mixed solution of $CaCl_2$ and $GaCl_3$ solution. It is evident from the XPS profiles of Ca2p and Ga2p in Figure 3 that all of the Ca ions and most of the Ga ions were located in interspaces of Ga–CHT, whereas some Ga ions were located at the $Ti^{4+}$ site. When the Ga–CHT with a Ga/Ca ratio of 0.26 was subjected to the heat treatment, it was transformed into gallium-containing calcium titanate (Ga–CT), rutile and anatase, as shown in Figure 2d. Interestingly, the metal treated thusly formed apatite on its surface in SBF within 3 days, as shown in Figure 7a, even without the final water treatment reported to be essential to induce apatite formation on Ti formed with calcium titanate (CT), Sr-containing calcium titanate (Sr–CT) and Mg-containing calcium titanate (Mg–CT) [17,18,34]. The apatite formation on Ti with Ga–CT is probably due to the relatively large capacity for Ca ion release, as shown in Figure 6a, compared to that of Ti with CT, Sr–CT, and Mg–CT. The large capacity of Ca ion release on Ga–CT might be due to the Ga ions located at the $Ti^{4+}$ site of Ga–CT, resulting in a distortion in the tight CT structure. Thus, most of the Ca ions were replaced with $H_3O^+$ ions when the treated metal was subsequently subjected to the final water treatment, as shown in Table 1. As a result, Ca ion release was significantly decreased, as shown in Figure 6b, which is in contrast to the slight decrease in Ca content and increase in Ca ion release on the Ti formed with CT [18]. Nevertheless, the apatite formation of Ga–CT was further increased by the final water treatment, as shown in Figure 7b. Thus, the increased apatite formation was not attributed to Ca ion release. This was also not attributed to the crystalline phases that were not obviously changed by the final water treatment, as shown in Figure 2d,e, but rather is probably due to the abundant Ti–OH groups that were preferable for apatite nucleation [42], as shown in Figure 4c. It was clear, however, as shown in Figure 6a,b, that the capacity of Ga ion release on Ga–CT was poor, even after the final water treatment.

When the Ti was soaked in 100 mM $GaCl_3$ solution after the NaOH treatment, 8.0% Ga was incorporated into the metal surface, forming gallium hydrogen titanate (GHT). Most of the Ga ions were located in the interspaces of GHT whereas some were at the $Ti^{4+}$ site, as seen in the case of Ga–CHT. The GHT became dehydrated and transformed into gallium titanate (GT), rutile and a small amount of anatase by the heat treatment. The treated metal did not form apatite in SBF within 3 days, but formed a marked amount apatite particles after the final water treatment, as shown in Figure 7c,d. The increased apatite formation by the final water treatment might be attributable to the increased Ga ion release as well as abundant Ti–OH groups, as shown in Figures 4c and 6d. The concentration of Ga in PBS become 1.88 ppm after 1 h, and slowly increased up to 3.75 ppm after 7 days, which remained there until after 14 days. When the treated metal was soaked in SBF, it is assumed that Ga ions were also released into SBF so as to increase the pH of the surrounding fluid via an exchange with $H_3O^+$ ions, and therefore formed Ti–OH groups on the metal surface, as in the case of the Na ions in the sodium titanate that was formed on Ti [43]. The Ti–OH groups become negatively charged in an alkaline environment [44], and combine with the positively charged Ca ions and, subsequently, negatively charged phosphate ions in the SBF. As a result, an amorphous calcium phosphate is formed, which is eventually transformed into crystalline, bone-like apatite [45]. Thus, the formed apatite becomes integrated with the nanostructured surface layer by growing into a uniform layer of about 4 μm in thickness, as shown in Figure 9. It was evident from the EDX line analysis in Figure 9 and XPS profiles on apatite far from the surface layer in Figure 10 that a small amount of Ga ions was taken into the apatite precipitated in SBF. It is reported that Ga ions in physiological saline solution supersaturated with Ca and $PO_4$ preferentially adsorbed on apatite crystal to reduce the rates of direct precipitation of apatite, transformation of amorphous calcium phosphate to apatite and growth of apatite seed crystals when the concentrations of Ga ions in the solution were more than 7, 17.5, and 1.75 ppm, respectively [46]. In the present study, no distinguishable differences were observed in terms of the crystallinity of the apatite formed on Ti with Ga–CT and GT regardless of the difference in their capacities for Ga release, as can be seen from Figures 7 and 8. This might be due to the relatively small Ga ion amounts (maximum at 3.75 ppm) compared to the reported value. The capacity of apatite formation on the metal, treated in this manner, was essentially maintained, even in 95% relative humidity at 80 °C for at least 1 week, as shown in Figure 7e.

It has been shown that a rod comprised of the NaOH- and heat-treated Ti was able to be pulled out of its site only by being accompanied with a bone fragment when it was implanted into an intramedullary canal of rabbit femur for 12 weeks [47]. A large number of bone fragments remained on the CT layer that formed on Ti by Ca-heat treatment when the metal was detached from the rabbit tibia after an implantation period of 4 to 26 weeks [48]. This indicates that the bonding strength between the substrate and the surface layer formed by the NaOH-heat or Ca-heat treatment was higher than the tensile strength of the bone. The strong bonding to bone was attributed to the capacity of apatite formation of the treated metals in both cases [47,48]. Thus, it is expected that Ti with Ga-CT or GT formed by the present treatments will also be strongly bonded to bone due to its capacity for apatite formation.

It was recently shown that the Ga-doped Ti produced by anodic spark deposition, releasing about 0.2 ppm of Ga ions into PBS, achieved better bacterial inhibition against one strain (DSM 30007) and two clinical isolates (AB1 and AB2) of *Acinetobacter baumannii* than Ag-doped Ti by the same method [24]. The antibacterial mechanism of Ga was suggested to be the "Trojan horse" mechanism in which $Ga^{3+}$ ions effectively compete with $Fe^{3+}$ ions for binding to siderophores, thus interrupting crucial Fe-dependent metabolic pathways in bacteria due to the biological similarity of $Ga^{3+}$ to $Fe^{3+}$ ions [24]. In the present study, Ti with an increased capacity for Ga ion release was prepared by the simple chemical and heat treatments. The Ti treated with Ga–CT or GT exhibited strong bacterial inhibition of *Acinetobacter baumannii* clinical isolates (MRAB12), as shown in Figures 11 and 12. The inhibitory effect was much higher for the latter, which released a greater amount of Ga. In addition, the effect of Ga ions on bone resorption and osteoclasts has been reported by many researchers. Hall et al. reported that 0.027 to 27 ppm of Ga ions derived from gallium nitrate produced a concentration-dependent inhibition of bone resorption by the osteoclasts isolated from neonatal rat long bones and cultured on slices of rat cortical bone [49] Verron et al. reported that 0.7 to 7 ppm of Ga ions derived from gallium nitrate inhibited the in vitro resorption activity of RBC and induced a significant decrease in the expression level of transcripts coding for osteoclastic markers in RAW 264.7 cells [27]. Studies have shown that gallium is adsorbed onto the surface of bone, where it is effective in blocking osteoclastic resorption. At antiresorptive concentrations, gallium does not appear to be cytotoxic to osteoclasts [28], or to act as a cellular metabolic inhibitor [29]. It is expected that the Ti that formed the Ga–CT or GT layer as the result of the NaOH-100Ca + 0.05Ga-heat-water or NaOH-100Ga-heat-water treatment in the present study would exhibit inhibitory effect on bone resorption and microbial activity so as to increase bone density and prevent infection in the living body, since it releases a concentration of Ga ions comparable to or greater than the reported values [27,49].

## 4. Materials and Methods

### 4.1. Surface Treatment

Pure Ti (Ti > 99.5%; medical grade, ISO5832-2, Nilaco Co., Tokyo, Japan) with dimensions of $10 \times 10 \times 1$ mm$^3$ was abraded with #400 diamond plates, washed with acetone, 2-propanol and ultrapure water in an ultrasonic cleaner for a period of 30 min, then dried at 40 °C. The Ti samples were soaked in 5 mL of 5 M NaOH (Reagent grade; Kanto Chemical Co., Inc., Tokyo, Japan) aqueous solution at 60 °C in an oil bath, then shaken at a speed of 120 strokes/min for a period of 24 h followed by gentle rinse with ultrapure water for a period of 30 s. They were subsequently soaked in 10 mL of a mixed solution of 100 mM CaCl$_2$ (Reagent grade; Kanto Chemical Co., Inc., Tokyo, Japan) and $X$ mM GaCl$_3$ (Reagent grade; Kanto Chemical Co., Inc., Tokyo, Japan) at 40 °C, where $X$ is a range from 0.01 to 0.1 and designated as "100Ca + XGa", shaken at a speed of 120 strokes/min for a period of 24 h, washed and dried in a similar manner. One metal soaked in 10 mL of 100 mM GaCl$_3$ after the NaOH treatment was also prepared and designated as "100Ga". They were subsequently heated to 600 °C at a rate of 5 °C/min and maintained at 600 °C for 1 h, then allowed to be cooled in a Fe–Cr electrical furnace. After the heat treatment, they were soaked in 10 mL of hot water at 80 °C, shaken at a speed of 120 strokes/min for a period of 24 h, washed and dried.

### 4.2. Surface Analysis

#### 4.2.1. Scanning Electron Microscopy

The chemical- and heat-treated samples were coated with a Pt–Pd thin film, then their surfaces and cross sections were observed under field emission scanning electron microscopy (FE-SEM: S-4300, Hitachi Co., Tokyo, Japan) with a voltage of 15 kV.

#### 4.2.2. Energy Dispersive X-ray Analysis

The surface chemical composition of the samples was analysed with an energy dispersive X-ray spectrometer (EDX: EMAX-7000, Horiba Ltd., Kyoto, Japan) by using 9 kV–K for the Ca, Ga, O and Ti in five areas. Their averaged value was used for analysis.

#### 4.2.3. Thin-Film X-ray Diffraction and Fourier Transform Confocal Laser Raman Spectrometry

The surfaces of the samples after the chemical and heat treatments were analysed by a thin-film X-ray diffractometer (TF-XRD: model RNT-2500, Rigaku Co., Tokyo, Japan) and Fourier transform confocal laser Raman spectrometer (FT-Raman: LabRAM HR800, Horiba Jobin Yvon, Longjumeau, France). In TF-XRD analysis, a CuKα X-ray source was used and the incident beam was fixed to an angle of 1° against the sample surface. The measurement was performed at 50 kV and 200 mA. In the FT-Raman measurements, an Ar laser with a wavelength of 514.5 nm was selected as the laser source and its laser power excitation was maintained to be 16 mW.

### 4.2.4. X-ray Photoelectron Spectroscopy

The distribution of elements with a near-surface depth in the metal samples that has been subjected to the chemical and heat treatments was analyzed using X-ray photoelectron spectroscopy (XPS, PHI 5000 Versaprobe II, ULVAC-PHI, Inc., Kanagawa, Japan) under Ar sputtering at a rate of 4 nm per min (SiO$_2$ conversion). An Al-Kα radiation line was used as the X-ray source with the XPS take-off angle at 45 degrees so that the system is able to detect photoelectrons to a depth of 1 to 5 nm from the surface. The calibration of the measured spectra was performed by reference to the C1s peak of the surfactant CH$_2$ groups on the substrate occurring at 284.8 eV in binding energy.

### 4.3. Ion Release

The samples after the chemical and heat treatments were soaked in 2 mL of phosphate-buffered saline (PBS) at a specific concentration (Na$^+$ 158.14, K$^+$ 1.06, Cl$^-$ 155.17, and HPO$_4{}^{2-}$ 4.03 mM) the pH adjusted to 7.4 at 36.5 °C. They were gently shaken at a speed of 50 strokes/min for predetermined periods of up to 7 days. After being removed from the PBS, the Ca and Ga ion concentrations in the PBS were measured by inductively coupled plasma emission spectroscopy (ICP, SPS3100, Seiko Instruments Inc., Chiba, Japan). Three samples were independently prepared for each soaking condition, and the values of average and standard deviation were calculated.

### 4.4. Soaking in SBF

Apatite formation of the samples after the chemical and heat treatments were examined by soaking in 24 mL of SBF that has ion concentrations (Na$^+$ 142.0, K$^+$ 5.0, Ca$^{2+}$ 2.5, Mg$^{2+}$ 1.5, Cl$^-$ 147.8, HCO$_3{}^-$ 4.2, HPO$_4{}^{2-}$ 1.0, and SO$_4{}^{2-}$ 0.5 mM) nearly equal to those of human blood plasma at 36.5 °C for various periods of up to 7 days. The SBF was prepared by dissolving reagent grade NaCl, NaHCO$_3$, KCl, K$_2$HPO$_4$·3H$_2$O, MgCl$_2$·6H$_2$O, CaCl$_2$, and Na$_2$SO$_4$ (Nacalai Tesque Inc., Kyoto, Japan) in ultrapure water, and then buffered at pH = 7.4 with tris(hydroxymethyl)aminomethane (CH$_2$OH)$_3$CNH$_2$ and 1 M HCl (Nacalai Tesque Inc., Kyoto, Japan) at 36.5 °C [50]. After being removed from the SBF, the samples were gently rinsed with ultrapure water and dried, then their surface apatite formation was analysed using TF-XRD, FE-SEM and EDX.

### 4.5. Antibacterial Activity Test

### 4.5.1. Bacterial Culture

Clinically isolated strains of Multidrug Resistant, *Acinetobacter baumannii* (MRAB12, isolated from patients' catheters, Concord Hospital, Sydney, Australia). The bacteria were grown in Tryptone Soy Broth (TSB) over a period of 16 h at 37 °C and 150 rpm. Following the growth period, a 1:10 dilution of bacterial suspension was made in TSB for inoculation of the surfaces.

### 4.5.2. Biofilm Formation

Ti samples as-polished or treated in the manner described in Section 2.1 were placed in 24-well cell culture plates. One millilitre of the diluted bacterial suspension was used to submerge the samples so as to allow adequate contact of the bacterial suspension with the surfaces. Biofilm was allowed to form over a period of 7 days, with media being replaced with 1 mL of fresh media on Day 3 and 5 to ensure sufficient nutrient availability for bacterial growth. On Day 7, media was removed and the samples rinsed with 1 × phosphate buffer saline (PBS) thrice, in order to remove any planktonic or loosely attached bacteria from the surface. Biofilms were then stained using the fluorescent nucleic acid stain SYTO-9 for detection of live cells, and with the cell-impairment nucleic acid stain Propidium iodide (PI) for detection of the dead cells (i.e., cells with compromised membranes) (the LIVE/DEAD Bac-Light Bacterial Viability Kit, Molecular Probes, Thermo Fisher Scientific, Waltham, Massachusetts, USA ). The stain mixture was prepared as per the manufacturer's instructions and added directly to the surfaces, then allowed to incubate under dark at room temperature for a period of 30 min.

### 4.5.3. Confocal Microscopy and Analysis

Live/dead stained biofilm samples were imaged using a Nikon C2 Confocal microscope (Nikon Corp., Tokyo, Japan) with laser settings of 473 nm and 559 nm for green (live) and red (dead) staining, respectively. Fiji ImageJ software (National Institutes of Health, USA) was employed to generate images and quantify the biovolume as well as the percentage of live and dead biofilm.

### 4.5.4. Statistical Analysis

Statistical analysis was performed using GraphPad Prism 7.02 software (GraphPad Software, Inc., La Jolla, CA, USA). The differences in the means between the sample groups were analysed using Tukey's multiple comparisons test, with a $p$-value $< 0.05$ indicating a statistically significant difference.

## 5. Conclusions

We have developed a simple and cost-effective technology to fabricate next-generation interfaces for titanium implants that:

- Notably enhance mineralisation of the matrix directly on the surface, thus supporting bone tissue formation.
- Sustainably release gallium ions, which have positive effects on bone formation and bone quality.
- Strikingly reduce biofilm formation on the surface.Figur

The developed technology is significant, because there are no efficacious treatment options available for patients living with infected implants, hence there is a huge unmet need. Furthermore, the proposed bioactive and antimicrobial surfaces represent a rare breakthrough towards the development of true multifunctional surfaces. We have also demonstrated that the release of the gallium can be tuned to adjust the strength of antimicrobial activity to individual patients/needs and it allows us to use our two-prong approach to eliminate infections and to promote tissue regeneration. This concept is fully translatable for other devices that suffer from high rate of infections and limited integration within bodily environment.

Specifically, our interfaces are composed of Ga-containing calcium titanate (Ga–CT) or gallium titanate (GT) accompanied by rutile and anatase. Ga–CT samples encouraged the apatite formation in SBF within 3 days, but released Ga ions at a low level: 0.23 ppm in PBS within 14 days. In contrast, GT interfaces released slowly 3.75 ppm of Ga ions up to 7 days and formed apatite within 3 days after the final water treatment. Although, a trace amount of Ga was incorporated into the apatite formed on GT samples, it had no effect on the apatite crystal structure. The inhibitory effect on micro and bioactivity was shown got both Ga–CT and GT, with some increase of ABC observed for GT samples. It can be expected that multifunctional interfaces will be particularly useful for orthopaedic and dental implants since it will facilitate direct disposition of the bone at the surface, thus it will promote desired implant integration, while the sustained release of Ga ions will prevent infections and inhibit bone resorption.

**Acknowledgments:** This work was partially supported by Chubu University Grant (AII).

**Author Contributions:** Tadashi Kokubo, Tomiharu Matsushita, Seiji Yamaguchi and Shekhar Nath conceived and designed the experiments; Seiji Yamaguchi, Yoko Sugawara, Shekhar Nath, Kamini Divakarla, Theerthankar Das and Jim Manos performed the experiments; Seiji Yamaguchi and Wojciech Chrzanowski wrote the paper.

**Conflicts of Interest:** The authors declare no conflict of interest.

## References

1. Yan, W.Q.; Nakamura, T.; Kobayashi, M.; Kim, H.M.; Miyaji, F.; Kokubo, T. Bonding of chemically treated titanium implants to bone. *J. Biomed. Mater. Res.* **1997**, *37*, 267–275. [CrossRef]
2. Leeuwenburgh, S.C.G.; Wolke, J.G.C.; Jansen, J.A.; de Groot, K. *Bioceramics and Their Clinical Applications*; Kokubo, T., Ed.; Woodhead Publishing: Cambridge, UK, 2008; Chapter 20; pp. 464–484, ISBN 978-1-84569-204-209.
3. Strange, D.G.T.; Oyen, M.L. Biomimetic bone-like composites fabricated through an automated alternate soaking process. *Acta Biomater.* **2011**, *7*, 3586–3594. [CrossRef] [PubMed]
4. Nayab, S.N.; Jones, F.H.; Olsen, I. Modulation of the human bone cell cycle by calcium ion-implantation of titanium. *Biomaterials* **2007**, *28*, 38–44. [CrossRef] [PubMed]
5. Tsutsumi, Y.; Niinomi, M.; Nakai, M.; Tsutsumi, H.; Doi, H.; Nomura, N.; Hanawa, T. Micro-arc oxidation treatment to improve the hard-tissue compatibility of Ti–29Nb–13Ta–4.6Zr alloy. *Appl. Surf. Sci.* **2012**, *262*, 34–38. [CrossRef]
6. Park, J.W.; Park, K.B.; Suh, J.Y. Effects of calcium ion incorporation on bone healing of Ti6Al4V alloy implants in rabbit tibiae. *Biomaterials* **2007**, *28*, 3306–3313. [CrossRef] [PubMed]
7. Park, J.W.; Kim, Y.J.; Jang, J.H.; Kwon, T.G.; Bae, Y.C.; Suh, J.Y. Effects of phosphoric acid treatment of titanium surfaces on surface properties, osteoblast response and removal of torque forces. *Acta Biomater.* **2010**, *6*, 1661–1670. [CrossRef] [PubMed]
8. Brammera, K.S.; Ohd, S.; Cobba, C.J.; Bjurstenb, L.M.; van der Heydec, H.; Jina, S. Improved bone-forming functionality on diameter-controlled TiO$_2$ nanotube surface. *Acta Biomater.* **2009**, *5*, 3215–3223. [CrossRef] [PubMed]
9. Kim, H.M.; Kokubo, T.; Fujibayashi, S.; Nishiguchi, S.; Nakamura, T. Bioactive macroporous titanium surface layer on titanium substrate. *J. Biomed. Mater. Res.* **2000**, *52*, 553–557. [CrossRef]
10. Takemoto, M.; Fujibayashi, S.; Neo, M.; Suzuki, J.; Matsushita, T.; Kokubo, T.; Nakamura, T. Osteoinductive porous titanium implants: Effect of sodium removal by dilute HCl treatment. *Biomaterials* **2006**, *27*, 2682–2691. [CrossRef] [PubMed]
11. Kokubo, T.; Miyaji, F.; Kim, H.M.; Nakamura, T. Spontaneous formation of bonelike apatite layer on chemically treated titanium metals. *J. Am. Ceram. Soc.* **1996**, *79*, 1127–1129. [CrossRef]
12. Kim, H.M.; Miyaji, F.; Kokubo, T.; Nakamura, T. Preparation of bioactive Ti and its alloys via simple chemical surface treatment. *J. Biomed. Mater. Res.* **1996**, *32*, 409–417. [CrossRef]
13. Kim, H.M.; Miyaji, F.; Kokubo, T.; Nakamura, T. Effect of heat treatment on apatite-forming ability of Ti metal induced by alkali treatment. *J. Mater. Sci. Mater. Med.* **1997**, *8*, 341–347. [CrossRef] [PubMed]
14. Nishiguchi, S.; Fujibayashi, S.; Kim, H.M.; Kokubo, T.; Nakamura, T. Biology of alkali- and heat-treated titanium implants. *J. Biomed. Mater. Res.* **2003**, *67*, 26–35. [CrossRef] [PubMed]
15. Kawanabe, K.; Ise, K.; Goto, K.; Akiyama, H.; Nakamura, T.; Kaneuji, A.; Sugimori, T.; Matsumoto, T. A new cementless total hip arthroplasty with bioactive titanium porous-coating by alkaline and heat treatment: average 4.8-year results. *J. Biomed. Mater. Res. Part B Appl. Biomater.* **2009**, *90*, 476–481. [CrossRef] [PubMed]
16. So, K.; Kaneuji, A.; Matsumoto, T.; Matsuda, S.; Akiyama, H. Is the Bone-bonding Ability of a Cementless Total Hip Prosthesis Enhanced by Alkaline and Heat Treatments? *Clin. Orthop. Relat. Res.* **2013**, *471*, 3847–3855. [CrossRef] [PubMed]
17. Yamaguchi, S.; Nath, S.; Matsushita, T.; Kokubo, T. Controlled release of strontium ions from a bioactive Ti metal with a Ca-enriched surface layer. *Acta Biomater.* **2014**, *10*, 2282–2289. [CrossRef] [PubMed]
18. Yamaguchi, S.; Matsushita, T.; Kokubo, T. A bioactive Ti metal with a Ca-enriched surface layer releases Mg ions. *RSC Adv.* **2013**, *3*, 11274–11282. [CrossRef]
19. Okutsu, Y.; Fujibayshi, S.; Otsuki, B.; Goto, K.; Yamamoto, K.; Yamaguchi, S.; Matsushita, T.; Kokubo, T.; Matsuda, S. *Abstract of the 35th Annual Meeting of the Research Society for Orthopaedic Biomaterials*; Orthopedics Biomaterial Bureau: Nara, Japan, 2015; p. 25.
20. Tian, Y.; Fujibayashi, S.; Yamaguchi, S.; Matsushita, T.; Koubo, T.; Matsuda, S. In vivo study of the early bone-bonding ability of Ti meshes formed with calcium titanate via chemical treatments. *J. Mater. Sci. Mater. Med.* **2015**, *26*, 271. [CrossRef] [PubMed]

21. Kizuki, T.; Matsushita, T.; Kokubo, T. Antibacterial and bioactive calcium titanate layers formed on Ti metal and its alloys. *J. Mater. Sci. Mater. Med.* **2015**, *25*, 1737–1746. [CrossRef] [PubMed]

22. Warrell, R.P., Jr. *Handbook of Metal Ligand Interactions in Biological Fluids, Bioinorganic Medicine*; Berthon, G., Ed.; Marcel Dekker: New York, NY, USA, 1995; Volume 2, pp. 1253–1265.

23. Valappil, S.P.; Ready, D.; Neel, E.A.A.; Pickup, D.M.; Chrzanowski, W.; O'Dell, L.A.; Nweport, R.J.; Smith, M.E.; Wilson, M.; Knowles, J.C. Antimicrobial gallium-doped phosphate-based glasses. *Adv. Funct. Mater.* **2008**, *18*, 732–741. [CrossRef]

24. Cochis, A.; Azzimonti, B.; Della, V.C.; De Giglio, E.; Bloise, N.; Visai, L.; Cometa, S.; Rimondini, L.; Chiesa, R. The effect of silver or gallium doped titanium against the multidrug resistant *Acinetobacter baumannii*. *Biomaterials* **2016**, *80*, 80–95. [CrossRef] [PubMed]

25. Kaneko, Y.; Thoende, M.; Olakanmi, O.; Britigan, B.E.; Singh, P.K. The transition metal gallium disrupts Pseudomonas aeruginosa iron metabolism and has antimicrobial and antibiofilm activity. *J. Clin. Investig.* **2007**, *117*, 877–888. [CrossRef] [PubMed]

26. Niesvizky, R. Gallium nitrate in multiple myeloma: prolonged survival in a cohort of patients with advanced-stage disease. *Semin. Oncol.* **2003**, *30*, 20–24. [CrossRef]

27. Verron, E.; Masson, M.; Khoshniat, S.; Duplomb, L.; Wittrant, Y.; Baud'huin, M.; Badran, Z.; Bujoli, B.; Janvier, P.; Scimeca, J.C.; et al. Gallium modulates osteoclastic bone resorption in vitro without affecting osteoblasts. *Br. J. Pharmacol.* **2010**, *159*, 1681–1692. [CrossRef] [PubMed]

28. Bockman, R.S. Studies on the mechanism of action of gallium nitrate. *Semin. Oncol.* **1991**, *18*, 21–25. [PubMed]

29. Schlesinger, P.H.; Teitelbaum, S.L.; Blair, H.C. Osteoclast inhibition by $Ga^{3+}$ contrasts with bisphosphonate metabolic suppression: competitive inhibition of $H^+$ ATPase by bone-bound gallium. *J. Bone Miner. Res.* **1991**, *6*, S127.

30. Yamaguchi, S.; Takadama, T.; Matsushita, T.; Nakamura, T.; Kokubo, T. Cross-sectional analysis of the surface ceramic layer developed on Ti metal by NaOH-heat treatment and soaking in SBF. *J. Ceram. Soc. Jpn.* **2009**, *117*, 1126–1130. [CrossRef]

31. Sun, X.; Li, Y. Synthesis and characterization of ion-exchangeable titanate nanotubes. *Chem. Eur. J.* **2003**, *9*, 2229–2238. [CrossRef] [PubMed]

32. Kawai, T.; Kizuki, T.; Takadama, H.; Matsushita, T.; Kokubo, T.; Unuma, H.; Nakamura, T. Apatite formation on surface titanate layer with different Na content on Ti metal. *J. Ceram. Soc. Jpn.* **2010**, *9*, 19–24. [CrossRef]

33. Kolwn'ko, Y.V.; Kovnir, K.A.; Gavrilov, A.I.; Garshev, A.V.; Frantti, J.; Lebedev, O.I.; Churagulov, B.R.; Tendeloo, G.V.; Yoshimura, M. Hydrothermal synthesis and characterization of nanorods of various titanates and titanium dioxide. *J. Phys. Chem. B* **2006**, *110*, 4030–4038. [CrossRef]

34. Kizuki, T.; Matsushita, T.; Kokubo, T. Preparation of bioactive Ti metal surface enriched with calcium ions by chemical treatment. *Acta Biomater.* **2010**, *6*, 2836–2842. [CrossRef] [PubMed]

35. Yamaguchi, S.; Takadama, H.; Matsushita, T.; Nakamura, T.; Kokubo, T. Apatite-forming ability of Ti–15Zr–4Nb–4Ta alloy induced by calcium solution treatment. *J. Mater. Sci. Mater. Med.* **2010**, *21*, 439–444. [CrossRef] [PubMed]

36. Yamaguchi, S.; Kizuki, T.; Takadama, H.; Matsushita, T.; Nakamura, T.; Kokubo, T. Formation of a bioactive calcium titanate layer on gum metal by chemical treatment. *J. Mater. Sci. Mater. Med.* **2012**, *23*, 873–883. [CrossRef] [PubMed]

37. Naumkin, A.V.; Kraut-Vass, A.; Gaarenstroom, S.W.; Powell, C.J. *NIST X-ray Photoelectron Spectroscopy Data Base Version 4.1*; the Measurement Services Division of the National Institute of Standards and Technology: Gaithersburg, MD, USA, 2012.

38. Liu, X.; Khan, M.; Liu, W.; Xiang, W.; Guan, M.; Jiang, P.; Cao, W. Synthesis of nanocrystalline Ga–$TiO_2$ powders by mild hydrothermal method and their visible light photoactivity. *Ceram. Int.* **2015**, *41*, 3075–3080. [CrossRef]

39. Deng, Q.R.; Gao, Y.; Xia, X.H.; Chen, R.S.; Wan, L.; Shao, G. V and Ga co-doping effect on optical absorption properties of $TiO_2$ thin films. *J. Phys. Conf. Ser.* **2009**, *152*, 012073. [CrossRef]

40. Hinkle, C.L.; Milojevic, M.; Brennan, B.; Sonnet, A.M.; Aguirre-Tostado, F.S.; Hughes, G.J.; Vogel, E.M.; Wallace, R.M. Detection of Ga suboxides and their impact on III-V passivation and Fermi-level pinning. *Appl. Phys. Lett.* **2009**, *94*, 162101. [CrossRef]

41. Sham, T.K. X-ray photoelectron spectroscopy (XPS) studies of clean and hydrated $TiO_2$ (rutile) surfaces. *Chem. Phys. Lett.* **1979**, *68*, 426–432. [CrossRef]

42. Li, P.; Ohtsuki, C.; Kokubo, T.; Nakanishi, K.; Soga, N.; de Groot, K. The role of hydrated silica, titania, and alumina in inducing apatite on implants. *J. Biomed. Mater. Res.* **1994**, *28*, 7–15. [CrossRef] [PubMed]

43. Kim, H.M.; Himeno, T.; Kawashita, M.; Lee, J.H.; Kokubo, T.; Nakamura, T. Surface potential change in bioactive titanium metal during the process of apatite formation in simulated body fluid. *J. Biomed. Mater. Res.* **2003**, *67*, 1305–1309. [CrossRef] [PubMed]

44. Textor, M.; Sitting, C.; Franchiger, V.; Tosatti, S.; Brunette, D.M. Properties and biological significance of natural oxide films on titanium and its alloys. In *Titanium in Medicine*; Brunette, D.M., Tengrall, P., Textor, M., Thomsen, P., Eds.; Springer: New York, NY, USA, 2001; Chapter 7; pp. 172–230, ISBN 978-3-642-63119-1.

45. Takadama, H.; Kim, H.M.; Kokubo, T.; Nakamura, T. TEM-EDX Study of mechanism of bonelike apatite formation on bioactive titanium metal in simulated body fluid. *J. Biomed. Mater. Res.* **2001**, *57*, 441–448. [CrossRef]

46. Blumenthal, N.C.; Cosma, V.; Levine, S. Effect of gallium on the in vitro formation, growth, and solubility of hydroxyapatite. *Calcif. Tissue Int.* **1989**, *45*, 81–87. [CrossRef] [PubMed]

47. Kokubo, T. Design of bioactive bone substitutes based on biomineralization process. *Mater. Sci. Eng. C* **2005**, *25*, 97–104. [CrossRef]

48. Fukuda, A.; Takemoto, M.; Saito, T.; Fujibayashi, S.; Neo, M.; Yamaguchi, S.; Kizuki, T.; Matsushita, T.; Niinomi, M.; Kokubo, T.; et al. Bone bonding bioactivity of Ti metal and Ti–Zr–Nb–Ta alloys with Ca ions incorporated on their surfaces by simple chemical and heat treatments. *Acta Biomater.* **2011**, *7*, 1379–1386. [CrossRef] [PubMed]

49. Hall, T.J.; Chambers, T.J. Gallium inhibits bone resorption by a direct effect on osteoclasts. *Bone Miner.* **1990**, *8*, 211–216. [CrossRef]

50. Kokubo, T.; Takadama, H. How useful is SBF in predicting in vivo bone bioactivity? *Biomaterials* **2006**, *27*, 2907–2915. [CrossRef] [PubMed]

© 2017 by the authors. Licensee MDPI, Basel, Switzerland. This article is an open access article distributed under the terms and conditions of the Creative Commons Attribution (CC BY) license (http://creativecommons.org/licenses/by/4.0/).

*Article*

# Evaluation of MC3T3 Cells Proliferation and Drug Release Study from Sodium Hyaluronate-1,4-butanediol Diglycidyl Ether Patterned Gel

**Sumi Bang [1], Dipankar Das [1,2], Jiyun Yu [2] and Insup Noh [1,2,\*]**

[1]   Convergence Institute of Biomedical Engineering and Biomaterials, Seoul National University of Science of Technology, Seoul 01811, Korea; bobosumi48@gmail.com (B.S.); dipankardas@seoultech.ac.kr (D.D.)

[2]   Department of Chemical and Biomolecular Engineering, Seoul National University of Science of Technology, 232 Gongneung-ro, Nowon-gu, Seoul 01811, Korea; jiyun_@seoultech.ac.kr

\*   Correspondence: insup@seoultech.ac.kr; Tel.: +82-2-970-6603; Fax: +82-2-977-8317

Received: 18 September 2017; Accepted: 5 October 2017; Published: 14 October 2017

**Abstract:** A pattern gel has been fabricated using sodium hyaluronate (HA) and 1,4-butanediol diglycidyl ether (BDDGE) through the micro-molding technique. The cellular behavior of osteoblast cells (MC3T3) in the presence and absence of dimethyloxalylglycine (DMOG) and sodium borate (NaB) in the pattern gel (HA-BDDGE) has been evaluated for its potential application in bone regeneration. The Fourier transform infrared spectroscopy (FTIR), $^{13}$C-nuclear magnetic resonance spectroscopy ($^{13}$C NMR), and thermogravimetric analysis (TGA) results implied the crosslinking reaction between HA and BDDGE. The scanning electron microscopy (SEM) analysis confirmed the formation of pattern on the surface of HA-BDDGE. The gel property of the crosslinked HA-BDDGE has been investigated by swelling study in distilled water at 37 °C. The HA-BDDGE gel releases DMOG in a controlled way for up to seven days in water at 37 °C. The synthesized gel is biocompatible and the bolus drug delivery results indicated that the DMOG containing patterned gel demonstrates a better cell migration ability on the surface than NaB. For local delivery, the pattern gel with 300 μM NaB or 300 μM DMOG induced cell clusters formation, and the gel with 150 μM NaB/DMOG showed high cell proliferation capability only. The vital role of NaB for bone regeneration has been endorsed from the formation of cell clusters in presence of NaB in the media. The in vitro results indicated that the pattern gel showed angiogenic and osteogenic responses with good ALP activity and enhanced HIF-1$\alpha$, and Runx2 levels in the presence of DMOG and NaB in MC3T3 cells. Hence, the HA-BDDGE gel could be used in bone regeneration application.

**Keywords:** cell proliferation; bone regeneration; hyaluronate; MC3T3 cell; pattern gel

---

## 1. Introduction

The development of size and shape specific biologically applicable hydrogels have opened inventive prospects in addressing challenges in tissue engineering like tissue architecture, vascularization, and cell seeding [1]. Hydrogels are water-swollen, physically or chemically crosslinked polymers which are remarkable in regenerative tissue engineering because of their capability to mimic the physical characteristics of tissues [2,3]. One of the precise advantages of hydrogels is their simple treating conditions, which allow directly for cell encapsulation in the gel [2]. The capacity to seed or encapsulate cells within a three-dimensional (3-D) network has distinct significance, because these substrates well replicate in vivo microenvironments than that of cell seeding on two-dimensional (2-D) materials [2,4,5].

For in vitro tissue engineering and scaffold designs, the precise spatial mechanism and association of cells are vital to characterize the appropriate microenvironment around the cells, simulating in vivo physical and chemical cues [6]. Cell–cell contact, biomolecules delivery, and tissue architecture are the main factors that regulate cell behaviors. Even though, in tissue engineered scaffolds, cells have self-assemble capacity to recover essential features of their cell-cell interactions, while, many of the interactions are eternally disappeared in the time of tissue isolation and seeding processes. Moreover, the homogeneous cell seeding throughout the scaffolds is quite difficult, because of the presence of a large number of cells in the border of scaffolds. The recent archetype of tissue engineering is concentrated on the application of cell-seeded scaffolds for in vitro generation of tissues and subsequent in vivo implantation [7]. Surface topography has been employed for stimulating the cellular orientation and monitoring the biological activity of cells within the structure of constructs [7]. However, the main drawbacks of in vitro tissue engineering is the failure to recapitulate cell dense paradigms with the required variety in cell types and structural complexity to mimic native tissues [8]. In vivo, cells are strongly organized in 3D architectures controlled by factors like cell-cell, cell-extracellular matrix (ECM), and cell-soluble factors [9]. Bioactive signals and stem cells that can respond to biomimetic morphogens and scaffolds are also important requisites in tissue engineering [10]. Bioactive signals demonstrate numerous vital roles on cell function and behavior. In most biological studies, soluble biochemical signals, for instance, growth factors or cytokines are added directly into the media to preserve and/or control cell behaviors in vitro. Conversely, these systems cannot precisely mimic convinced in vivo biological signaling motifs, which are regularly immobilized to ECM and also exhibit spatial gradients, which are critical for tissue morphology. In addition, biochemical cues and biophysical characteristics, for example, material hardness can influence cell activities but is not simple to control in conventional cell culturing practices [11]. One of the biggest challenges in tissue engineering is mass transfer limitations. This is the limiting factor in the size of any tissue developed in vitro, in addition to the successive incorporation of these constructs in vivo [12]. In vitro viable tissue-like systems frequently display dimensions beyond convenient perfusion limits, and have no functional blood vessels with flowing blood to supply nutrients and oxygen, and to remove waste products [13].

Currently, tissue engineering has turned into one of the most universally exploited approaches for cartilage and bone tissue reconstruction and regeneration [14]. The commonly used techniques for bone repair, like autografting and allografting are restricted due to the risks of donor-site morbidity, potential infection, and a high non-union rate with host tissues. Bone defects are one of the principal reasons of morbidity and disability in aged patients. Consequently, developing a technique to perfectly and enduringly repair the damaged cartilage and bone tissue is of noteworthy clinical attention for patients with cartilage lesions and bone defects. Preferably, for clinical application, the scaffolds of both cartilage and bone tissue engineering should be porous, highly biocompatible, nontoxic, and competent of endorsing cell differentiation and new tissue formation. They should also have stable mechanical properties, degrade in response to the formation of new tissue, facilitate the diffusion of nutrients and metabolites, adhere and integrate with the surrounding native tissue, and properly fill the injured site. Till date, a variety of biopolymers based hydrogels and hydrogels composites were employed in vitro and in vivo tissue engineering, especially for bone regeneration. For example, Han et al. developed alginate-based hybrid hydrogel that efficiently promotes the adhesion, proliferation, and differentiation of osteogenic and angiogenic cells [15]. Kook et al. prepared collagen-based sponge, which simulated natural bone tissue and supports cellular activity by enhancing cell adhesion and proliferation [16]. Vo et al. designed $N$-isopropylacrylamide/gelatin-based composite hydrogel, which enhanced bony bridging and mineralization within the defect and direct bone-implant contact [17]. Fu et al. designed PEG–PCL–PEG, collagen, and nanohydroxyapatite-based hydrogel that demonstrated a good biocompatibility and in vivo studies showed better performance in bone regeneration [18]. Dhivyaet al. reported chitosan/nanohydroxyapatite/β-glycerophosphate-based hydrogel for in vivo studies in a rat bone defect, where the hydrogel accelerated bone formation at molecular and cellular levels [19]. Vishnu Priya et al. developed chitin and poly(butylene succinate) based hydrogel, which enhanced the initiation of differentiation and expression of alkaline phosphatase and osteocalcin, thus indicating its promise for regenerating irregular bone defects [20]. In vivo study of bone regeneration using alginate-bone ECM hydrogels was also reported by Gothard et al. [21]. A clinical study for bone generation was performed by Laino et al. [22]. They reported a comparative study of the histological aspects of bone formation in atrophic posterior mandibles augmented by autologous bone block from chin area with corticocancellous bone block allograft used as inlays with the sandwich technique [22]. Biomimetic mineralization on a macroporous cellulose-based matrix for bone regeneration was reported by Petrauskaite et al. [23] The porous cellulose matrix was non-cytotoxic, allowed the adhesion and proliferation of human osteoblastic cells, while both properties were improved on the mineralized cellulose matrices [23].

Our aim was to develop a shape-specific, hyaluronate-based patterned hydrogel for its application to bone regeneration. Sodium hyaluronate was chosen owing to its unique properties like abundant as natural ECM in human body, and its physico-chemical and immune-neutral characteristics [24,25]. The valuable properties led to development of various hyaluronic acid (HA) hydrogels for biomedical applications such as dermal fillers [26,27], cartilage regeneration [28,29], nucleus pulposus regeneration [30], and wound healing [31,32]. The microfabrication method was employed to pattern gel because of its noteworthy significance in tissue engineering as it can be employed to replicate structures (0.1–10 μm), to regulate the microenvironment of individual cells (10–400 μm), to control the structure of clusters of cells (>400 μm), and to control the interactions between multiple cell clusters [1]. In this aspect, soft lithography method has been established as an economical and effective process for patterning of bare glass [33] or metal-coated glass [34], polystyrene materials to flexible poly(dimethyl siloxane) (PDMS) materials [35], and biomaterials [36]. Soft lithography includes stamps fabricated from an elastomer or soft material like PDMS [6]. The PDMS stamp can mark ECM, self-assembled monolayer (SAM), and hydrogel to print on materials [6,24]. The well-defined ECM micro-patterns showed a significant effect on numerous imperative cell behaviors, such as cell adhesion and spreading [37], cell proliferation and differentiation [38], cell polarity [39], and migration [40].

In this study, a pattered hydrogel has been fabricated using sodium hyaluronate (HA) as biopolymer, 1,4-butanediol diglycidyl ether (BDDGE) as crosslinker, and sodium hydroxide (NaOH) as base by the micro-molding technique. BDDGE has been selected because of the presence of two epoxy rings, where, neucleophilile can attack on both ends and crosslinking will take place in HA. Scanning electron microscopy (SEM) analysis confirmed the formation of uniform pattern on the surface of gel. The hydrogel is biocompatible against MC3T3 cells. The HA-BDDGE gel releases dimethyloxalylglycine (DMOG) in a controlled manner for up to seven days in distilled water at 37 °C. It is observed that the hydrogel with more than 100 μM NaB showed MC3T3 cell clusters formation at day seven. In the cell proliferation study, the system with bolus drug delivery showed the best cell proliferations at the concentrations of 100 μM NaB and DMOG, individually. The presence of NaB helps the formation of MC3T3 cell clusters, supporting the vital role of NaB in bone regeneration. It has been also noticed that when drugs were delivered locally, the HA-BDDGE patterned gel showed higher intensity of ALP and Runx2, indicating a better bone regeneration ability. Although the experiment results demonstrated significant bone regeneration characteristics of the HA-BDDGE gel, still this study has some limitations that are associated with the in vitro study along with the lack of porosity in hydrogel and absence of adequate mechanical strength. Thus, to get better efficiency and the perfect ability for clinical and surgical experiments further modification would be appreciated in future. With the variation of amount of crosslinker and by mineralization with inorganic particles or ceramics, hydrogels scaffolds with different porosity and adequate mechanical property could be achieved. Furthermore, with the incorporation of biological matrix, such as gelation or collagen in the hydrogel network cell adhesiveness and tissue regeneration could be improved in future for use in a clinical study. Finally, the HA-BDDGE gel, with well micro-patterned architecture, biocompatibility, controlled release ability of DMOG drug, clusters formation ability of MC3T3 cells and higher intensity in alkaline phosphatase activity (ALP), hypoxia induced factor (HIF)-1α, and runt-related transcription factor 2 (Runx2) studies signified that the pattern gel could be used in bone regeneration application.

## 2. Materials and Methods

### 2.1. Materials

Hyaluronic acid (HA) (MW: 575 kDa) was received as gift from Hanmi Pharmaceutical Co. (Pyeongtaek, Korea), 1,4-butanediol diglycidyl ether (BDDGE, MW: 202 Da), α-MEM and sodium butyrate (NaB) were purchased from Sigma-Aldrich (St. Luis, MO, USA). Sodium hydroxide (MW: 40 Da) was purchased from Yakuri Pure Chemical Co. (Kyoto, Japan). Poly(dimethyl siloxane) (PDMS) (184 Sylgard) was purchased from Dow Corning (Auburn, MI, USA). Penicillin-streptomycin was purchased Lonza Korea (Basel, Switzerland). Cell counting kit-8 (CCK-8) solution was bought from Dojindo Laboratories (Kumamoto, Japan). Live & dead viability/cytotoxicity kit for mammalian cells was purchased from Invitogen (Carlsbad, CA, USA). Dimethyloxalylglycine (DMOG) was procured from Cyman Chemical (Ann Arbor, MI, USA). Anti-RUNX2 antibody, anti-HIF-1-alpha antibody, anti-beta actin antibody, anti-osteocalcin antibody and Goat Anti- Mouse IgG H&L (HRP) were bought from Abcam (Cambridge, UK). RIPA buffer, protease inhibitor, and phosphatase inhibitor were purchased from Sigma-Aldrich (USA).

### 2.2. Fabrication of Sodium Hyaluronate-BDDGE Patterned Gel (HA-BDDGE)

Fabrication of sodium hyaluronate-BDDGE patterned gel (HA-BDDGE) was performed using the method reported in our previous paper [24]. Briefly, sodium hyaluronate (HA, 0.18 g) was homogeneously mixed in 1 mL of 1% NaOH solution ($w/v$ %) using centrifuge at room temperature with 10,000 rpm speed for 2 h. Then, 72 μL of BDDGE was added and mixed with a spatula. After that, the mixture was transported into a 10 mL syringe, injected on the PDMS mold supported with teflon-glass slide, and kept 24 h for crosslinking and pattern formation. Afterwards, the crosslinked patterned gel was taken out from the PDMS mold and put in 100% ethanol for 24 h to remove the unreacted reagents. Then, the patterned gel was immersed in phosphate buffered solution (PBS) solution for three days by exchanging the PBS solution after every 12 h. The gel was dried in lyophilizer at −75 °C for further characterizations.

### 2.3. Characterizations

The attenuated total reflectance Fourier transform infrared (ATR-FTIR) spectra of HA, BDDGE, and dried HA-BDDGE patterned gel were recorded using ATR-FTIR spectrometer (Model: Travel IR, Smiths Detection, Edgewood, MD, USA) in the wavelength range of 650–4000 cm$^{-1}$. The $^{13}$C NMR analyses of HA and dried HA-BDDGE gel were executed in solid state, while BDDGE was carried out in liquid state with 700 MHz nuclear magnetic resonance (NMR) spectrometer (Model: DD2 700, Agilent Technologies-Korea, Santa Clara, CA, USA). The TGA analyses of HA and dried HA-BDDGE gel were carried out using thermogravimetric analyzer (Model: DTG-60, Shimadzu, Kyoto, Japan) under nitrogen atmosphere. The scan rate was 5 °C/min. The surface morphology of HA and dried patterned HA-BDDGE gel were observed by SEM (Model: SEM, TESCAN VEGA3, Tescan, Seoul, Korea).

### 2.4. Swelling Study

The % swelling of the HA-BDDGE patterned gel was evaluated gravimetrically. In brief, the pre-weighed dried patterned gel ($2r$ = 1 cm) was immersed in 100 mL distilled water at room temperature (25 °C) for 6 h. After a regular interval (1 h), the patterned gel was taken out from distilled water and the surface water was blotted off by tissue paper. Then, the patterned gel was reweighed until equilibrium of their weight was achieved. The % swelling was calculated by the Equation (1):

$$\text{Swelling } (\%) = \frac{\text{Weight of gel} - \text{Initial dried weight of gel}}{\text{Initial dried weight of gel}} \times 100 \tag{1}$$

*2.5. DMOG Loading in HA-BDDGE Patterned Gel and In Vitro Release Study*

For DMOG loading inside the HA-BDDGE patterned gel, dried gel samples were immersed into 2 mL of 25, 50 and 100 μM DMOG solutions at room temperature for 2 days in a 12 well plate. After absorption of drug solutions, the gels were dried in lyophilizer at −75 °C.

The release study was performed in distilled water (pH: 7). After 1, 3, 6, 12, 24, 72, 120 and 144 h, aliquots were taken out and absorptions were recorded by UV-Vis spectrophotometer (Model: BioMATE 3, Thermo Scientific, Madison, WI, USA).

*2.6. In Vitro MC3T3 Cell Culture on the Surface of HA-BDDGE Patterned Gel*

An osteoblast precursor cell line derived from Mus musculus (mouse) calvaria (MC3T3, Sigma Aldrich Co., St. Louis, MO, USA) was used after 10 passage for in vitro cell study. The HA-BDDGE patterned gel was sterilized by autoclave (AC-02, Jeio Tech, Daejeon, Korea) for 24 h. Then, the MC3T3 cells at the density of 10,000 cells/cm$^2$ were cultured on the surfaces of patterned gel with/without drugs for 7 days. The α-MEM media containing both 10% fetal bovine serum (Gibco Life Science, Waltham, MA, USA) and penicillin-streptomycin (100 unit/mL) was added in the 24 well plate and incubated with 5% CO$_2$ at 37 °C.

Cell adhesion and proliferation were evaluated with the CCK-8 after seeding MC3T3 cells on the surface of the hydrogel. The cell number was counted by the CCK-8 assay with a microplate reader (Tecan, Port Melbourne VIC, Australia). In brief, 100 μL CCK-8 solution was mixed with 900 μL of α-MEM medium in a 15 mL tube. Afterwards, the culture media was removed and the mixed CCK-8 solution was put in the 24 well plate and incubated for 2 h with 5% CO$_2$ at 37 °C. After 2 h, 100 μL medium was transferred into a 96 well plate and the optical density was measured at the wavelength of 450 nm by the microplate reader.

*2.7. Live & Dead Assay*

In vitro cell viability and adhesions of the HA-BDDGE patterned gel were observed with MC3T3 cells in the 24-well culture plate. Live & dead viability/cytotoxicity kit for mammalian cells was prepared according to the protocol suggested by the vendor (Invitrogen, Carlsbad, CA, USA). The 1.2 μL of 2 mM ethidium homodimer-1 (EthD-1) and 0.3 μL of 4 mM calcein AM were added into 600 μL PBS and used for live and dead assay. The solution was put in the well plate and incubated for 30 min with 5% CO$_2$ at 37 °C. The images of the MC3T3 cells on the HA-BDDGE patterned gel were captured by a fluorescence microscope (Leica DMLB, Wetzlar, Germany).

*2.8. Alkaline Phosphatase (ALP) Activity Assay*

ALP activities were determined by measuring the amount of *p*-nitrophenol produced using *p*-nitrophenol phosphate substrate. Cell lysates were mixed with alkaline buffer solution and gently shaken for 10 min. ALP substrate was added at room temperature for 30 min. After that, the reaction was stopped with the addition of 0.05 (N) NaOH, and the absorbance at 405 nm was read and compared with a standard curve prepared with *p*-nitrophenol standard solution.

*2.9. Western Blot Analysis*

The in vitro protein expressions of HIF-1α and Runx2 of MC3T3 cells on the patterned hydrogel with/without drugs by using western blot assay. After loading MC3T3 cells at a density of 10,000 cells/cm$^2$, cell culture lasted for seven days by employing medium with/without DMOG, NaB. Tris buffered solution (TBS) washing was performed on the cell cultured samples and then radio-immunoprecipitation assay (RIPA) buffer with protease and phosphatase inhibitors was loaded on each well and patterned gels. Cells were harvested from the surfaces by using cell scraper, transferred to a 1 mL microtube in cold and stored at 4 °C in a refrigerator for 30 min. Centrifuge was performed at by using 16,000 rpm at 4 °C for 20 min and then surface layer was transferred to a new microtube. Cell lysate was obtained by heating the cell solutions with Lammeli sample buffer at 95 °C for 5 min, and then transferred into PVDF membrane after loading into DS-PAGE gel. The PVDF membrane was blocked with 5% skim milk solution. Primary antibodies were grafted by incubating with anti-HIF-1α antibody, anti-RUNX-2 antibody, anti-osteocalcin antibody, and then secondary antibodies were done by incubating with goat anti-mouse IgG connected with horse radish protein (HRP). After loading ECL solution in the PVDF membrane with secondary antibody, excitation of drugs was measured with X-ray film by using β-actin as a loading control.

*2.10. Statistical Analysis*

Data were expressed as mean ± standard deviation. Statistical significance was assessed with one-way and multi-way ANOVA by employing the SPSS 18.0 program (ver. 18.0, SPSS Inc., Chicago, IL, USA). The comparisons between two groups were carried out using a *t*-test. The samples were considered as significantly different when $p < 0.05$.

## 3. Results and Discussion

*3.1. Fabrication of Sodium Hyaluronate-BDDGE Patterned Gel (HA-BDDGE)*

The HA-BDDGE gel was synthesized using HA as biopolymer, BDDGE as crosslinking agent, and NaOH as base. It is assumed that the base (NaOH) abstracts hydroxyl proton from HA and form negative charge over oxygen atom, which act as neucleophilile in the reaction media. The nucleophile attacks on the less hindered electrophilic center of the epoxide rings of BDDGE and opens the epoxide rings. Thus, two HA moieties react with two epoxide rings of BDDGE and form a covalent bond between HA and BDDGE (Scheme 1). Hence, it is supposed that one BDDGE molecule covalently crosslinked two HA molecules through the nucleophilic addition reaction. For the formation of pattern gel, the micro-molding technique was used, where PDMS mold acted as fabrication chamber. The pattern formation in the gel was confirmed by SEM analysis, which is described in the characterization section.

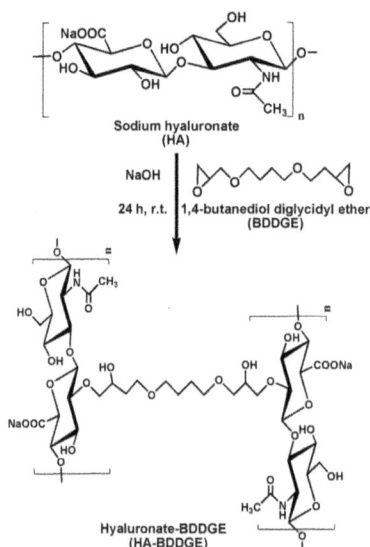

**Scheme 1.** Schematic representation of the reaction between sodium hyaluronate (HA) and 1,4-butanediol diglycidyl ether (BDDGE).

## 3.2. Characterizations

Figure 1 represents the ATR-FTIR spectra of HA, BDDGE, and dried HA-BDDGE gel. In the FTIR spectrum of sodium hyaluronate (HA, Figure 1a), the peaks at 3301, 2898, 1610, 1592, 1407, and 1038 cm$^{-1}$ are because of the stretching vibrations of O–H/N–H bond, C–H bond, C=O bond, amide-II, C–O bond of –COONa group, and C–O–C bond, respectively [41]. The peaks at 1376 and 947 cm$^{-1}$ are due to the vibrations of C–H bending and C–O–H deformation, respectively [41]. In the FTIR spectrum of BDDGE (Figure 1b), the characteristics peaks at 2927, 2865, 1253, 1100, and 908 cm$^{-1}$ are responsible for C–H stretching of epoxy ring, C–H stretching –CH$_2$ bond, C–C bond, C–O–C stretching, and C–O stretching vibrations of epoxy ring, respectively [42]. While, in the FTIR spectrum of HA-BDDGE gel (Figure 1c), the peaks at 3317, 2924, 1608, 1562, 1405, 1374, and 946 cm$^{-1}$ are because of the stretching vibrations of O–H/N–H bond, C–H bond, C=O bond, amide-II, C–O bond of –COONa group, vibration C–H bending, and C–O–H deformation, respectively. These peaks suggest the presence of HA moiety in the HA-BDDGE gel. Again, the peaks for C–O–C stretching vibrations of HA and BDDGE moieties merged and gave a peak in the spectrum of HA-BDDGE gel with high intensity at 1036 cm$^{-1}$ (Figure 1c). Most importantly, the disappearance of the peak for C–O bond (908 cm$^{-1}$) of epoxy ring indicates the reaction between –OH groups of HA and epoxy rings of BDDGE (Figure 1c). Whereas, the increase of peak intensity at 3317 cm$^{-1}$ suggest the appearance of new free –OH groups due to the reaction between HA and BDDGE (Figure 1c and Scheme 1).

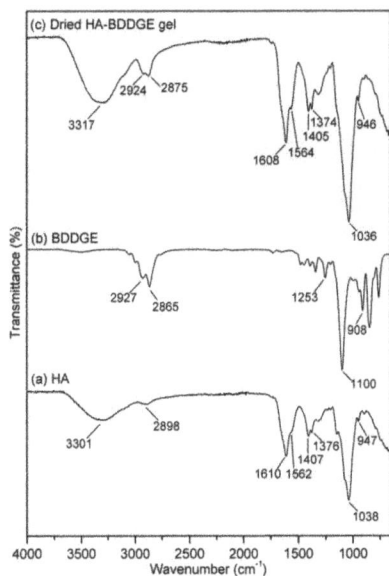

**Figure 1.** ATR-Fourier transform infrared spectroscopy (FTIR) spectra of (a) HA, (b) BDDGE, and (c) dried sodium hyaluronate-BDDGE patterned gel (HA-BDDGE) gel.

Figure 2 describes the $^{13}$C-nuclear magnetic resonance (NMR) spectra of HA (solid state), BDDGE (liquid state), and dried HA-BDDGE gel (solid state). In the NMR spectrum of HA (Figure 2a), the chemical shifts at δ = 174.1, 103.8, 76.2, and 24.6 ppm are due to the presence of carbon atoms of C=O groups (C6, C7), anomeric position (C1, C1'), polysaccharide rings (C2–C5, C2'–C6'), and –CH$_3$ group (C8), respectively [43]. In the NMR spectrum of BDDGE (Figure 2b), the sharp chemical shifts at δ = 44.8 and 51.5 ppm are owing to the carbon atoms of epoxy rings (C1, C1') and (C2, C2'), respectively. While, the chemical shifts at δ = 25.2 and 70.9 ppm are because of the chain carbons (C5, C5') and (C3–4, C3'–4'), respectively (Figure 2b). In the NMR spectrum, HA-BDDGE gel showed the chemical shifts at δ = 174.1, 101.9, 74.8, 71.4, 26.7, and 23.8, which are responsible for the carbon atoms of –C=O (C6, C7), anomeric position (C1, C1′), polysaccharide rings (C2–C5, C2′–C6′), C9–C12 positions, C13, and –CH₃ group (C8), respectively (Figure 2c). The presence of characteristics peaks for C=O group and anomeric carbon (C1, C1′) and –CH₃ group (C8) imply the presence of HA in the gel network (Figure 2c). While, the peaks for the carbons of C13 and between C9–C12 confirmed the presence of BDDGE in the gel network and successful formation of HA-BDDGE compound (Figure 2c).

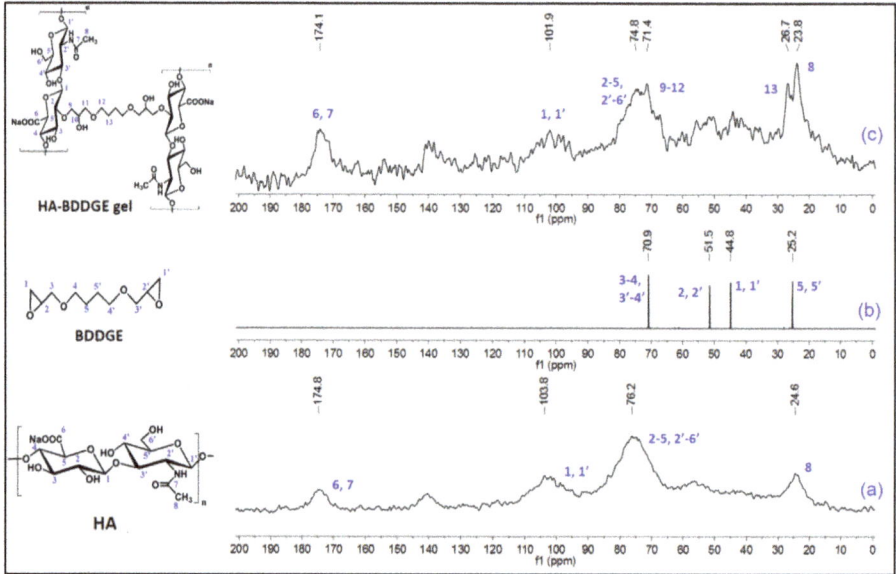

**Figure 2.** $^{13}$C NMR spectra of (**a**) HA (solid state), (**b**) BDDGE (liquid state), and (**c**) dried HA-BDDGE gel (solid state).

The thermal properties of HA and dried HA-BDDGE gel were analyzed by thermogravimetric analysis (TGA) and results are shown in Figure 3.

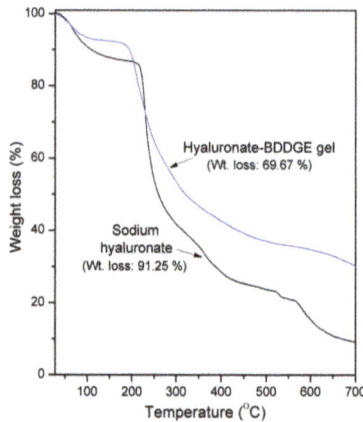

**Figure 3.** Thermogravimetric analysis (TGA) plots of HA and dried HA-BDDGE patterned gel.

A significant weight loss has been noticed for two samples between the temperature range of 50–520 °C because of the thermal breakdown of HA [44]. The initial weight loss for HA between 28 and 100 °C is owing to moisture evaporation (Figure 3). A sharp weight loss zone is seen between 200–260 °C, then relatively slow decompositions are noticed between 260–400 °C, and 400–700 °C, which are owing to the complete breakdown of polysaccharide residue. The HA showed ~91.25% weight loss in the TGA analysis (Figure 3). In the TGA plot of HA-BDDGE gel, the first weight loss region (28–100 °C) is because of the evaporation of moisture (Figure 3). The steady second (160–310 °C), third (310–500 °C), and fourth (500–700 °C) weight loss zones are due to the decomposition of HA, BDDGE unit and completely breakdown of crosslinked network. The weight loss of HA-BDDGE gel was found as 69.67%, which indicates that the covalent attachment between HA and BDDGE increased the thermal stability of the HA-BDDGE gel.

Figure 4 depicts the SEM images of HA and dried HA-BDDGE patterned gel. From Figure 4a, it is observed that HA exhibits an aggregated pod-like structure. After modification with BDDGE, the morphology of HA totally changed and the gel appeared with a relatively smooth surface (Figure 4b). While, the pattern shape is clearly observed from the both surface (Figure 4b) and cross-section images (Figure 4c). It has also been noticed that the width of individual pattern is approximately 5 µm, whereas, the gap between two pattern (valley) is 15 µm (Figure 4b). It is expected that the regular distribution of pattern in the gel structure could be assisted for high adhesion and proliferation of cells.

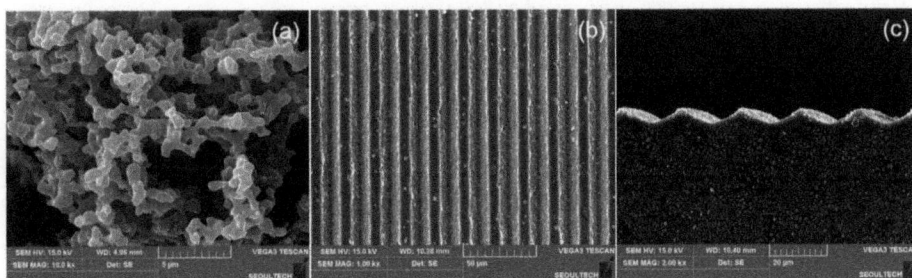

**Figure 4.** SEM images of (**a**) HA powder, (**b**) surface of HA-BDDGE patterned gel, and (**c**) cross-section of HA-BDDGE patterned gel.

### 3.3. Swelling Study

To confirm the gel characteristics of the crosslinked HA-BDDGE polymer, the swelling study was performed with distilled water at 37 °C. Figure 5a represents the swelling study result of HA-BDDGE compound in distilled water at 37 °C.

From the Figure 5a, it is obvious that the rate of water absorption by the HA-BDDGE polymer was higher at the initial stage, then it decreased and finally attained equilibrium swelling state around at 5 h. The HA-BDDGE patterned gel demonstrated the % of swelling as $439 \pm 21\%$. The swelling property confirmed the formation of hydrogel of crosslinked HA-BDDGE polymer in distilled water at 37 °C.

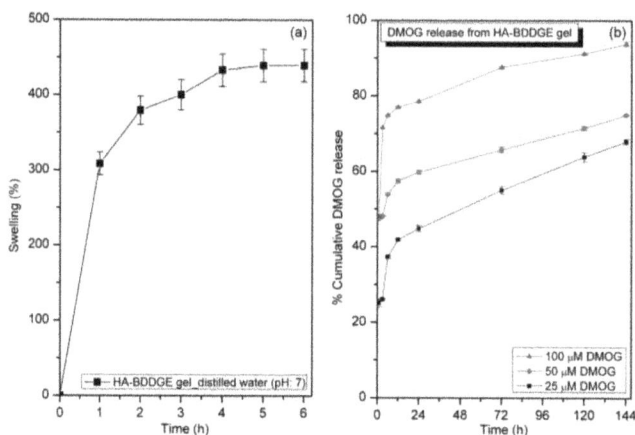

**Figure 5.** (a) Graph of % swelling vs. time for HA-BDDGE patterned gel, and (b) in vitro release profile of dimethyloxalylglycine (DMOG) from loaded HA-BDDGE patterned gel over time.

### 3.4. In Vitro DMOG Release from Loaded HA-BDDGE Patterned Gel

The in vitro DMOG release behavior of DMOG-loaded HA-BDDGE patterned gel with different concentrations of DMOG (25, 50 and 100 μM) is shown in Figure 5b. Release study was performed after the absorption of 25, 50 and 100 μM DMOG. Distilled water (pH 7) was used as medium, while temperature was 37 °C. After 1, 3, 6, 12, 24, 72, 120 and 144 h, absorptions of the aliquots were measured by UV-Vis spectrophotometer (Model: BioMATE 3, Thermo Scientific, Madison, WI, USA). From the release profile (Figure 5b), it is apparent that the initial rate of DMOG release is higher as compared to later stage. This is may be because of the higher rate of swelling of the HA-BDDGE patterned gel and the release of DMOG molecules present on the surface of gel. However, after reaching the equilibrium swelling, the rate of drug diffusion decreased. Among three loaded HA-BDDGE patterned gels (25, 50, and 100 μM), the gel containing 100 μM DMOG exhibited the highest rate of DMOG diffusion (Figure 5b). This is because of the existence of a higher dose of DMOG than that of other grades. In case of 100 μM system, the lesser amount of the % polymer presents in the formulation as compared to 25 and 50 μM systems. Thus, the interaction with HA-BDDGE gel and DMOG will be to certain extent weaker. Besides, the gel layer from which drug molecules diffuse will be weaker and the rate of DMOG release will be faster. After seven days, the % DMOG releases from the HA-BDDGE patterned gel are 93.8 ± 0.6% (for 100 μM), 74.9 ± 0.1% (for 50 μM), and 67.9 ± 0.9% (for 25 μM).

### 3.5. In Vitro MC3T3 Cells Study

#### 3.5.1. Effect of Drug Molecules of Cell Proliferation

To observe the effects of DMOG and NaB release on cell response, 100 μM of NaB, DMOG and NaB/DMOG were bolus-delivered in 24 well plate after seeding $1 \times 10^5$ no. of MC3T3 cells/well with media and the cell culture was lasted for seven days. Cell counting was performed by CCK-8 assay (Figure 6a), while cells images were captured with light microscopy (Figure 6b) at day 1, 3, 5, and 7.

It has been observed that the overall cell proliferation improved when DMOG and NaB were delivered individually or in a combined way (Figure 6a,b). When the proliferation values were normalized with the day 1, the cell proliferation rates were 115 ± 2% and 321 ± 34% at day 3 and 7, respectively (Figure 6a). Importantly, when NaB were bolus-delivered, cell proliferation was improved as 121 ± 2% and 346 ± 33% at day 3 and 7, respectively (Figure 6a). On the other hand, when DMOG were bolus-delivered, cell proliferation was improved as 136 ± 2% and 354 ± 20%, which are similar to those of NaB cases (Figure 6a). In the case of simultaneous delivery of NaB and DMOG, degrees of cell proliferation were $129 \pm 2\%$ and $332 \pm 14\%$ after day 3 and 7 (Figure 6a). No significant differences in cell proliferations were observed between two drugs.

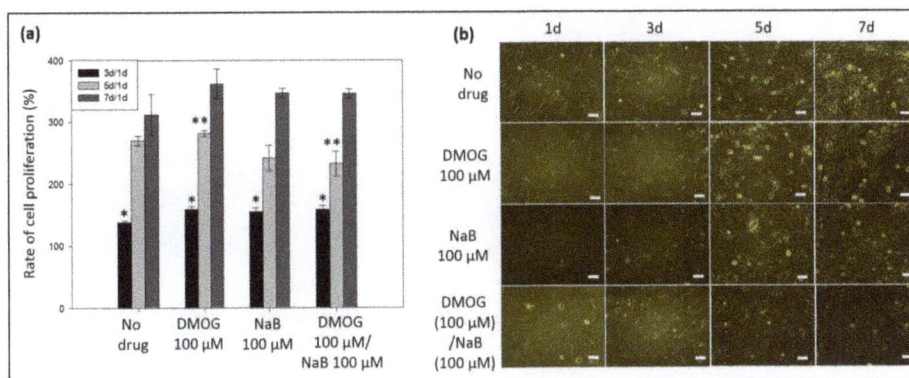

**Figure 6.** (a) Effect of MC3T3 cell proliferation with bolus delivery of drugs (DMOG 100 μM, NaB 100 μM, NaB 100 μM + DMOG 100 NaB 100 μM) on well plate, (b) images of MC3T3 cells on well plate by light microscopy (scale bar = 50 μm).

### 3.5.2. Effect of DMOG or NaB to MC3T3 Cell Proliferation Cultured on the Surface of HA-BDDGE Patterned Gel

To observe the effect of DMOG and NaB delivery to MC3T3 cells cultured on the surface of HA-BDDGE patterned gel, cells were culture in native α-MEM medium and the medium with 25 μM, 100 μM, 400 μM DMOG, and NaB drugs. The MC3T3 cells were seeded and *in vitro* cultured on the HA-BDDGE patterned gel at a density of $1 \times 10^5$ cells/cm² for seven days. While, cellular behaviors and cell proliferation were observed with light microscopy and CCK-8 assay, respectively.

According to both LM and L/D assays, it is observed that all of the cells were alive after day 7 (Figure 7A–N). The NaB containing hydrogel showed cells clusters formation. Specifically, the patterned gel with 25 μM NaB showed cell migration on surface (Figures 5L and 7E), whereas, the hydrogel with 100 μM and higher than 100 μM NaB showed cells clusters at day (Figure 7F–N). The patterned gel with DMOG demonstrated better cell migration on surface than NaB (Figure 7B–K). The cell proliferation was measured using CCK-8 assay by normalizing the data of day 3, 5 and 7 by day 1 (Figure 8). The patterned gel with DMOG showed initial quick cell proliferation at day 3, but their proliferations rate decreased when the concentrations of DMOG increased (Figure 8). The patterned gel with NaB showed lower cell proliferation over time (Figure 8).

### 3.5.3. Effect of DMOG and NaB to MC3T3 Cell Proliferation on the Surface of HA-BDDGE Patterned Gel Depending on the Way of Delivery

To detect the cell behaviors on the HA-BDDGE patterned gel, 12.5, 50, 100, 150, 200, 300 and 600 μM of DMOG and NaB were used. The $1 \times 10^5$ MC3T3 cells/cm² were employed for in vitro study for seven days. The LM, LD, and CCK-8 tests were performed. For the effects of drugs on cell behaviors drugs were loaded in gel and add in the medium. According to LD assay results at day 7, the medium with both NaB and DMOG induced cell clusters on the patterned HA gel (Figure 9A–D,H–K). For local delivery, the HA gel with either 300 μM or more than 300 μM NaB/DMOG induced cell clusters. But those with 150 μM showed cell high proliferation and cell-cell contact was found on the pattern architecture (Figure 9E–G,L–N).

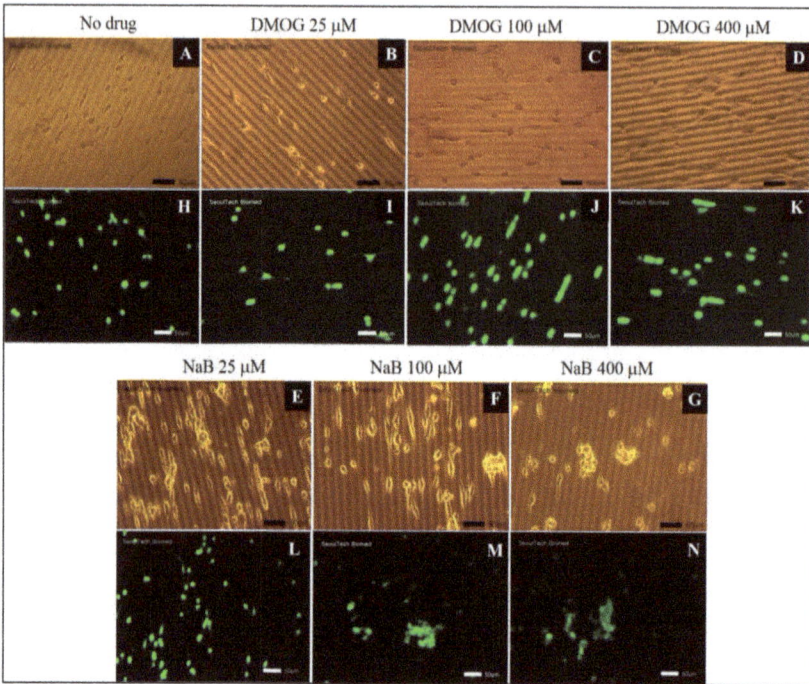

**Figure 7.** Cellular behavior of MC3T3 cells on the surface of HA-BDDGE patterned gel in absence or presence of DMOG and NaB in medium after 7 days: (**A,H**) No drug, (**B,I**) DMOG-25 μM, (**C,J**) DMOG-100 μM, (**D,K**) DMOG-400 μM, (**E,L**) NaB-25 μM, (**F,M**) NaB-100 μM, (**G,N**) NaB-400 μM, (scale bar = 50 μm).

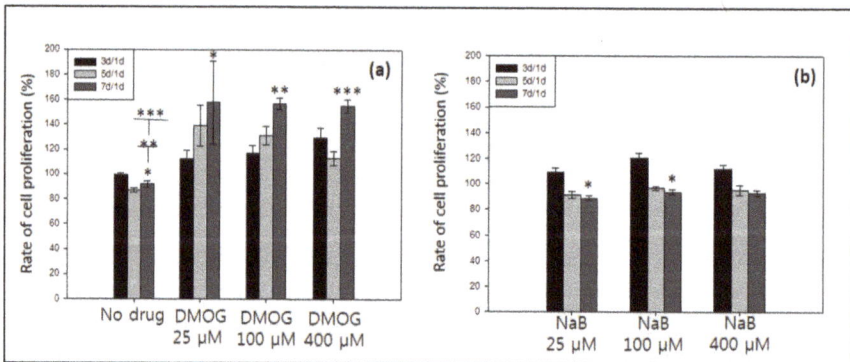

**Figure 8.** Cell proliferation of MC3T3 cells in absence or presence 25, 100 and 400 μM of (**a**) DMOG, and (**b**) NaB, measured by CCK-8 assay.

For cell proliferation assays, the system with bolus drug delivery showed the best cell proliferations at the concentrations of 100 μM NaB and DMOG individually (Figure 10a). When the drugs are locally delivered in the patterned gel, cell proliferations lasted longer over time. When both drugs having concentrations of 150 and 300 μM individually, cell proliferations were more effective

(Figures 10b and 9E,F,L,M). However, the patterned gel with 600 μM DMOG and NaB low proliferation and induced cell clusters (Figures 10b and 9G,N).

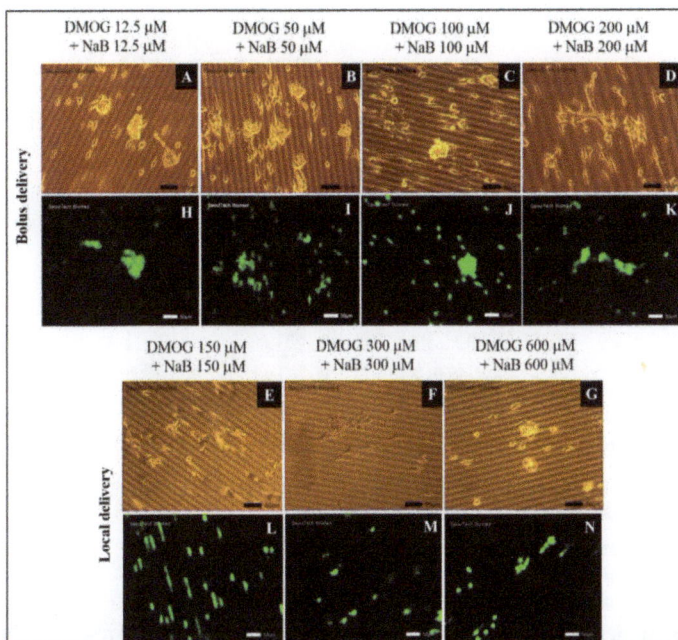

**Figure 9.** Cellular behavior of MC3T3 cells on the surface of HA-BDDGE patterned gel in presence of both DMOG and NaB after 7 days: bolus delivery (**A,H**) NaB/DMOG 12. 5 μM, (**B,I**) NaB/DMOG 50 μM, (**C,J**) NaB/DMOG 100 μM, (**D,K**) NaB/DMOG 200 μM, and local delivery (**E,L**) NaB/DMOG 150 μM, (**F,M**) NaB/DMOG 300 μM, (**G,N**) NaB/DMOG 600 μM.

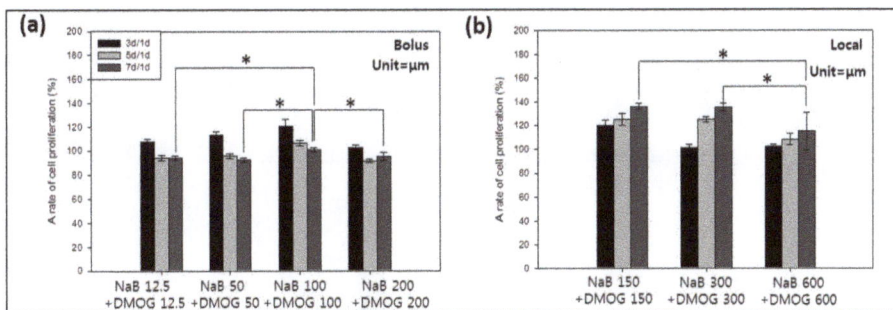

**Figure 10.** Cell proliferation results of MC3T3 cells in (**a**) bolus, and (**b**) local delivery with 12.5, 50, 100, 150, 200, 300 and 600 μM of DMOG and NaB, measured by CCK-8 assay.

### 3.5.4. Cell Cluster Formation on the Surface of HA-BDDGE Patterned Gel by Bolus and Local Deliveries of DMOG and NaB

The cell clusters formation on the patterned HA gel were observed as shown in Figure 11.

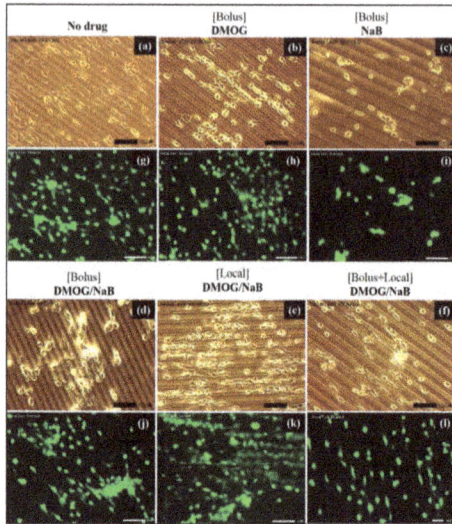

**Figure 11.** Cell cluster behaviours results: (**a–l**) live and dead assay of cells on the HA-BDDGE patterned gel with and without the effects of local and bolus DMOG/NaB delivery.

For bolus study, 100 μM DMOG and NaB was added three times (100 μM × 3) during seven days. The same amount of drug loaded gel was used for local delivery. The patterned gel either no drugs or DMOG showed no cell clusters formation (Figure 11). While, the presence of NaB helps the formation of clusters, indicating NaB has vital role for bone regeneration (Figure 11c–l). The NaB either in the medium or inside the patterned gel induced cell cluster (Figure 11c–l). When NaB was introduced in medium, the cell area reduced and the area of clusters increased as the concentration increased. When both NaB and DMOG were added in the medium, surface area of cells decreased significantly. It is also observed that cells were considerably grown on the pattern shape of the gel. The bright field and stain images confirmed that the pattern architecture assisted to form cell-cell contact and association (Figure 11). On the other hand, when DMOG/NaB were added inside the patterned gel, did not showed any significant trends. However, the clusters formation were highest when 300 μM DMOG and NaB were added individually. The presence of NaB, showed lower cell proliferation during cluster formation (Figure 12).

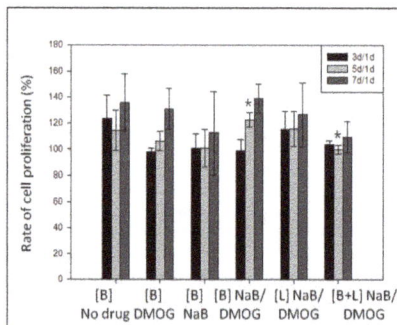

**Figure 12.** Cell proliferation of MC3T3 cells on the patterned HA gel during cluster formation, measured by CCK-8 assay.

### 3.6. Effect of DMOG and NaB on Osteogenic or Angiogenic Responses of MC3T3 Cells Cultured on Pattern Gel

The degree of protein expression was analyzed using MC3T3 cell lines by western blot assay. The results are shown in Figure 13. Combined treatment of 100 μM of both DMOG and NaB through local delivery increased ALP activity (Figure 13a). While, DMOG and NaB individually did not significantly alter the ALP activity. It was well established that NaB increased ALP activity and act as a good marker for the differentiation of MC3T3 osteoblasts [45]. When compared to experiments without drugs, DMOG increased the level of HIF-1α protein (Figure 13b). Combined treatment with DMOG and NaB in both delivery modes also elevated the level of HIF-1α protein. Similar types of results were observed in other reports [46]. Generally, angiogenic activity of cells are interrelated with the HIF-1 pathway [47]. HIF-1 is an oxygen-sensitive compound, which contains HIF-1α and HIF-1β. Under normoxic conditions, HIF-1α is hydroxylated by the enzyme HIF-PH, resulting in ubiquitylation and the degradation of HIF-1α [47]. DMOG is a cell permeable prolyl-4-hydroxylase inhibitor, which can upregulate the protein level of HIF-1α post-transcriptionally under normoxic conditions [47]. In this study, western blotting test showed that the expression level of HIF-1α in MC3T3 enhanced when DMOG was used either individually or through combined treatment with NaB. After MC3T3 cells exposing to DMOG, HIF-1α will be accumulated in cells, and activates HIF-1 complex and shows the expression of various angiogenic genes.

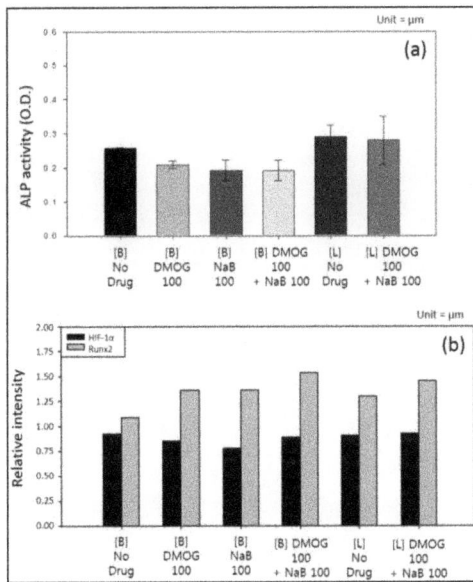

**Figure 13.** (a) Assays of alkaline phosphatase activity, (b) and expression of HIF-1α and Runx2 by the delivery of with and without DMOG/NaB through bolus and local modes.

On the other hand, DMOG, NaB and combined treatments showed high intensities of runt-related transcription factor 2 (Rnx2) expression than that of experiment without drug (Figure 13b). Runx2 is a runt family of transcription factors (Runx1–3) that control the growth and differentiation of various cell lineages [48]. Runx2 protein comprises 128 amino acid Runt domain, which is responsible for the DNA binding. The Runx2 can interact with several proteins including co-regulatory proteins and chromatin remodeling factors, leading to complex role in regulating bone specific genes and differentiation. Mouse MC3T3 cell lines represent Runx2 positive pre-osteoblasts committed to osteogenic lineage [48]. However, for clinical implementation, treatment with rh-BMP2 is essential. As it is reported that the addition of rh-BMP2 to the sites of distraction osteogenesis improves soft tissue healing, and reduced graft exposure and protecting the bone tissue healing [49]. The above in vitro results indicate that DMOG and NaB exerted their angiogenic and osteogenic responses in MC3T3 cells, respectively. Hence, the HA-BDDGE gel could be used in bone regeneration application.

## 4. Conclusions

The HA-BDDGE gel has been successfully fabricated through nucleophilic addition reaction using NaOH as base. The chemical analyses such as ATR-FTIR, $^{13}$C NMR, and TGA analyses implied the formation of crosslinked networks. SEM analysis confirmed the pattern architecture on the surface of the gel. The width of pattern was approximately 5 μm, while the width of valley between two patterns was 15 μm. The crosslinked polymer attained equilibrium swelling state after ~5 h in distilled water at 37 °C, which confirmed the hydrogel nature of HA-BDDGE. In vitro release study from DMOG loaded HA-BDDGE patterned gel showed a controlled release nature. The hydrogel is biocompatible against MC3T3 cell lines. In cell proliferation study, the bolus delivery results indicated that the DMOG containing patterned gel demonstrates better cell migration ability on surface than NaB. For local delivery, the pattern gel with 300 μM NaB or 300 μM DMOG induced cell clusters formation, while, the gel with 150 μM NaB/DMOG showed high cell proliferation capability only. The vital role of NaB in bone regeneration has been endorsed from the formation of cell clusters in presence of NaB. The bright field and stain images confirmed that the pattern architecture assisted to form cell-cell contact and association. Through local delivery mode of DMOG and NaB, the HA-BDDGE patterned gel showed higher intensity of ALP, HIF-1α, and Runx2 activities, signify angiogenic and osteogenic responses in MC3T3 cells. However, for clinical application, in vivo study also needs to be explored in the future by modifying some characteristics of gel like porosity, cell adhesiveness, tissue regeneration, and mineralization with inorganic particles or ceramics to improved mechanical properties. Finally, the HA-BDDGE gel, with well micro-patterned architecture, biocompatibility, controlled release ability of DMOG drug, clusters formation ability of MC3T3 cells, and higher intensity in ALP, HIF-1α, and Runx2 studies implied that it could be used in bone regeneration application.

**Acknowledgments:** This work was supported by the National Research Foundation of Korea (NRF) Grant (2014K2A2A7060928 and 2015R1A2A1A10054592).

**Author Contributions:** Sumi Bang worked most of experiments with Dipankar Das and Jiyun Yu. Dipankar Das analyzed the chemical data and wrote down the paper. Jiyun Yu did both osteogenic and angiogenic response works. Prof. Noh supervised the research. All authors discussed the results and commented on the manuscript.

**Conflicts of Interest:** The authors declare no conflict of interest.

## References

1.	Cha, S.H.; Lee, H.J.; Koh, W.G. Study of myoblast differentiation using multi-dimensional scaffolds consisting of nano and micropatterns. *Biomater. Res.* **2017**, *21*, 1. [CrossRef] [PubMed]
2.	Khetan, S.; Burdick, J.A. Patterning hydrogels in three dimensions towards controlling cellular interactions. *Soft Matter* **2011**, *7*, 830–838. [CrossRef]
3.	Tibbitt, M.W.; Anseth, K.S. Hydrogels as extracellular matrix mimics for 3D cell culture. *Biotechnol. Bioeng.* **2009**, *103*, 655–663. [CrossRef] [PubMed]
4.	Engler, A.J.; Sen, S.; Sweeney, H.L.; Discher, D.E. Matrix Elasticity Directs Stem Cell Lineage Specification. *Cell* **2006**, *126*, 677–689. [CrossRef] [PubMed]
5.	Shin, Y.M.; Kim, K.S.; Lim, Y.M.; Nho, Y.C.; Shin, H. Modulation of Spreading, Proliferation, and Differentiation of Human Mesenchymal Stem Cells on Gelatin-Immobilized Poly(l-lactide-co-ε-caprolactone) Substrates. *Biomacromolecules* **2008**, *9*, 1772–1781. [CrossRef] [PubMed]
6.	Tang, X.; Ali, M.Y.; Saif, M.T.A. A novel technique for micro-patterning proteins and cells on polyacrylamide gels. *Soft Matter* **2012**, *8*, 7197–7206. [CrossRef] [PubMed]
7.	Fathi, A.; Lee, S.; Breen, A.; Shirazi, A.N.; Valtchev, P.; Dehghani, F. Enhancing the mechanical properties and physical stability of biomimetic polymer hydrogels for micro-patterning and tissue engineering applications. *Eur. Polym. J.* **2014**, *59*, 161–170. [CrossRef]

8.  Akintewe, O.O.; DuPont, S.J.; Elinen, K.K.; Cross, M.C.; Toomey, R.G.; Gallant, N.D. Microcontact printing of tissue precursors via geometrically patterned shape-changing hydrogel stamps preserves cell viability and organization. *Bioprinting* **2017**. [CrossRef]
9.  Khademhosseini, A.; Eng, G.; Yeh, J.; Fukuda, J.; Blumling, J., III; Langer, R.; Burdick, J.A. Micromolding of photocrosslinkable hyaluronic acid for cell encapsulation and entrapment. *J. Biomed. Mater. Res. A* **2006**, *79*, 522–532. [CrossRef] [PubMed]
10. Alaribe, F.N.; Manoto, S.L.; Motaung, S.C.K.M. Scaffolds from biomaterials: Advantages and limitations in bone and tissue engineering. *Biologia* **2016**, *71*, 353–366. [CrossRef]
11. Dorsey, T.B.; Grath, A.; Wang, A.; Xu, C.; Dai, G.; Hong, Y. Evaluation of photochemistry reaction kinetics to pattern bioactive proteins on hydrogels for biological applications. *Bioact. Mater.* **2017**. [CrossRef]
12. Lovett, M.; Lee, K.; Edwards, A.; Kaplan, D.L. Vascularization Strategies for Tissue Engineering. *Tissue Eng. Part B Rev.* **2009**, *15*, 353–370. [CrossRef] [PubMed]
13. Sakaguchi, K.; Shimizu, T.; Horaguchi, S.; Sekine, H.; Yamato, M.; Umezu, M.; Okano, T. In Vitro Engineering of Vascularized Tissue Surrogates. *Sci. Rep.* **2013**, *3*, 1316. [CrossRef] [PubMed]
14. Liu, M.; Zeng, X.; Ma, C.; Yi, H.; Ali, Z.; Mou, X.; Li, S.; Deng, Y.; He, N. Injectable hydrogels for cartilage and bone tissue Engineering. *Bone Res.* **2017**, *5*, 17014. [CrossRef] [PubMed]
15. Han, Y.; Zeng, Q.; Li, H. The calcium silicate/alginate composite: Preparation and evaluation of its behavior as bioactive injectable hydrogels. *Acta Biomater.* **2013**, *9*, 9107–9117. [CrossRef] [PubMed]
16. Kook, Y.J.; Lee, D.H.; Song, J.E.; Tripathy, N.; Jeon, Y.S.; Jeon, H.Y.; Oliveira, J.M.; Reis, R.L.; Khang, G. Osteogenesis evaluation of duck's feet-derived collagen/hydroxyapatite sponges immersed in dexamethasone. *Biomater. Res.* **2017**, *21*, 2. [CrossRef] [PubMed]
17. Vo, T.N.; Shah, S.R.; Lu, S. Injectable dual-gelling cell-laden composite hydrogels for bone tissue engineering. *Biomaterials* **2016**, *83*, 1–11. [CrossRef] [PubMed]
18. Fu, S.; Ni, P.; Wang, B. Injectable and thermo-sensitive PEG-PCL-PEG copolymer/collagen/n-HA hydrogel composite for guided bone regeneration. *Biomaterials* **2012**, *33*, 4801–4809. [CrossRef] [PubMed]
19. Dhivya, S.; Saravanan, S.; Sastry, T.P. Nanohydroxyapatite-reinforced chitosan composite hydrogel for bone tissue repair in vitro and in vivo. *J. Nanobiotechnol.* **2015**, *13*, 40. [CrossRef] [PubMed]
20. Vishnu Priya, M.; Sivshanmugam, A.; Boccaccini, A.R. Injectable osteogenic and angiogenic nanocomposite hydrogels for irregular bone defects. *Biomed. Mater.* **2016**, *11*, 035017. [CrossRef] [PubMed]
21. Gothard, D.; Smith, E.L.; Kanczler, J.M.; Black, C.R.; Wells, J.A.; Roberts, C.A.; White, L.J.; Qutachi, O.; Peto, H.; Rashidi, H.; et al. In Vivo Assessment of Bone Regeneration in Alginate/Bone ECM Hydrogels with Incorporated Skeletal Stem Cells and Single Growth Factors. *PLoS ONE* **2015**, *10*, e0145080. [CrossRef] [PubMed]
22. Laino, L.; Iezzi, G.; Piattelli, A.; Muzio, L.L.; Cicciu, M. Vertical Ridge Augmentation of the Atrophic Posterior Mandible with Sandwich Technique: Bone Block from the Chin Area versus Corticocancellous Bone Block Allograft—Clinical and Histological Prospective Randomized Controlled Study. *BioMed Res. Int.* **2014**, *2014*, 982104. [CrossRef] [PubMed]
23. Petrauskaite, O.; Gomes, P.S.; Fernandes, M.H.; Juodzbalys, G.; Stumbras, A.; Maminskas, J.; Liesiene, J.; Cicciu, M. Biomimetic Mineralization on a Macroporous Cellulose-Based Matrix for Bone Regeneration. *BioMed Res. Int.* **2013**, *2013*, 452750. [CrossRef] [PubMed]
24. Park, H.S.; Lee, S.Y.; Yoon, H.; Noh, I. Biological evaluation of micro-patterned hyaluronic acid hydrogel for bone tissue engineering. *Pure Appl. Chem.* **2014**, *86*, 1911–1922. [CrossRef]
25. Lee, K.Y.; Mooney, D.J. Hydrogels for Tissue Engineering. *Chem. Rev.* **2001**, *101*, 1869–1880. [CrossRef] [PubMed]
26. Sidwell, R.U.; Dhillon, A.P.; Butler, P.E.M.; Rustin, M.H.A. Localized granulomatous reaction to a semi-permanent hyaluronic acid and acrylic hydrogel cosmetic filler. *Clin. Exp. Dermatol.* **2004**, *29*, 630–632. [CrossRef] [PubMed]
27. Lowe, N.J.; Maxwell, C.A.; Lowe, P.; Duickb, M.G.; Shah, K. Hyaluronic acid skin fillers: Adverse reactions and skin testing. *J. Am. Acad. Dermatol.* **2011**, *45*, 930–933. [CrossRef] [PubMed]
28. Mohabatpour, F.; Karkhaneh, A.; Sharifi, A.M. A hydrogel/fiber composite scaffold for chondrocyte encapsulation in cartilage tissue regeneration. *RSC Adv.* **2016**, *6*, 83135–83145. [CrossRef]
29. Park, H.; Lee, H.J.; Ana, H.; Lee, K.Y. Alginate hydrogels modified with low molecular weight hyaluronate for cartilage regeneration. *Carbohydr. Polym.* **2017**, *162*, 100–107. [CrossRef] [PubMed]

30. Zhu, Y.; Tan, J.; Zhu, H.; Lin, G.; Yin, F.; Wang, L.; Song, K.; Wang, Y.; Zhou, G.; Yi, W. Development of kartogenin-conjugated chitosan–hyaluronic acid hydrogel for nucleus pulposus regeneration. *Biomater. Sci.* **2017**, *5*, 784–791. [CrossRef] [PubMed]
31. Hubbell, J.A. Hydrogel systems for barriers and local drug delivery in the control of wound healing. *J. Control. Release* **1996**, *39*, 305–313. [CrossRef]
32. Bose, R.; Lee, S.H.; Park, H.S. Lipid-based surface engineering of PLGA nanoparticles for drug and gene delivery applications. *Biomater. Res.* **2016**, *20*, 34. [CrossRef] [PubMed]
33. Nishizawa, M.; Takoh, K.; Matsue, T. Micropatterning of HeLa Cells on Glass Substrates and Evaluation of Respiratory Activity Using Microelectrodes. *Langmuir* **2002**, *18*, 3645–3649. [CrossRef]
34. Falconnet, D.; Csucs, G.; Grandin, H.M.; Textor, M. Surface engineering approaches to micropattern surfaces for cell-based assays. *Biomaterials* **2006**, *27*, 3044–3063. [CrossRef] [PubMed]
35. Ahmed, W.W.; Wolfram, T.; Goldyn, A.M.; Bruellhoff, K.; Rioj, B.A.; Moller, M.; Spatz, J.P.; Saif, T.A.; Groll, J.; Kemkemer, R. Myoblast morphology and organization on biochemically micro-patterned hydrogel coatings under cyclic mechanical strain. *Biomaterials* **2009**, *31*, 250–258. [CrossRef] [PubMed]
36. Chen, C.S.; Jiang, X.; Whiteside, G.M. Microengineering the Environment of Mammalian Cells in Culture. *MRS Bull.* **2005**, *30*, 194–201. [CrossRef]
37. Poellmann, M.J.; Harrell, P.A.; King, W.P.; Johnson, A.J.W. Geometric microenvironment directs cell morphology on topographically patterned hydrogel substrates. *Acta Biomater.* **2010**, *6*, 3514–3523. [CrossRef] [PubMed]
38. Nelson, C.M.; Jean, R.P.; Tan, J.L.; Liu, W.F.; Sniadecki, N.J.; Spector, A.A.; Chen, C.S. Emergent patterns of growth controlled by multicellular form and mechanics. *Proc. Natl. Acad. Sci. USA* **2005**, *102*, 11594–11599. [CrossRef] [PubMed]
39. James, J.; Goluch, E.D.; Hu, H.; Liu, C.; Mrksich, M. Subcellular curvature at the perimeter of micropatterned cells influences lamellipodial distribution and cell polarity. *Cell. Motil. Cytoskelet.* **2008**, *65*, 841–852. [CrossRef] [PubMed]
40. Mahmud, G.; Campbell, C.J.; Bishop, K.J.M.; Komarova, Y.A.; Chaga, O.; Soh, S.; Huda, S.; Kandere-Grzybowska, K.; Grzybowski, B.A. Directing cell motions on micropatterned ratchets. *Nat. Phys.* **2009**, *5*, 606–612. [CrossRef]
41. Gilli, R.; Kacurakova, M.; Mathlouthi, M.; Navarini, L.; Paoletti, S. FTIR studies of sodium hyaluronate and its oligomers in the amorphous solid phase and in aqueous solution. *Carbohydr. Res.* **1994**, *263*, 315–326. [CrossRef]
42. Kim, D.; Park, H.J.; Lee, K.Y. Study on Curing Behaviors of Epoxy Acrylates by UV with and without Aromatic Component. *Macromol. Res.* **2015**, *23*, 944–951. [CrossRef]
43. Cowman, M.K.; Hittner, D.M.; Feder-Davis, J. $^{13}$C-NMR Studies of Hyaluronan: Conformational Sensitivity to Varied Environments. *Macromolecules* **1996**, *29*, 2894–2902. [CrossRef]
44. Sheu, C.; Shalumon, K.T.; Chen, C.H.; Kuo, C.Y.; Fong, Y.T.; Chen, J.P. Dual crosslinked hyaluronic acid nanofibrous membranes for prolonged prevention of post-surgical peritoneal adhesion. *J. Mater. Chem. B* **2016**, *4*, 6680–6693. [CrossRef]
45. Iwami, K.; Moriyama, T. Effects of short chain fatty acid, sodium butyrate, on osteoblastic cells and osteoclastic cells. *Int. J. Biochem.* **1993**, *25*, 1631–1635. [CrossRef]
46. Woo, K.M.; Jung, H.M.; Oh, J.H.; Rahman, S.; Kim, S.M.; Baek, J.H.; Ryoo, H.M. Synergistic effects of dimethyloxalylglycine and butyrate incorporated into α-calcium sulfate on bone regeneration. *Biomaterials* **2015**, *39*, 1–14. [CrossRef] [PubMed]
47. Ding, H.; Chen, S.; Song, W.Q.; Gao, Y.S.; Guan, J.J.; Wang, Y.; Sun, Y.; Zhang, C.Q. Dimethyloxaloylglycine Improves Angiogenic Activity of Bone Marrow Stromal Cells in the Tissue Engineered Bone. *Int. J. Biol. Sci.* **2014**, *10*, 746–756. [CrossRef] [PubMed]
48. Tarkkonen, K.; Hieta, R.; Kytola, V.; Nykter, M.; Kiviranta, R. Comparative analysis of osteoblast gene expression profiles and Runx2 genomic occupancy of mouse and human osteoblasts in vitro. *Gene* **2017**, *626*, 119–131. [CrossRef] [PubMed]
49. Herford, A.S.; Cicciu, M.; Eftimie, L.F.; Miller, M.; Signorino, F.; Fama, F.; Cervino, G.; Giudice, G.L.; Bramanti, E.; Lauritano, F.; et al. rhBMP-2 applied as support of distraction osteogenesis: A split-mouth histological study over nonhuman primates mandibles. *Int. J. Clin. Exp. Med.* **2016**, *9*, 17187–17194.

© 2017 by the authors. Licensee MDPI, Basel, Switzerland. This article is an open access article distributed under the terms and conditions of the Creative Commons Attribution (CC BY) license (http://creativecommons.org/licenses/by/4.0/).

*nanomaterials*

MDPI

Article

# Room Temperature Tunable Multiferroic Properties in Sol-Gel-Derived Nanocrystalline Sr(Ti$_{1-x}$Fe$_x$)O$_{3-\delta}$ Thin Films

Yi-Guang Wang [1], Xin-Gui Tang [1,*], Qiu-Xiang Liu [1], Yan-Ping Jiang [1] and Li-Li Jiang [2]

[1]  School of Physics & Optoelectric Engineering, Guangdong University of Technology,
     Guangzhou Higher Education Mega Centre, Guangzhou 510006, China;
     wangyiguang2011@gmail.com (Y.-G.W.); liuqx@gdut.edu.cn (Q.-X.L.); ypjiang@gdut.edu.cn (Y.-P.J.)
[2]  Laboratory Teaching Center, Guangdong University of Technology, Guangzhou Higher Education
     Mega Center, Guangzhou 510006, China; jianglili@gdut.edu.cn
*   Correspondence: xgtang@gdut.edu.cn; Tel./Fax: +86-20-3932-2265

Received: 17 August 2017; Accepted: 5 September 2017; Published: 8 September 2017

**Abstract:** Sr(Ti$_{1-x}$Fe$_x$)O$_{3-\delta}$ ($0 \leq x \leq 0.2$) thin films were grown on Si(100) substrates with LaNiO$_3$ buffer-layer by a sol-gel process. Influence of Fe substitution concentration on the structural, ferroelectric, and magnetic properties, as well as the leakage current behaviors of the Sr(Ti$_{1-x}$Fe$_x$)O$_{3-\delta}$ thin films, were investigated by using the X-ray diffractometer (XRD), atomic force microscopy (AFM), the ferroelectric test system, and the vibrating sample magnetometer (VSM). After substituting a small amount of Ti ion with Fe, highly enhanced ferroelectric properties were obtained successfully in SrTi$_{0.9}$Ti$_{0.1}$O$_{3-\delta}$ thin films, with a double remanent polarization ($2P_r$) of 1.56, 1.95, and 9.14 $\mu$C·cm$^{-2}$, respectively, for the samples were annealed in air, oxygen, and nitrogen atmospheres. The leakage current densities of the Fe-doped SrTiO$_3$ thin films are about $10^{-6}$–$10^{-5}$ A·cm$^{-2}$ at an applied electric field of 100 kV·cm$^{-1}$, and the conduction mechanism of the thin film capacitors with various Fe concentrations has been analyzed. The ferromagnetic properties of the Sr(Ti$_{1-x}$Fe$_x$)O$_{3-\delta}$ thin films have been investigated, which can be correlated to the mixed valence ions and the effects of the grain boundary. The present results revealed the multiferroic nature of the Sr(Ti$_{1-x}$Fe$_x$)O$_{3-\delta}$ thin films. The effect of the annealing environment on the room temperature magnetic and ferroelectric properties of Sr(Ti$_{0.9}$Fe$_{0.1}$)O$_{3-\delta}$ thin films were also discussed in detail.

**Keywords:** SrTi$_{1-x}$Fe$_x$O$_3$ thin films; sol-gel; multiferroic; leakage current; conduction mechanism

## 1. Introduction

Strontium titanate SrTiO$_3$ has been widely applied in electronically tunable microwave devices for its high dielectric, low dielectric losses and high tunability [1,2]. Pure SrTiO$_3$ is known as an incipient ferroelectric or paraelectric, since its remaining paraelectric is down to the 0 K under a stress-free condition and it has the instability of ferroelectric at a low temperature [3,4]. A ferrodistortive phase transition temperature from cubic to tetragonal for SrTiO$_3$ is as low as 105 K [5,6], which means the ferroelectric properties of the SrTiO$_3$ are unavailable in most cases. To obtain the room temperature ferroelectric properties of SrTiO$_3$, many efforts have been taken, such as introducing by the strain [7–9], substituting the O$^{16}$ with O$^{18}$, or doping with other elements [10,11]. For example, the ferroelectric property has been obtained in the epitaxial SrTiO$_3$ film [12] and the large epitaxial strain induced by the lattice mismatch. Besides that, the doping of aliovalent ionic may also be used to provide the strain in the material to promote ferroelectric properties. Among these, the Fe-doped SrTiO$_3$ has been proven to be a ferromagnetic [13] with resistive switching characteristics [14], and can also be used as electrode materials [15]. Such characteristics inspire us to investigate the possible multiferroic properties of the Fe-doped $SrTiO_3$, which may have potential applications in memory devices, sensors, and actuators [16,17]. Owing to the obtained ferromagnetic property in the Fe-doped $SrTiO_3$ [13,18–20], and the excellent dielectric properties of $SrTiO_3$, the study of the ferroelectric properties will play a significant role on the multiferroic application of the $SrTi_{1-x}Fe_xO_3$.

In fact, besides the strain, the decreasing crystal symmetry and the defects such as oxygen vacancies induced by the Fe ion substitution may also promote the ferroelectric properties in $Sr(Ti_{1-x}Fe_x)O_3$. Thus, the Fe substituting is a quite feasible approach to promote the ferroelectric properties in $SrTiO_3$. However, we can find only a little research about the ferroelectric properties of the $Sr(Ti_{1-x}Fe_x)O_3$ films [21–23], and the influence of Fe substitution concentration on the ferroelectric and leakage current behavior of the sol-gel derived $Sr(Ti_{1-x}Fe_x)O_3$ thin films has not been reported still now.

In this work, a series of Fe-doped $Sr(Ti_{1-x}Fe_x)O_{3-\delta}$ (STF, $x = 0, 0.05, 0.1, 0.15$ and $0.2$; abbreviated as STO, STF05, STF10, STF15, and STF20, respectively) thin films were synthesized on the $LaNiO_3$ (LNO) coated Si(100) substrates by the sol-gel method. The multiferroic properties, leakage current behaviors, and conduction mechanism of the STF thin films were investigated. The effect of annealing environment on the room temperature magnetic and ferroelectric properties of $Sr(Ti_{0.9}Fe_{0.1})O_3$ thin films were also discussed in detail.

## 2. Results and Discussion

The X-ray Diffraction (XRD) patterns of the $Sr(Ti_{1-x}Fe_x)O_3$ (STF) thin films grown on the $LaNiO_3$ (LNO) buffered Si(100) substrates were shown in Figure 1. The (110) peak responsible for the perovskite structure was observed in all film samples, showing a typical polycrystalline perovskite nature of the STF films with a highly preferential (110) orientation. The orientation can be ascribed to the quite similar lattice constant and crystal structure of the LNO and STF thin films, where the LNO can be used as a seeding layer for favoring the nucleation and growth of the STF films [24]. The XRD peaks shift to a lower diffraction angle with increasing Fe concentration, which is the result of increased lattice parameters. The lattice parameters are 3.855, 3.860, 3.876, 3.882, and 3.908 Å, respectively, for $Sr(Ti_{1-x}Fe_x)O_3$ thin films with $x = 0$ to 20. This increase of lattice parameters with $x$ is attributed to the increasing of low-valence-state Fe ions [25]. The various ratios of the lattice parameters $a$ with respect to the bulk STO material (cubic: $a_0 = 3.905$ Å), i.e., $(a - a_0)/a_0$, are presented in the inset of Figure 1. The compressed lattice parameter of thin film compared to the bulk material (ratio < 0) indicates the existence of strain in the film. The magnified plot of the peaks was also presented in the inset of Figure 1, which indicates the superposed peak of the LNO (110) and STF (110) clearly.

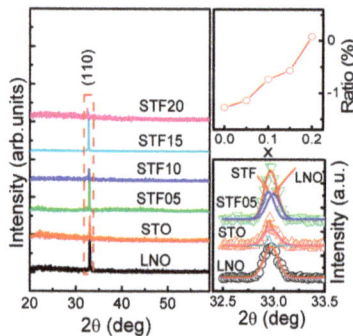

**Figure 1.** X-ray diffraction (XRD) patterns of the $Sr(Ti_{1-x}Fe_x)O_3$(STF) thin films grown on the $LaNiO_3$ (LNO) buffered Si(100) substrates. The top inset shows the variation ratio of the in-plane lattice parameters $a$ with respect to the bulk STO material; $(a - a_0)/a_0$, the bottom inset, shows the magnified plot of the (110) peaks for the LNO and STF.

The valence states of the Fe ions in STF10 thin film characterized by X-ray photoelectron spectroscopy (XPS) (Thermo Fisher Scientific Inc., Waltham, MA, USA) were shown in Figure 2. The composition of STF10 thin film was analyzed by XPS as well. The atomic ratio of Sr:Ti:Fe was found to be 10.35:10.49:2.35, which slightly deviated from the theoretical value of 10:9:1 of the stoichiometric thin film. Besides the analytical error, the deviation may be relevant with the presence of divalent Fe ions. Since the ions can be incorporated in the SrO sublattice, higher Fe concentration may indicate its migration towards extended defects and the surface [26], which is, however, not sufficient to kill the long range atomic order. The Fe 2p 2/3 and 2p 1/2 doublets of STF10 were seen in the vicinity of 706 eV (705.5, 707.3 eV) and 721.76 eV, respectively. These peaks appear at a lower binding energy compared to measurements on $Fe_2O_3$ with $Fe^{3+}$ [13], implying the existence of $Fe^{2+}$ and $Fe^{3+}$ in STF10. The clear $Fe^{3+}$ satellite peak was present at 718.48 eV. The peak at 711.6 eV is about 0.9 eV higher than the Fe 2p 2/3 of $Fe_2O_3$, indicating the possible existence of $Fe^{4+}$. Thus, the XPS result shows the coexistence of $Fe^{2+}$, $Fe^{3+}$, and $Fe^{4+}$ mixed valence states in STF10 thin film, with dominance of the $Fe^{3+}$ states, which is consistent with the previous works and corresponds to the existence of oxygen vacancies [13].

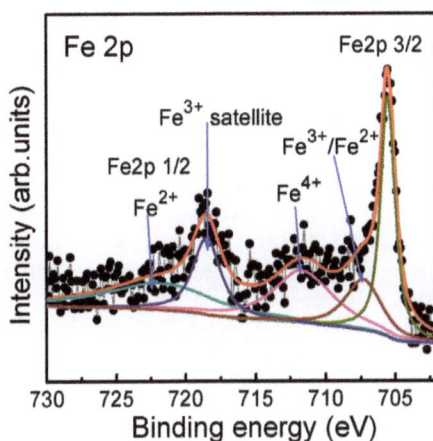

**Figure 2.** Fe 2p X-ray photoelectron spectroscopy (XPS) splitting spectrum of the STF10 thin film on LNO/Si(100) substrate.

The morphology of the STF10 thin film was displayed in the inset of Figure 3a, which exhibits a dense micro-structure with no cracks. The calculated results of the atomic force microscopy (AFM) (Being Nano-Instruments Ltd. Beijing, China) image showed that the average grain size is about 86 nm and the root mean square roughness of the STF10 thin film is 5.8 nm. Such nanoscale grains were the result of the rapid thermal annealing (RTA) process, which can restrain the grain growth effectively.

The electric-filed-induced polarization (*P-E*) switching behavior measurements under 1 kHz at room temperature were shown in Figure 3a. It is observed that all samples showed an almost perfect symmetrical *P-E* loop along both the electric field axes and the polarization axes, indicating the existence of ferroelectric properties. The rather low values of saturation polarization ($P_s$) and remnant polarization ($P_r$) of STO imply a lack of obvious ferroelectric signal in the STO thin film, which is consistent with the nature of incipient ferroelectric, and the weak signal can be interrupted as the result of internal strain. When substituting the Ti ion with the Fe ion, the ferroelectric properties of the thin film improved significantly. The variation values of $P_r$, $P_s$, and the coercive field ($E_c$) with the concentration of Fe are shown in Figure 3b. With increasing Fe concentration $x$ (when $x \leq 0.1$), the value of $2P_s$ increases and reaches a maximum of 12.34 $\mu C \cdot cm^{-2}$ when $x = 0.1$. However, upon a further increase in Fe concentration (when $x > 0.1$), the $2P_s$ decreases instead. Slightly different, the maximum of $2P_r$ was obtained with $2P_r$ of 1.71 $\mu C \cdot cm^{-2}$ for $x = 0.05$.

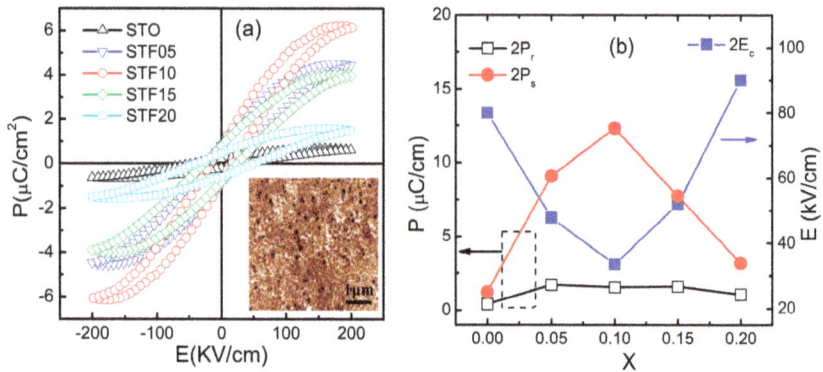

**Figure 3.** Ferroelectric properties and morphology of the STF thin films: (**a**) electric polarization ($P$) as a function of the electric field ($E$) for the STF thin films. The inset is the atomic force microscopy (AFM) image of the STF10 thin film; (**b**) the variation values of the remnant polarization ($P_r$), saturation polarization ($P_s$), and the coercive field ($E_c$) as a function of the Fe concentration $x$.

Figure 4 shows the $P$-$E$ loops of the STF10 thin films annealed at various atmospheres. When the thin film samples are annealed at air and oxygen atmospheres, the values of the $2P_r$ and $2E_c$ are 1.56 $\mu C \cdot cm^{-2}$ and 32.0 kV$\cdot cm^{-1}$, and 1.95 $\mu C \cdot cm^{-2}$ and 65.1 kV$\cdot cm^{-1}$, respectively. The ferroelectric properties of the thin films were annealing at oxygen atmosphere (which is better than that of the samples that were annealed at air atmosphere). When the sample was annealing in a nitrogen atmosphere, the values of the $2P_r$ and $2E_c$ were 9.14 $\mu C \cdot cm^{-2}$ and 265.4 kV$\cdot cm^{-1}$, respectively. The $2P_r$ value of the nitrogen atmosphere annealing sample was the best; the major source was from the leakage current. The $2P_r$ value (9.14 $\mu C \cdot cm^{-2}$) of the STF films is higher than that of tetragonal strontium titanate thin films on SrTiO$_3$ (001) substrates (2.5 $\mu C \cdot cm^{-2}$) [27], higher than that of tensile-strained SrTiO$_3$ thin films on the GdScO$_3$ (110) substrate (0.5 $\mu C \cdot cm^{-2}$) [28], and compares with the etragonally strained SrTiO$_3$ thin films on the single crystal Rh substrate (8 C/cm$^2$) [29], respectively, at room temperature.

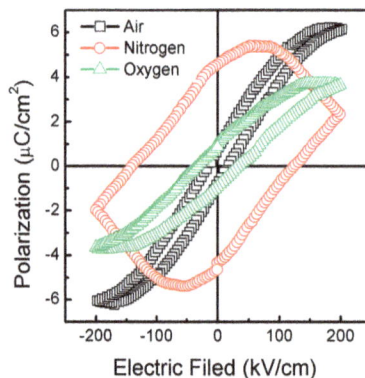

**Figure 4.** Ferroelectric properties of the STF10 thin films were annealed at various atmospheres.

Although the cause of improved ferroelectric properties in STF thin films is complicated, there are primarily three possible factors that influence the ferroelectric properties in this system. One is the internal strain induced by the misfit of the substrate and ionic substitution of Fe, the other is the decrease of crystal symmetry induced by the aliovalent Fe ion and oxygen distributions; both of them are beneficial for the enhancement of spontaneity [30]. The last is defect-induced by the $Fe^{3+}$ or $Fe^{2+}$ ion substitution. For instance, in the typical case of STF10, the existence of $Fe^{3+}$ has been confirmed by the XPS results, and the possible $Fe^{2+}$ also been observed. The Ti ion usually presents a valence of +4 in STO, when substituting that ion with the Fe ion in the STO lattice, oxygen vacancies ($V_{\ddot{O}}$) are generally created in order to maintain the charge balance [31]. When the concentration of Fe is above a degree, the defect dipoles of the cations (i.e. Fe ions)-$V_{\ddot{O}}$ complex could exist in the unit cells by binding cations and oxygen vacancies [32,33], and therefore increase the total polarization [34]. However, meanwhile, the doping of Fe ions can also weaken the ferroelectric of the STF thin films because of the strong magnetic coupling between the doped Fe ions [18]. Therefore, the polarization did not increase linearly with the increasing concentration of Fe. The $2E_c$ showed a completely opposite trend with the variation of $2P_s$, which is consistent with the general variation trend of ferroelectric properties. The change of $E_c$ could arise from the change in domain switching dynamics, and can similarly be explained as the result of changed crystal symmetry and defect dipoles induced by the Fe-doping.

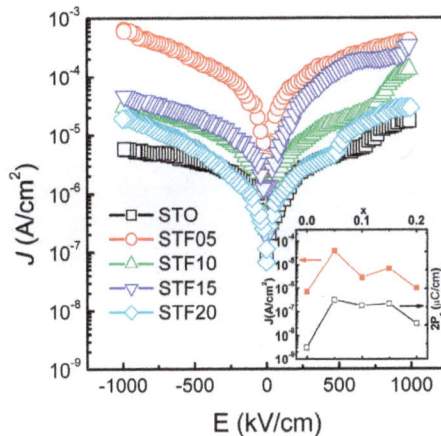

**Figure 5.** Leakage current densities (*J*) as a function of electric field (*E*) for STF thin films. Note the negative voltage axis was used. The inset shows *J* at applied field of 100 kV/cm and the values of double remnant polarization ($2P_r$) varying with the concentration of Fe.

The leakage current densities (*J*) as a function of the electric field (*E*) for the STF thin films were shown in Figure 5. The Inset illustrates the influence of the Fe doping on the leakage current densities of the STF thin films. After substituting the Ti with Fe, the leakage current densities of the STO thin film increased from $7.22 \times 10^{-7}$ A·cm$^{-2}$ to about $10^{-6}$–$10^{-5}$ A·cm$^{-2}$ at an applied field of 100 kV·cm$^{-1}$. The effect could be attributed to the oxygen defect induced by the mixed-valence Fe substitution, where Fe ions can shape the energy-band structure of STF, resulting in a decrease in band-gap energy, reduction enthalpy, and increasing levels of disorder in the oxygen sublattice [35], therefore increasing the free carriers in the films. On the other hand, since oxygen vacancies can be used as trapping centers for electrons, shallow trap energy levels may be generated within the band gap for the mobility of activated electrons. This is similar with the case in Ni-doped $BiFeO_3$, where the increased leakage was caused by the oxygen vacancies [36]. However, due to the possible changes in the oxidation states of the Fe ions and the decrease of crystal asymmetry, the oxygen vacancies will not always increase with the increasing Fe concentrations [37]. In addition to the bulk limited conduction, the significant difference in leakage when the bias reversed indicates the effect of interface limited conduction [38]. Thus the leakage current density did not exhibit a linear increasing trend with the increasing Fe concentration. The consistent variation of the $P_r$ and $J$ indicates the possible contribution of $J$ to the ferroelectric properties of STF thin films, most probably through the oxygen vacancies and the defect dipoles. Though the difference between the pure STO and STF thin films is about a 1–2 order of magnitude, the $J$ in the films is low enough compared to that of the atomic layer deposition pure STO films [39], which indicates a potential application in dielectric devices.

Several models were applied in order to investigate the conduction mechanism of the films in detail, and the typical fitted curves were shown in Figure 6. Figure 6a,b shows the log$J$ versus log$E$ (or $E^{1/2}$) plot of the STF thin film with the positive and negative bias on the Au electrode, respectively. In the case of the STO thin films with the positive bias on the Au electrode, see Figure 6a, the log$J$ versus log$E$ plot in low field region ($E < 700$ kV·cm$^{-1}$) was fitted well by the linear segment with slope ~1, which indicates the Ohmic conduction behavior [40,41]. The same behavior was also observed for many other samples, especially in the low field regions. In the high field region ($E > 700$ kV·cm$^{-1}$) of the STO thin film, as seen in Figure 6a, the slope of the log$J$ versus log$E$ plots is larger than 2, which reveals a space-charge-limited conduction (SCLC) [42]:

$$J = \frac{9\varepsilon_0\varepsilon_r\mu\theta E^2}{8d} \tag{1}$$

where $\varepsilon_0$ is the permittivity of free space, $\varepsilon_r$ is the low-frequency permittivity of the film, $\mu$ is the charge carrier mobility, $\theta$ is the ratio of the free carriers to the total carriers, and $d$ is the thickness of the film. The conduction mechanism also predominated in the high field region ($E > 490$ kV·cm$^{-1}$) of the STF10, STF15, and STF20 thin films. Different from that, for the negative bias region of the STF20 thin films, the log$J$ showed a linear relationship with the $E^{1/2}$ (Figure 6c,d), indicating the Schottky emission conduction mechanism [43]:

$$J = A^*T^2\exp\left[-\frac{1}{k_BT}\left(\Phi_B - \left(\frac{q^3E}{4\pi\varepsilon\varepsilon_0}\right)^{\frac{1}{2}}\right)\right] \tag{2}$$

where $A^*$ is the Richardson constant, and $\Phi_B$ is the Schottky barrier height. Here, barriers are induced by the contact of Au and STF20.

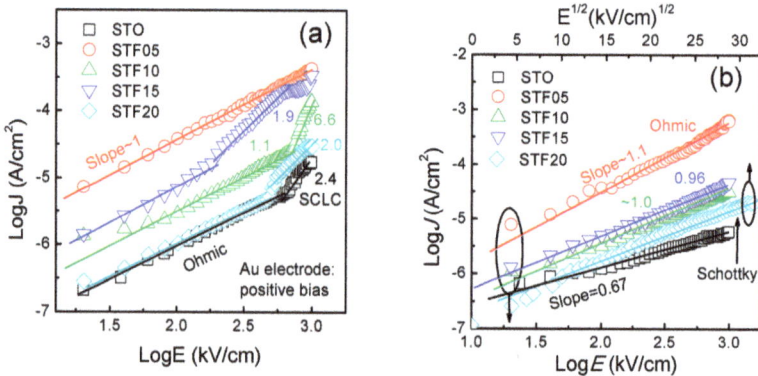

**Figure 6.** log$J$ versus log$E$ (or $E^{1/2}$) plot of the STF thin films when (**a**) positive bias (**b**) negative bias is applied on the Au electrode, respectively. The red lines show the fitted linear segments of the plots.

The conduction mechanisms for all samples in the full region were listed in the Table 1. In the low field regions, most thin films showed Ohmic conduction behavior, and the SCLC mechanism was observed in higher field regions. This is because the SCLC will not be observed until the injected free-carrier density exceeds the volume-generated, free-carrier density [44]. In this way, the transition field can indicate the level of volume-generated, free-carrier density indirectly. We thus can speculate that the STF10 and STF15 thin films possess a higher volume free-carrier than the STO thin film, because the transition field from Ohmic to the SCLC mechanism for STO is lower, and the three thin films have the same conduction mechanism in other regions. This is exactly consistent with the experiment result, where the $J$ in STF10 is higher than that of STO and STF15 possess a higher $J$ than that of STF10. As discussed above, after Fe substitution, oxygen vacancies were generated and the energy-band structure of STF was shaped with a reduced band-gap energy and reduction enthalpy, resulting in the increase of the free carriers in the films. A further increase of $x$ ($x = 0.20$) leads to the significant effect of interface-limited Schottky emission. The mechanism arises from the difference in Fermi levels between the electrode and the thin films, which will create a potential barrier to limit the $J$. It is likely that the doped Fe ion has influenced the barrier through a changing concentration of defects and therefore affects the Fermi level, which is consistent with the pervious study that Fe substitution can change the energy-band structure of STF [31]. Also, based on the study, the defect concentrations of bulk STF were determined by the balance between the intrinsic electronic and ionic disorder and the redox reaction. Moreover, owing to the existence of the interface in the thin films, the defects and carrier concentrations are more difficult to study, and more works are needed to test the specific details.

**Table 1.** The conduction mechanism of the STF thin films annealed at 650 °C in air.

| Samples | STF/LNO Interface | | | | Au/STF Interface |
|---|---|---|---|---|---|
| | Low Field (kV/cm) | Mechanism | High Field (kV/cm) | Mechanism | Mechanism |
| STO | <700 | Ohmic | >700 | SCLC | Ohmic |
| STF05 | - | Ohmic | - | Ohmic | Ohmic |
| STF10 | <710 | Ohmic | >710 | SCLC | Ohmic |
| STF15 | <120 | Ohmic | >120 | SCLC | Ohmic |
| STF20 | <490 | Ohmic | >490 | SCLC | Schottky |

Figure 7 shows the magnetization-magnetic field ($M$-$H$) curves of the various thin films on LNO/Si(100) substrates at room temperature by applying an in-plane magnetic field. The top inset shows the magnified plot of the vicinity "0" magnetic field. In the cases of STO, STF10, and STF20, the thin films exhibits weak ferromagnetism with average remnant magnetization ($M_r$), saturated magnetization ($M_s$), and a coercive magnetic field ($H_c$) of $1.46 \times 10^{-2}$ emu·cm$^{-3}$, $1.62 \times 1^{-1}$ emu·cm$^{-3}$, and 123 Oe for STO, $3.74 \times 10^{-2}$ emu·cm$^{-3}$, $9.22 \times 10^{-2}$ emu·cm$^{-3}$, and 1231 Oe for STF10, and $2.43 \times 10^{-2}$ emu·cm$^{-3}$, $2.33 \times 10^{-1}$ emu·cm$^{-3}$, and 136 Oe for STF20, respectively. The weak room temperature ferromagnetism in SrTiO$_3$ thin films was usually attributed to the surface defects, such as oxygen vacancies, cation vacancies, oxygen-ended polar terminations, and the effects of the grain boundary [45–51]. For the STF thin films, magnetoelastic effects may be an important contributor to the obviously enhanced ferromagnetic properties. In transition-metal-substituted oxides thin films, lattice mismatch, thermal mismatch, and coalescence during growth typically lead to strain. Also, magnetoelastic effects can be associated with the magnetic behaviours in the strain thin films, which have been investigated in some transition-metal-substituted STO films [52]. Based on the preceding discussions, the surface defects such as oxygen vacancies, strontium vacancies, iron vacancies, titanium vacancies, and oxygen-ended polar terminations [51] are perhaps the most probable magnetic sources, all of them being mainly located at the surface of the nanograins. A series of papers have proposed the grain boundaries as the controlling factor for the ferromagnetic behavior of oxides [47–51]. The bottom inset of Figure 7 (lower right) shows the variation of average $M_r$ and $H_c$ as a function of $x$. The $M_r$ and

$H_c$ values of the thin films decrease when the Fe concentration $x$ is higher than 10%. The phenomenon may be caused by the antiferromagnetic coupling between the Fe ions, which can only be exhibited as the Fe concentration increases to a certain degree [18].

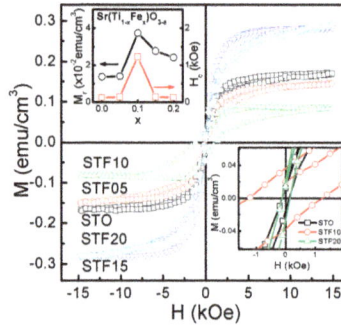

**Figure 7.** Magnetization-magnetic field (*M-H*) curves of STF thin films measured at room temperature. The inset shows the magnified plot of the vicinity "0" magnetic field (**bottom**) and the variation values of the remanent magnetization ($M_r$) and coercive magnetic field ($H_c$) as a function of the Fe concentration $x$ (**top**).

Figure 8 Room temperature *M-H* curves of the STF10 thin films were annealed at various atmospheres. In the case of the various atmospheres annealing, the thin films exhibit weak ferromagnetism with average values of the $M_r$, $M_s$, and $H_c$, which are $1.21 \times 10^{-2}$ emu·cm$^{-3}$, $8.83 \times 10^{-2}$ emu·cm$^{-3}$, and 203 Oe for air atmosphere, $1.09 \times 10^{-2}$ emu·cm$^{-3}$, $4.01 \times 10^{-2}$ emu·cm$^{-3}$, and 613 Oe for nitrogen atmospheres, and $7.40 \times 10^{-3}$ emu·cm$^{-3}$, $3.77 \times 10^{-2}$ emu·cm$^{-3}$, and 359 Oe for oxygen atmosphere, respectively. Compared with the ferroelectric properties, the sample shows the largest remanent magnetization when it is annealing in air atmosphere and showing good ferromagnetism. The samples show relatively weak ferromagnetism when it was annealed in oxygen or nitrogen atmosphere. Among them, the sample of nitrogen annealing also has less magnetic susceptibility in the case of the high magnetic field, showing a certain diamagnetism. Therefore, the air annealed sample with the higher oxygen vacancy concentration and grain boundaries exhibits good ferromagnetism. However, when the concentration of the oxygen vacancy is too high, the antiferromagnetic phase appears in the sample, which shows certain diamagnetism in the sample. Similar to ferroelectricity, the ferromagnetism shows a significant change in vacancy concentration, which further verifies that the ferromagnetic properties of the samples are closely related to the mixed valence ions and the effects of the grain boundary in the samples [47–51].

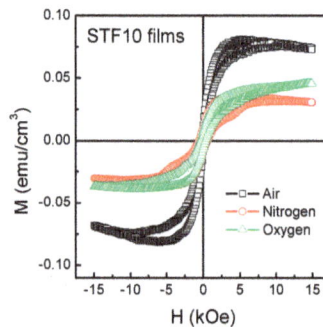

**Figure 8.** Room temperature *M-H* curves of the STF10 thin films were annealed at various atmospheres.

## 3. Materials and Methods

$Sr(Ti_{1-x}Fe_x)O_3$ (STF, $x$ = 0, 0.05, 0.1, 0.15, and 0.2; abbreviated as STO, STF05, STF10, STF15, and STF20, respectively) thin films were synthesized on the $LaNiO_3$ (LNO) coated Si(100) substrates by a sol-gel route with a spin-coating process [23]. To form the precursor, reagent-grade strontium acetate, iron nitrate nonahydrate, and titanium butyrate were dissolved under continuous stirring at 60 °C in solvents acetic acid, 2-methoxyethanol, and acetyl acetone, respectively. Here, acetyl acetone was also used to stabilize titanium butyrate. The three solutions were mixed and then stirred together at 60 °C for 1.5 h, forming a complete homogeneous transparent solution. The concentration of the final solution was adjusted to 0.25 M with a pH value of 2–3 by adding 2-methoxyethanol and acetic acid. Prior to spin-coating, the solution was filtered to avoid particulate contamination. The LNO thin layer prepared by chemical precursor solutions was described in previous literatures [24]. The STF layers were spin-coated onto the LNO films at a speed of 4000 rpm for 60 s. After each spin coating process, samples were heat-treated at 300 °C for 1 h on the hot plate. The step is repeated twice to obtain the desired thickness of the STF thin films. The STF films on LNO coated Si(100) substrates were finally annealed at 650 °C for 15 min by rapid thermal annealing (RTA) in air. To compare the multiferroic properties of the STF thin films annealing at various atmospheres, the STF10 thin films samples were selected annealing at 650 °C for 15 min by RTA in air, oxygen, and nitrogen atmospheres, respectively. As measured by a surface profiler (KLA-Tencor P-10, ClassOne Equipment, Inc. Decatur, GA, USA), the mean thickness of the annealed LNO and STF films was about 80 nm and 100 nm, respectively.

The crystalline phase and micrograph of the STF thin films were identified by the X-ray diffractometer (XRD, Pgeneral XD-2, PERSEE, Beijing, China) using CuKα radiation and atomic force microscopy (AFM, BenYuan CSPM-5500, Being Nano-Instruments Ltd. Beijing, China), respectively. The surface chemical states of the thin film were characterized by X-ray photoelectron spectroscopy (XPS, Thermo Scientific ESCALAB 250, Thermo Fisher Scientific Inc., Waltham, MA, USA) with the Al Kα radiation source. To investigate the electrical properties of the STF thin films, the gold (Au) top electrode with a diameter of 0.2 mm were deposited on the surface of the STF films by a vacuum evaporation apparatus (KYKY SB-12, KYKY Technology Co., Ltd. Beijing, China) through a shadow metal mask. The ferroelectric properties and the leakage current characteristics were measured by a ferroelectric test system (Radiant Precision Premier II, Albuquerque, NM, USA). Magnetic properties of the films were measured by a vibrating sample magnetometer (VSM, PPMS-9, Quantum Design, San Diego, CA, USA) at room temperature.

## 4. Conclusions

$SrTi_{1-x}Fe_xO_3$ ($0 \leq x \leq 0.2$) thin films have been synthesized on the LNO-coated Si(100) substrates by the sol-gel technique. The ferroelectric properties of the STO thin film has been improved through Fe doping, with a maximum double saturated polarization ($2P_s$) of 12.34 $\mu C \cdot cm^{-2}$ when doped with 10% Fe. The strain, crystal asymmetry, and defect dipoles induced by the ion substitution were ascribed to the possible origin of the enhanced ferroelectric properties. The leakage current densities of the Fe-doped STO thin films are about $10^{-5}$–$10^{-6}$ $A \cdot cm^{-2}$ at an applied field of 100 kV/cm, about 1–2 orders of magnitude larger than that of the pure STO thin film. The conduction mechanism of the thin films with various Fe concentrations has been discussed in detail. The oxygen vacancies are concluded to play a significant role on the conduction properties of the thin films. The ferromagnetic properties of the STF thin films have been investigated. The mixed valence ions and effects of the grain boundary were used to explain the room temperature ferromagnetism. The obtained low current densities in the films allow the possible application in multiferroic and electronic devices.

**Acknowledgments:** This work was supported by the National Natural Science Foundation of China (Grant No. 11574057), the Guangdong Provincial Natural Science Foundation of China (Grant No. 2016A030313718), and the Science and Technology Program of Guangdong Province of China (Grant No. 2016A010104018).

**Author Contributions:** X.G. Tang conceived and designed the experiments, and analyzed and interpreted the data. Y.G. Wang prepared the samples, preformed the structural, ferroelectric, and magnetic measurements and, together with Q. X. Liu, Y.P. Jiang, and L.L. Jiang, primarily analyzed the experimental data. Y.G. Wang and X.G. Tang wrote the manuscript. All authors commented on the manuscript.

**Conflicts of Interest:** The authors declare no conflict of interest.

## References

1.  Fuchs, D.; Schneider, C.W.; Schneider, R.; Rietschel, H. High dielectric constant and tunability of epitaxial SrTiO$_3$ thin film capacitors. *J. Appl. Phys.* **1999**, *85*, 7362–7369. [CrossRef]
2.  Huang, X.X.; Zhang, T.F.; Tang, X.G.; Jiang, Y.P.; Liu, Q.X.; Feng, Z.Y.; Zhou, Q.F. Dielectric relaxation and pinning phenomenon of (Sr,Pb)TiO$_3$ ceramics for dielectric tunable device application. *Sci. Rep.* **2016**, *6*, 31960. [CrossRef] [PubMed]
3.  Müller, K.A.; Burkard, H. SrTiO$_3$: An intrinsic quantum paraelectric below 4 K. *Phys. Rev. B* **1979**, *19*, 3593–3602. [CrossRef]
4.  Neville, R.C.; Hoeneisen, B.; Mead, C.A. Permittivity of strontium titanate. *J. Appl. Phys.* **1972**, *43*, 2124–2131. [CrossRef]
5.  Zhong, W.; Vanderbilt, D. Effect of quantum fluctuations on structural phase transitions in SrTiO$_3$ and BaTiO$_3$. *Phys. Rev. B* **1996**, *53*, 5047–5050. [CrossRef]
6.  Fleury, P.A.; Scott, J.F.; Worlock, J.M. Soft phonon modes and the 110°K phase transition in SrTiO$_3$. *Phys. Rev. Lett.* **1968**, *21*, 16–19. [CrossRef]
7.  Li, Y.L.; Choudhury, S.; Haeni, J.H.; Biegalski, M.D.; Vasudevarao, A.; Sharan, A.; Ma, H.Z.; Levy, J.; Gopalan, V.; Trolier-McKinstry, S.; et al. Phase transitions and domain structures in strained pseudocubic (100) SrTiO$_3$ thin films. *Phys. Rev. B* **2006**, *73*, 184112. [CrossRef]
8.  Verma, A.; Raghavan, S.; Stemmer, S.; Jena, D. Ferroelectric transition in compressively strained SrTiO$_3$ thin films. *Appl. Phys. Lett.* **2015**, *107*, 192908. [CrossRef]
9.  Haismaier, R.C.; Engel-Herbert, R.; Gopalan, V. Stoichiometry as key to ferroelectricity in compressively strained SrTiO$_3$ films. *Appl. Phys. Lett.* **2016**, *109*, 032901. [CrossRef]
10. Itoh, M.; Wang, R.; Inaguma, Y.; Yamaguchi, T.; Shan, Y.J.; Nakamura, T. Ferroelectricity induced by oxygen isotope exchange in strontium titanate perovskite. *Phys. Rev. Lett.* **1999**, *82*, 3540–3543. [CrossRef]
11. Ranjan, R.; Hackl, R.; Chandra, A.; Schmidbauer, E.; Trots, D.; Boysen, H. High-temperature relaxor ferroelectric behavior in Pr-doped SrTiO$_3$. *Phys. Rev. B* **2007**, *76*, 224109. [CrossRef]
12. Haeni, J.H.; Irvin, P.; Chang, W.; Uecker, R.; Reiche, P.; Li, Y.L.; Choudhury, S.; Tian, W.; Hawley, M.E.; Craigo, B.; et al. Room-temperature ferroelectricity in strained SrTiO$_3$. *Nature* **2004**, *430*, 758–761. [CrossRef] [PubMed]
13. Kim, D.H.; Aimon, N.M.; Bi, L.; Dionne, G.F.; Ross, C.A. The role of deposition conditions on the structure and magnetic properties of SrTi$_{1-x}$Fe$_x$O$_3$ films. *J. Appl. Phys.* **2012**, *111*, 07A918. [CrossRef]
14. Menke, T.; Meuffels, P.; Dittmann, R.; Szot, K.; Waser, R. Separation of bulk and interface contributions to electroforming and resistive switching behavior of epitaxial Fe-doped SrTiO$_3$. *J. Appl. Phys.* **2009**, *105*, 066104. [CrossRef]
15. Fagg, D.P.; Kharton, V.V.; Kovalevsky, A.V.; Viskup, A.P.; Naumovich, E.N.; Frade, J.R. The stability and mixed conductivity in La and Fe doped SrTiO$_3$ in the search for potential SOFC anode materials. *J. Eur. Ceram. Soc.* **2001**, *21*, 1831–1835. [CrossRef]
16. Shukla, V.K.; Mukhopadhay, S. Anomalous ferroelectric switching dynamics in single crystalline SrTiO$_3$. *Appl. Phys. Lett.* **2016**, *120*, 154102. [CrossRef]
17. Khanbabaee, B.; Mehner, E.; Richter, C.; Hanzig, J.; Zschornak, M.; Pietsch, U.; Stocker, H.; Leisegang, T.; Meyer, D.C.; Gorfman, S. Large piezoelectricity in electric-field modified single crystals of SrTiO$_3$. *Appl. Phys. Lett.* **2016**, *109*, 222901. [CrossRef]
18. Guo, Z.G.; Pan, L.Q.; Bi, C.; Qiu, H.M.; Zhao, X.D.; Yang, L.H.; Rafique, M.Y. Structural and multiferroic properties of Fe-doped Ba$_{0.5}$Sr$_{0.5}$TiO$_3$ solids. *J. Magn. Magn. Mater.* **2013**, *325*, 24–28. [CrossRef]
19. Hu, J.; Lv, X.; Zhu, W.; Hou, Y.; Huang, F.; Lu, X.; Xu, T.T.; Su, J.; Zhu, J. Induction and control of room-temperature ferromagnetism in dilute Fe-doped SrTiO$_3$ ceramics. *Appl. Phys. Lett.* **2015**, *107*, 012409.

20. Kim, H.S.; Bi, L.; Kim, D.H.; Yang, D.J.; Choi, Y.J.; Lee, J.W.; Kang, J.K.; Park, Y.C.; Dionne, G.F.; Ross, C.A. Ferromagnetism in single crystal and nanocomposite Sr(Ti,Fe)O$_3$ epitaxial films. *J. Mater. Chem.* **2011**, *21*, 10364–10369. [CrossRef]

21. Kumar, A.S.; Suresh, P.; Kumar, M.M.; Srikanth, H.; Post, M.L.; Sahner, K.; Moos, R.; Srinath, S. Magnetic and ferroelectric properties of Fe doped SrTiO$_{3-\delta}$ films. *J. Phys. Conf. Ser.* **2010**, *200*, 092010. [CrossRef]

22. Kim, K.T.; Kim, C.; Fang, S.P.; Yoon, Y.K. Room temperature multiferroic properties of (Fe$_x$,Sr$_{1-x}$)TiO$_3$ thin films. *Appl. Phys. Lett.* **2014**, *105*, 102903. [CrossRef]

23. Wang, Y.G.; Tang, X.G.; Liu, Q.X.; Jiang, Y.P.; Feng, Z.Y. Ferroelectric and ferromagnetic properties of SrTi$_{0.9}$Fe$_{0.1}$O$_{3-\delta}$ thin films. *Solid State Commun.* **2015**, *202*, 24–27. [CrossRef]

24. Tang, X.G.; Chan, H.L.W.; Ding, A.L. Electrical properties of (Pb$_{0.76}$Ca$_{0.24}$)TiO$_3$ thin films on LaNiO$_3$ coated Si and fused quartz substrates prepared by a sol–gel process. *Appl. Surf. Sci.* **2003**, *207*, 63–68. [CrossRef]

25. Shannon, R.D. Revised effective ionic radii and systematic studies of interatomic distances in halides and chacogenides. *Acta Cryst. A* **1976**, *32*, 751–767.

26. Szade, J.; Szot, K.; Kulpa, M.; Kubacki, J.; Lenser, C.; Dittmann, R.; Waser, R. Electronic structure of epitaxial Fe doped SrTiO$_3$ thin films. *Phase Transit.* **2011**, *84*, 489–500. [CrossRef]

27. Kim, Y.S.; Kim, D.J.; Kim, T.H.; Noh, T.W.; Choi, J.S.; Park, B.H.; Yoon, J.G. Observation of room-temperature ferroelectricity in tetragonal strontium titanate thin films on SrTiO$_3$ (001) substrates. *Appl. Phys. Lett.* **2007**, *91*, 042908. [CrossRef]

28. Kim, Y. S.; Choi, J.S.; Kim, J.; Moon, S.J.; Park, B.H.; Yu, J.; Kwon, J.H.; Kim, M.; Chung, J.S.; Noh, T.W.; et al. Defect-related room-temperature ferroelectricity in tensile-strained SrTiO$_3$ thin films on GdScO$_3$ (110) substrates. *Appl. Phys. Lett.* **2010**, *97*, 242907. [CrossRef]

29. Maeng, W.J.; Jung, I.; Son, J.Y. Room temperature ferroelectricity of tetragonally strained SrTiO$_3$ thin films on single crystal Rh substrates. *Solid State Commun.* **2012**, *152*, 1256–1258. [CrossRef]

30. Hill, N.A. Why Are There so Few Magnetic Ferroelectrics? *J. Phys. Chem. B* **2000**, *104*, 6694–6709. [CrossRef]

31. Denk, I.; Münch, W.; Maier, J. Partial Conductivities in SrTiO$_3$: Bulk Polarization Experiments, Oxygen Concentration Cell Measurements, and Defect-Chemical Modeling. *J. Am. Ceram. Soc.* **1995**, *78*, 3265–3272. [CrossRef]

32. Merle, R.; Maier, J. How is oxygen incorporated into oxides? A comprehensive kinetic study of a simple solid-state reaction with SrTiO$_3$ as a model material. *Angew. Chem. Int. Ed. Engl.* **2008**, *47*, 3874–3894. [CrossRef] [PubMed]

33. Kim, J.K.; Kim, S.S.; Kim, W.J.; Ha, T.G.; Kim, I.S.; Song, J.S.; Guo, R.; Bhalla, A.S. Improved ferroelectric properties of Cr-doped Ba$_{0.7}$Sr$_{0.3}$TiO$_3$ thin films prepared by wet chemical deposition. *Mater. Lett.* **2006**, *60*, 2322–2325. [CrossRef]

34. Warren, W.L.; Pike, G.E.; Vanheusden, K.; Dimos, D.; Tuttle, B.A.; Robertson, J. Defect-dipole alignment and tetragonal strain in ferroelectrics. *J. Appl. Phys.* **1996**, *79*, 9250–9257. [CrossRef]

35. Rothschild, A.; Menesklou, W.; Tuller, H.L.; Ivers-Tiffée, E. Electronic structure, defect chemistry, and transport properties of SrTi$_{1-x}$Fe$_x$O$_{3-y}$ solid solutions. *Chem. Mater.* **2006**, *18*, 3651–3659. [CrossRef]

36. Qi, X.D.; Dho, J.; Tomov, R.; Blamire, M.G.; MacManus-Driscoll, J.L. Greatly reduced leakage current and conduction mechanism in aliovalent-ion-doped BiFeO$_3$. *Appl. Phys. Lett.* **2005**, *86*, 062903. [CrossRef]

37. Makhdoom, A.R.; Akhtar, M.J.; Rafiq, M.A.; Hassan, M.M. Investigation of transport behavior in Ba doped BiFeO$_3$. *Ceram. Int.* **2012**, *38*, 3829–3834. [CrossRef]

38. Tang, X.G.; Wang, J.; Wang, X.X.; Chan, H.L.W. Preparation and electrical properties of highly (111)-oriented (Na$_{0.5}$Bi$_{0.5}$)TiO$_3$ thin films by a sol-gel process. *Chem. Mater.* **2004**, *16*, 5293–5296. [CrossRef]

39. Mojarad, S.A.; Goss, J.P.; Kwa, K.S.K.; Zhou, Z.; Al-Hamadany, R.A.S.; Appleby, D.J.R.; Ponon, N.K.; O'Neill, A. Leakage current asymmetry and resistive switching behavior of SrTiO$_3$. *Appl. Phys. Lett.* **2012**, *101*, 173507. [CrossRef]

40. Tang, X.G.; Wang, J.; Zhang, Y.W.; Chan, H.L.W. Leakage current and relaxation characteristics of highly (111)-oriented lead calcium titanate thin films. *J. Appl. Phys.* **2003**, *94*, 5163–5166. [CrossRef]

41. Wu, J.G.; Wang, J.; Xiao, D.Q.; Zhu, J.G. Valence-driven electrical behavior of manganese-modified bismuth ferrite thin films. *J. Appl. Phys.* **2011**, *109*, 124118. [CrossRef]

42. Scott, J.F. *Ferroelectric Memories*; Springer: Berlin/Heidelberg, Germany, 2000; pp. 85–90. ISBN 978-3-662-04307-3.

43. Pintile, L.; Vrejoiu, I.; Hesse, D.; LeRhun, G.; Alexe, M. Ferroelectric polarization-leakage current relation in high quality epitaxial Pb(Zr,Ti)O$_3$ Films. *Phys. Rev. B* **2007**, *75*, 104103. [CrossRef]

44. Yang, B.; Li, Z.; Gao, Y.; Lin, Y.H.; Nan, C.W. Multiferroic properties of $Bi_{3.15}Nd_{0.85}Ti_3O_{12}$–$CoFe_2O_4$ bilayer films derived by a sol–gel processing. *J. Alloys Compd.* **2011**, *509*, 4608–4612. [CrossRef]
45. Khalid, M.; Setzer, A.; Ziese, M.; Esquinazi, P.; Spemann, D.; Pöppl, A.; Goering, E. Ubiquity of ferromagnetic signals in common diamagnetic oxide crystals. *Phys. Rev. B* **2010**, *81*, 214414. [CrossRef]
46. Zhang, Y.J.; Hu, J.F.; Cao, E.S.; Sun, L.; Qin, H.W. Vacancy induced magnetism in $SrTiO_3$. *J. Magn. Magn. Mater.* **2012**, *324*, 1770–1775. [CrossRef]
47. Hsu, H.S.; Huang, J.C.A.; Chen, S.F.; Liu, C.P. Role of grain boundary and grain defects on ferromagnetism in Co:ZnO films. *Appl. Phys. Lett.* **2007**, *90*, 102506. [CrossRef]
48. Straumal, B.B.; Protasova, S.G.; Mazilkin, A.A.; Goering, E.; Schütz, G.; Straumal, P.B.; Baretzky, B. Ferromagnetic behaviour of ZnO: The role of grain boundaries. *Beilstein J. Nanotechnol.* **2016**, *7*, 1936–1947. [CrossRef] [PubMed]
49. Straumal, B.B.; Mazilkin, A.A.; Protasova, S.G.; Stakhanova, S.V.; Straumal, P.B.; Bulatov, M.F.; Schütz, G.; Tietze, T.; Goering, E.; Baretzky, B. Grain boundaries as a source of ferromagnetism and increased solubility of Ni in nanograined ZnO. *Rev. Adv. Mater. Sci.* **2015**, *41*, 61–71.
50. Tietze, T.; Audehm, P.; Chen, Y.C.; Schütz, G.; Straumal, B.B.; Protasova, S.G.; Mazilkin, A.A.; Straumal, P.B.; Prokscha, T.; Luetkens, H.; et al. Interfacial dominated ferromagnetism in nanograined ZnO: A µSR and DFT study. *Sci. Rep.* **2015**, *5*, 8871. [CrossRef] [PubMed]
51. Beltrán, J.J.; Barrero, C.A.; Punnoose, A. Combination of defects plus mixed valence of transition metals: A strong strategy for ferromagnetic enhancement in ZnO nanoparticles. *J. Phys. Chem. C* **2016**, *120*, 8969–8978. [CrossRef]
52. Kim, D.H.; Bi, L.; Jiang, P.; Dionne, G.F.; Ross, C.A. Magnetoelastic effects in $SrTi_{1-x}M_xO_3$ (M = Fe, Co, or Cr) epitaxial thin films. *Phys. Rev. B* **2011**, *84*, 014416. [CrossRef]

© 2017 by the authors. Licensee MDPI, Basel, Switzerland. This article is an open access article distributed under the terms and conditions of the Creative Commons Attribution (CC BY) license (http://creativecommons.org/licenses/by/4.0/).

*nanomaterials*

MDPI

*Review*

# Mesoporous Silica Nanoparticles as Drug Delivery Vehicles in Cancer

Anna Watermann and Juergen Brieger *

Department of Otorhinolaryngology, Head and Neck Surgery, University Medical Center Mainz,
Langenbeckstraße 1, 55131 Mainz, Germany; awaterma@uni-mainz.de
* Correspondence: brieger@uni-mainz.de; Tel.: +49-6131-17-3354

Received: 13 June 2017; Accepted: 18 July 2017; Published: 22 July 2017

**Abstract:** Even though cancer treatment has improved over the recent decades, still more specific and effective treatment concepts are mandatory. Surgical removal is not always possible, metastases are challenging and chemo- and radiotherapy can not only have severe side-effects but also resistances may occur. To cope with these challenges more efficient therapies with fewer side-effects are required. One promising approach is the use of drug delivery vehicles. Here, mesoporous silica nanoparticles (MSN) are discussed as biodegradable drug carrier to improve efficacy and reduce side-effects. MSN excellently fulfill the criteria for nanoparticulate carriers: their distinct structure allows high loading capacity and a plethora of surface modifications. MSN synthesis permits fine-tuning of particle and pore sizes. Moreover, drug release can be tailored through various gatekeeper systems which are for example pH-sensitive or redox-sensitive. Furthermore, MSN can either enter tumors passively by the enhanced permeability and retention effect or can be actively targeted by various ligands. PEGylation prolongs circulation time and availability. A huge advantage of MSN is their explicitly low toxic profile in vivo. Yet, clinical translation remains challenging. Overall, mesoporous silica nanoparticles are a promising tool for innovative, more efficient and safer cancer therapies.

**Keywords:** mesoporous silica nanoparticles; drug delivery; tumor targeting; biocompatibility

## 1. Introduction

Although cancer therapy has improved over the past decades and survival rates increased [1], the heterogeneity of cancer still demands new therapeutic strategies. Especially solid tumors at anatomical crucial sites e.g., glioblastoma, head and neck squamous cell carcinoma, lung adenocarcinoma are sometimes limited to radiotherapy and/or chemotherapy. Nonetheless, detrimental effects of these therapies are chemo- and radioresistance which promote locoregional recurrences, distant metastases and second primary tumors. Besides, severe side-effects reduce the patients' quality of life. Therefore, it is of utmost importance to develop new therapeutic strategies to overcome resistances and to reduce side-effects by targeted therapy. One possibility is to embrace the enhanced permeability and retention (EPR) effect of solid tumors: Due to a leaky vasculature and the lack of lymphatic drainage small structures such as nanoparticles can accumulate in the tumor [2]. Therefore, exploiting nanoparticles as drug delivery vehicles is a promising approach.

Research in nanomedicine prospered over the last decades and yielded several prerequisites for drug delivery systems. Nanoparticles should have a high loading capacity and the cargo should be protected until it reaches the side of action. Moreover, nanoparticles should be taken up predominantly and efficiently by cancer cells and evade the mononuclear phagocytic system (MPS). Once drug carriers are incorporated by the cells, endosomal escape and drug release is crucial. Good tumor accumulation and deep tumor penetration are also favorable. Importantly, nanoparticles need a good biocompatibility which is dependent on the used material but also influenced by degradation and excretion.

Over the past decades a plethora of different nanoparticles for drug delivery, organic and inorganic, were developed. Organic nanoparticles are represented for example by liposomes, polymer micelles, dendrimers and poly lactid-co-glycolic acid (PLGA)-based nanoparticles. In fact, some liposomal formulations are already approved by the US Food and Drug Administration (FDA), e.g., liposomal doxorubicin (Doxil®/Caelyx™) for treatment of Karposi's sarcoma, ovarian cancer and multiple myeloma [3]. Yet, the advantage of liposomes compared to the free drug is mostly limited to longer half-life and reduced toxicity [4]. Furthermore, several polymeric and micelle based vehicles for cancer therapy were or are in clinical trials, respectively [3].

Drug delivery systems can also be based on inorganic materials, e.g., gold nanoparticles, metal oxide such as iron oxide particles, carbon nanotubes, quantum dots and mesoporous silica nanoparticles (MSN) [5–9]. Particularly, iron oxide nanoparticles are already approved for glioblastoma therapy in Europe and as contrast enhancers for magnetic resonance imaging [3]. So far, no clinical trials were performed with MSN but an early phase I study (NCT02106598) is conducted with targeted silica nanoparticles for image-guided operative sentinel lymph node mapping [10]. However, MSN exhibit several superior features in comparison to other inorganic nanoparticles: MSN possess a unique structure with a tunable pore and particle size, resulting in a high specific surface area which can be easily functionalized, and most importantly are highly biocompatible. Silica is classified as "Generally Recognized as Safe" (GRAS) by the FDA and is used in cosmetics and as a food-additive [11]. The MSNs' porous structure allows a high drug loading capacity and a time-dependent drug release. But, the cargo can also be absorbed to the particle's surface. The pores are usually sealed by a gatekeeper system which is often also used for additional functionalization and improvement of pharmacodynamical characteristics.

In the following paragraphs we will discuss MSN synthesis, characteristics and surface modifications with regard to cancer cell targeting, controlled drug release and endosomal escape. Finally, MSN biocompatibility in vitro and in vivo will be reviewed and challenges of MSN application in cancer therapy will be discussed.

## 2. MSN Synthesis and Characteristics

First, MSN synthesis will be discussed briefly with regard to nanoparticle diameter and pore size. Then, the influence of the nanoparticles' characteristics is described with regard to drug delivery vehicles.

### 2.1. MSN Synthesis

Several different approaches are used for mesoporous silica nanoparticle synthesis resulting in a variety of engineered particle and pore sizes. For instance, MSN are synthesized based on a modified Stöber synthesis, using e.g., tetraethyl orthosilicate (TEOS) as precursor for silica condensation and different additives as templates such as surfactants like cetyltrimethylammonium bromide (CTAB), polymers, micelle forming agents or other dopants [12,13]. In brief, surfactants are stirred in a mixture of water and alcohol under basic conditions and TEOS or other silicates are added under agitation. Concentrations and compositions of silica sources, template-agents and stirring conditions determine particle size, pore size and shape. When the surfactant concentration is above the critical micelle concentration, CTAB self-aggregates into micelles and the silica precursor condensates at the surface. A silica structure is formed around the surface of the micelles. Then, the surfactants have to be completely removed to obtain biocompatible mesoporous silica nanoparticles which are usually further modified [14]. Another approach was first introduced by Zhao et al. who used triblock copolymers as templating agents for well-ordered hexagonal mesoporous silica structures with up to 30 nm pores [15]. The common pore size of MSN ranges between 2 and 5 nm but larger pore sizes of 23 nm can be generated e.g., by adding swelling agents such as trimethylbenzene [16]. Also, hollow-structured MSN were examined as drug carriers by Wu et al. who employed a stability difference-based selective bond breakage strategy. In brief, this strategy relies on the fact that a Si–C bond is weaker than a Si–O bond and can be degraded by hydrothermal treatment. By applying different temperatures, pore sizes were increased gradually up to 24 nm [17]. A greater variation can be found in the particle diameter which is also dependent on surface modifications. While some silica nanoparticles are 100–120 nm in diameter others are larger than 200 nm, yet pore sizes are similar (2.5 nm or 3.0 nm, respectively) [18–20]. However, the denoted particle diameter is also dependent on surface modifications such as coatings and the suspension medium. A more detailed description of MSN synthesis strategies can be found here [14].

## 2.2. Influence of MSN Characteristics on Biological Systems

The influence of nanoparticle characteristics including size, shape, surface area and chemistry on biological systems play an important role for efficient drug delivery and was extensively reviewed by Albanese, Tang and Chan [21]. MSN exhibit a high specific surface area of up to 1000 m$^2$/g which is decreased by surface modifications such as amination or coating [22,23]. Accordingly, large-pore nanoparticles (10 nm) exhibit a smaller specific surface area [24]. Yet, a large surface area increases loading efficiency for small molecule drugs and siRNA. For example, a nearly 1000 fold higher amount of Doxorubicin could be loaded in MSN compared to the FDA-approved liposomal formulation Doxil® [25]. MSN uniformity is important for quality assurance and can be determined by dynamic light scattering. Analyzing the Brownian motion reveals the polydispersity index (PDI) as indication for the colloidal dispersion size range and a low PDI is favorable [26]. Also, particle shape and size are analyzed with transmission electron microscopy (TEM). The nanoparticle characteristics such as shape, size and charge have an influence on particle uptake. Cellular entry is also dependent on the applied targeting strategy. In general, several possible endocytic pathways for cellular nanoparticle uptake were proposed, namely caveolae or clathrin-dependent endocytosis, caveolae or clathrin-independent endocytosis and micropinocytosis [27]. The most prominent cellular entry strategy is receptor-mediated endocytosis of targeted nanoparticles. After the MSNs' ligands bound the corresponding receptors on the cellular membrane, the endocytic process is initiated and particles are incorporated in endosomes [28]. Yet, untargeted MSN can also interact with the plasma membrane through their surface modifications by non-specific binding forces and then are endocytosed or penetrate the cellular membrane [29]. Regarding the nanoparticle shape, the best cellular uptake was achieved by rods, followed by spheres, cylinders and cubes when particles were larger than 100 nm [21,30]. Yet, spherical MSN of 50 nm showed a notable better incorporation by HeLa cells than 110, 170 or 280 nm particles, respectively [31]. The membrane-wrapping process and ligand-receptor interactions influence the uptake efficiency of different particle sizes. A smaller nanoparticle of 50 nm is able to induce membrane-wrapping by binding a sufficient number of receptors. While larger nanoparticles interact with a higher number of receptors and the uptake is limited by the receptors' redistribution on the cellular membrane through diffusion to compensate for local receptor shortage [21,29]. Since endosomes exhibit an acidic pH and pH decreases along the endocytic pathway from late endosomes to lysosomes [32], this acidic environment is used for a controlled release strategy, which is reviewed later. The surface charge also influences nanoparticle uptake. Positively charged particles have been found to be taken up faster than neutral or negatively charged particles by human cancer cells [28]. The cellular membrane has a slightly negative charge and favors binding of positively charged nanoparticles by electrostatic interaction. Yet, in a physiological environment nanoparticles are coated by a protein corona consisting of serum proteins, opsonins and ions which changes the in vitro determined parameters such as size and charge and thereby also influences cellular uptake and toxicity [29]. The absorbed proteins facilitate clearance by the MPS and agglomeration, but this can be prevented by coating the nanoparticles with poly(ethylene glycol) (PEG) resulting also in longer blood-circulation times [16,33]. So, MSN size and shape have a great influence on the nanoparticles' in vitro and in vivo behavior. Yet, surface modifications have an even greater impact on the drug delivery vehicles properties and will be discussed next.

### 3. Modifications to Control Cellular Uptake, Drug Release and Endosomal Escape

MSN surface modifications are necessary for several purposes: targeting moieties are supposed to direct the drug carrier to the desired destination, different capping systems ensure controlled drug release at the site of action and endosomal escape is not only crucial for efficacy but can also be influenced by certain alterations. The silanol groups present within the interior of the pores and on the outer surface can be modified with various functional molecules. These alterations will be discussed in the following sections.

#### 3.1. Passive and Active Targeting of MSNs

Scientists imagine a site-directed cancer therapy to lower toxic side-effects, enhance efficacy and reduce required drug doses. In general, three different strategies are exploited for this purpose, namely passive targeting, active targeting and magnetic-field directed targeting. These approaches were profoundly reviewed by Yang and Yu in 2015 [34], thus only an up to date summary will be given here.

#### 3.1.1. Passive Targeting

As mentioned above, nanoparticles accumulate favorably in solid tumor tissue due to the EPR effect, which is considered as passive targeting. Generally, tumors exceeding about one cubic millimeter in size require oxygen and nutrient supply to proliferate further [35]. Therefore, they rapidly form a highly abnormal vasculature by angiogenesis. The blood vessels are lined by a single, thin layer of flattened endothelial cells, the basement membranes have fenestrations varying in size and little or no pericytes cover the vessels [36]. Hence, macromolecules larger than 40 kDa, which is the threshold of renal clearance, can leave the blood vessels and accumulate in the adjacent tumor tissue but not in normal tissue. Also, solid tumors commonly lack effective lymphatic drainage, so accumulated macromolecules or nanoparticles remain longer in the tumor tissue without being cleared by the immune system [37]. To achieve efficient passive targeting, so far the focus laid on prolonging circulation time which is dependent on renal clearance and MPS escape. Phagocytic cells such as monocytes and macrophages are mainly located in liver, spleen, bone marrow and lymph nodes [38]. Hence, nanoparticles also tend to accumulate in these organs.

For efficient passive targeting several nanoparticle characteristics have to be considered such as particle size, morphology and surface modifications. To avoid renal clearance particles have to be at least 10 nm in diameter and a size of 100–200 nm seems to be optimal to also evade the MPS [34]. Besides, the nanoparticle shape also plays a role in passive targeting based on the EPR effect and was examined by Huang et al. in vivo. Using short-rod and long-rod MSN the main accumulation was found in liver, spleen and lung, which is no surprise considering the high blood flow rate of these organs. Yet, short-rod MSN tended to preferably accumulate in the liver with a fast clearance rate while long-rod MSN were distributed in the spleen with relatively slow clearance [39]. However, this study was performed without tumors and therefore no passive targeting was shown. Lu and colleagues could demonstrate enhanced tumor accumulation of MSN in comparison to normal tissue in vivo while MSN also exhibited good biocompatibility [40].

Surface modifications also have a major influence on nanoparticle tumor accumulation. As mentioned above, PEG is used to minimize opsonization and thereby evade the MPS. However, it has been implicated that PEGylation reduces cellular nanoparticle uptake in cancer cells but also in macrophages [33,41,42]. Nevertheless, Zhu and colleagues reported improved uptake of PEGylated hollow MSN in comparison to naked particles in cervical cancer cells and mouse embryonic fibroblasts [43]. Another considerable aspect with regard to passive targeting is the elevated interstitial fluid pressure in solid tumors which can be 10 to 40 fold higher compared to normal tissue [44]. This can create pressure gradients and heterogeneous flow in the interstitium which influences the distribution of nanoparticles and can lead to reduced particle concentrations in the tumor. Nonetheless, larger tumors and metastases often have necrotic tissue or highly hypovascular areas in the center because angiogenesis was slower than tumor growth. For this reason, nanoparticles can barely reach these regions by passive targeting.

Moreover, based on the data collected and analyzed by Wilhelm and colleagues [36] only $0.4 \pm 0.2\%$ of the administered untargeted MSN dose (7 data sets) could be found in the tumor tissue. However, $0.8 \pm 0.5\%$ of injected targeted MSNs (6 data sets) were found in tumors supporting the advantage of active targeting which will be discussed in the following paragraphs.

### 3.1.2. Active Targeting

In order to enhance drug delivery with nanocarriers and drug efficacy, active targeting is conducted to membrane receptors predominantly expressed in tumors, in vascular structures or in the nuclear membrane. In case of leukemic diseases nanoparticle targeting is inevitable because the EPR effect does not apply. So, different targeting moieties can be added to the MSNs' surface such as small molecules, short peptides, aptamers and whole antibodies or antibody fragments. Usually, the MSN are then taken up by receptor-mediated endocytosis. An overview of the described targeting ligands is given in Figure 1.

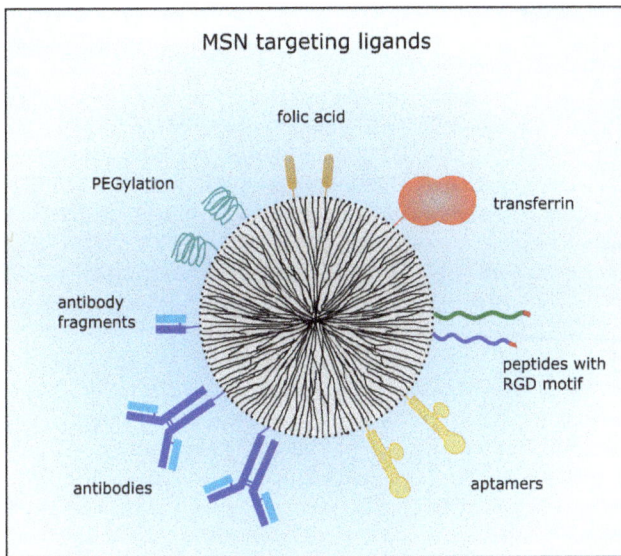

**Figure 1.** Ligands for active tumor targeting. MSN can be coated with poly (ethylene glycol) (PEG) to prolong circulation time. Small molecules such as folic acid are often used for active targeting. Different peptides with the arginine-glycine-aspartic acid (RGD) motif or proteins such as transferrin were also employed for tumor targeting. Moreover, aptamers, antibodies or antibody fragments are utilized to target membrane-receptors which are commonly overexpressed in cancer cells.

A prominently used tumor cell target is the folate receptor which is overexpressed in many tumors in comparison to healthy tissue [45]. Qi et al. targeted laryngeal carcinoma with folic acid-modified MSN. They successfully delivered commonly used chemotherapeutic drugs (paclitaxel, cisplatin, 5-fluoruracil) and siRNA targeting ABCG2, a drug efflux pump involved in multidrug-resistance, to CD133[+] positive laryngeal cancer cells in vitro and in vivo [46]. Before, the group showed a greater reduction in laryngeal tumor size in a mouse model by using cisplatin-loaded and folate-conjugated MSN compared to untargeted MSN [19]. Zhang and colleagues also utilized folate as targeting ligand on MSN to improve the radioenhancer effect of valproic acid in glioblastoma cells [47]. Moreover,

PEG-conjugated folate was applied by Cheng et al. as targeting ligand on pH-sensitive polydopamine coated MSN in vitro and in vivo. Doxorubicin delivery via folate-targeted MSN had improved efficacy compared to the free drug and untargeted MSN with doxorubicin in a xenograft tumor model. Also, distinctly higher tumor accumulation of folate-targeted MSN in comparison to untargeted nanoparticles was observed [48].

Using another concept, the glycoprotein transferrin was applied as targeting-ligand and redox-responsive gatekeeper by Chen et al. who could show the same toxicity of the free drug doxorubicin and doxorubicin in transferrin-targeted-MSN in hepatocellular carcinoma cells [49]. Furthermore, Chen and colleagues exploited the fucose-binding lectin UEA1 for colorectal adenocarcinoma, adenoma and polyposis coli targeting and detection. Fluorescently labeled and UEA1 carrying MSN were successfully tested in a mouse colon cancer model as a contrast agent to visualize malignant lesions in the colon [50].

Not only proteins can be utilized for targeting but also short peptides. For instance, Sweeney and coworkers attached a bladder-cancer specific peptide named Cyc6 to $Gd_2O_3$-MSN and thereby improved the detection of tumor boundaries in magnetic resonance imaging (MRI) scans in a mouse bladder cancer model [51]. The arginine-glycine-aspartic acid (RGD) motif is a prominent peptide sequence targeting integrin $\alpha v \beta 3$ which is overexpressed in certain tumors [52,53]. Therefore, peptides including the RGD motif have been used for targeting MSN to tumors in vivo by Pan et al. who showed good tumor accumulation and efficacy of doxorubicin loaded MSN. Even better tumor accumulation and reduction in tumor size were found when the cell-penetrating and nuclear-targeting peptide TAT was also coupled to the MSN besides RGD. In addition, bare MSN accumulation in liver and spleen was distinctly greater than RGD/TAT-MSN accumulation in those organs while untargeted MSN were found only in small concentration in the tumor tissue [54]. A similar approach was conducted by Ashley and colleagues who used the peptide SP-94 and a nuclear localization signal (NLS) for cancer cell and nuclear targeting, respectively. The MSN were loaded with siRNA and different chemotherapeutic drugs and then coated with a lipid bilayer which conveyed the targeting moieties, a fusogenic peptide for endosomal escape and PEG. Doxorubicin-loaded and targeted MSN significantly decreased cellular viability of hepatocellular carcinoma cells in comparison to hepatocytes which were barely affected [25].

Apart from small molecules and peptides, aptamers can be used for tumor cell targeting. Aptamers are synthetic single-stranded DNA or RNA oligonucleotides that show high affinity and specificity toward different targets. They are polyanionic and larger than small peptides but smaller than antibodies [55]. An aptamer binding to epithelial cell adhesion molecule (EpCAM) was employed by Babaei and colleagues for hepatocellular carcinoma targeting in vitro and in vivo. They encapsulated 5-Fluorouracil in MSN with citrate-modified gold nanoparticles as gatekeeper which were PEGylated and conjugated with the EpCAM aptamer. Targeted nanoparticles showed a greater reduction of cellular viability than untargeted nanoparticles. Moreover, in vivo the system was tested as a theranostic device and profoundly better tumor accumulation was observed after Rhodamine-6G loaded-MSN injection in in vivo imaging [56]. Another receptor for aptamer targeting is nucleolin which is expressed on cancer cells. Tang et al. developed a photoresponsive drug delivery system based on graphene oxide wrapped MSN for light-mediated drug release and a conjugated nucleolin-targeting aptamer. However, in vitro no difference between targeted and untargeted doxorubicin loaded MSN on cellular viability of breast cancer cells was recognized [57].

Finally, whole antibodies or antibody-fragments are used for tumor targeting of drug delivery vehicles. For instance, antibodies already approved for cancer therapy are utilized for this purpose including cetuximab (targeting (Epidermal Growth Factor) EGF receptor), trastuzumab (targeting (Human Epidermal Growth Factor Receptor 2) HER2/neu receptor) and bevacizumab (targeting (Vascular Endothelial Growth Factor) VEGF receptor) or related antibodies with similar targets. The group of Jeffery Brinker developed a drug nanocarrier named "protocell" which consists of a MSN core for drug loading and a lipid bilayer as gatekeeper and platform for surface modifications. They availed an epithelial growth factor receptor (EGFR)-antibody for targeting leukemic cells efficiently in vitro and in vivo [58]. Moreover, Zhou and colleagues conjugated rituximab to MSN and evaluated the drug delivery vehicle in vitro and in vivo [59]. Rituximab is a chimeric monoclonal antibody targeting the CD20 antigen on B cells and is approved amongst others for B-cell non-Hodgkin lymphoma therapy [60]. In a murine xenograft lymphoma model a pronounced effect on tumor volume reduction was observed for rituximab-targeted doxorubicin-loaded MSN, while the mice constitution remained better in comparison to mice treated with free doxorubicin [59]. Furthermore, for tumor vasculature targeting anti-CD105 antibody (TRC105) has been employed by Chen et al. in a murine breast cancer model. Tumor uptake of antibody-conjugated MSN was significantly larger compared to untargeted nanoparticles but still, liver accumulation 24 h after injection was witnessed [61]. The same group also used a TRC105 antibody fragment (Fab) to target dual-labeled MSN for in vivo targeted positron emission tomography (PET) imaging/near-infrared fluorescent dye (NIRF) imaging of the tumor vasculature in a mouse model [62].

In conclusion, several strategies are available for nanoparticle targeting and some have already been employed successfully in murine models. However, high accumulation in organs such as liver, spleen, lungs and kidneys still poses a problem for application in cancer therapy and regulatory approval.

## 3.2. Systems for Controlled Drug Release

A plethora of different approaches have been used to control MSN drug release and are reviewed in detail by Mekaru, Lu and Tamanoi [63]. The various gatekeeper systems are categorized by internal and external stimuli responses and an overview of the here described examples is shown in Figure 2. Internal stimuli include decreasing pH, reducing environment and enzymes. As mentioned before, nanoparticles are often engulfed via endocytosis, so a system responding to low pH is frequently applied using different concepts. Besides, the tumor microenvironment exhibits a low pH due to hypoxia [64] and therefore, drug release can be facilitated at the target site. Examples for low pH activated capping systems include pH-sensitive nanovalves such as pseudorotaxane encircled by β-cyclodextrin [65], tannic acid [66], polymer and lipid coatings as applied by Popat et al. and Durfee et al., respectively [58,67]. Another pH-sensitive system consisted of a block copolymer containing positively charged artificial amino acids and oleic acid blocks, which acted simultaneously as capping and endosomal release agents [24]. Upon protonation the pore blocking agents were removed or degraded and the cargo could be released. Furthermore, Liu et al. developed a cascade pH-responsive system using the weak acidic pH of the tumor microenvironment and the acidic endolysosomal pH. First, β-cyclodextrin was conjugated to hollow MSN with a boronic-acid-catechol ester bond for sealing the pores which was degraded in the endosomes or lysosomes at pH 4.5 to 6.5. Second, PEG was grafted to adamantine via a weak pH sensitive benzoic-imine bond which was degraded at pH 6.8 and PEGylated adamantine reacted with the sealed nanoparticles via host-guest interactions. Therefore, PEG was released in the tumor microenvironment facilitating nanoparticle uptake and more efficient drug delivery [68]. Another dual-responsive drug carrier was developed by Liu and coworkers who induced drug release at high temperature and low pH. The polymer poly[(*N*-isopropylacrylamide)-*co*-(methacrylic acid)] was grafted onto MSN to seal the pores and control the diffusion of the cargo in and out of the pore channels depending on temperature and pH [19].

MSN drug release can also be modulated by a redox-sensitive system. As intracellular glutathione concentration can be up to 10 mM, disulfide bonds linking the capping system to the MSN are reduced upon entering the cytoplasm and cargo can be released [69]. For example, Kim et al. used β-cyclodextrin directly linked to the MSN with a disulfide bond to seal the pores and efficient doxorubicin toxicity in lung adenocarcinoma cells was shown [70]. Also, polymers cross-linked by cystamine were utilized to close the MSNs' pores and the polymeric network was degraded in a reducing environment [71]. Besides, Wu et al. sealed their hollow structured MSN with poly-(β-amino-esters) via a disulfide-linker which was also reduced intracellularly [17]. Furthermore, a redox- and pH-sensitive dual response system was developed by Li and colleagues who utilized ammonium salt to seal the MSNs' pores. The ammonium salt was connected via an amide and a disulfide linker to the MSN. Hence, the disulfide bond was reduced glutathione-dependently and the amide bond was degraded at low pH upon cellular uptake [72].

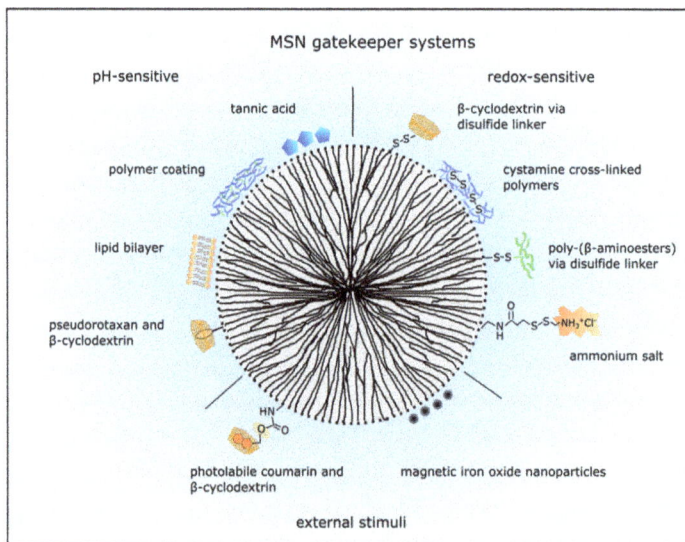

**Figure 2.** MSN gatekeeper systems to control drug release. Drug release can be regulated by internal stimuli such as pH decrease or reduction by glutathione or by external stimuli. PH-sensitive systems respond to acidic pH in the tumor microenvironment and in the endolysosomal system. Several examples are presented here such as pseudorotaxan encircled by β-cyclodextrin, tannic acid, polymer and lipid coatings. Several capping structures are linked to the MSN via disulfide bonds which are reduced by glutathione intracellularly. Then the pore blocking agents such as β-cyclodextrin, cystamine, poly-(β-aminoesters) and ammonium salt are released and the drugs can escape the nanoparticle. External stimuli such as light and magnetism are utilized to control drug release, too. Photolabile coumarin encircled by β-cyclodextrin is cleaved from the MSN by light or a magnetic field stimulates iron oxide nanoparticles to release the encapsulated drugs.

Using a biomolecule activated system, Mondragón et al. encapsulated camptothecin in MSN with a protease cleavable ε-poly-L-lysine and in human cervix epitheloid carcinoma cells viability was reduced after camptothecin-loaded nanoparticle incubation [73]. The same group also magnetic several hydrolyzed starch products as saccharides for enzyme-responsive drug release [74].

Apart from internal stimuli also external stimuli such as light or magnetic fields are utilized to control gatekeepers. These systems can generate more precise and local drug release, hence reducing toxicity towards normal cells. With regard to light activated drug release, the best wavelengths for adequate tissue penetration are within the biological spectra, typically 800–1100 nm [63]. In an in vitro model, Guardado-Alvarez et al. used two-photon excitation at 800 nm to cleave the nanoparticles' cap which consisted of photolabile coumarine-molecules bound to the nanoparticle surface and non-covalently conjugated β-cyclodextrin molecules [75]. Moreover, Croissant and colleagues also used two-photon light to control drug release via a photo-transducer from mesoporous silica nanoimpellers in human cancer cells [76]. However, tissue penetration of light is still limited, so using a magnetic field for external stimulated cargo release is more advantageous even though a magnetic component is necessary. Therefore, a magnetic iron oxide core is coated with mesoporous silica or MSN are capped with iron oxide nanoparticles [77,78]. The iron oxide core has superparamagnetic properties and can be heated up by an oscillating magnetic field which in turn can be used to open a nanovalve and for example release doxorubicin [78]. Several superparamagnetic iron oxide nanoparticles (SPION) are already FDA-approved imaging agents (endorem®/umirem® AMAG pharmaceuticals, Waltham, MA, USA) and iron oxide nanoparticles (Nanotherm®, MagForce, Berlin, Germany) are also approved in the European Union for glioblastoma therapy [3].

### 3.3. Endosomal Escape of MSN and Their Cargo

Once MSN entered the cancer cells by endocytosis, an endosomal escape of the nanoparticles or the delivered drug is mandatory for efficacy. The endosomal pH ranges from 6.0 to 6.5, but along the endocytic pathway acidity increases and late endosomes and lysosomes exhibit a pH from 4.5 to 5.5 [32]. Thus, the MSNs' cargo could be degraded or inactivated by lysosomal enzymes. To avoid this, several concepts are applied to enable drug release in the cell based on different theories (Figure 3). For example, endosomal escape can be achieved by the so-called "proton sponge effect", which relies on an increase of proton concentrations during hydrolysis. This leads to an increase in membrane potential and influx of counter-ions resulting in osmotic swelling and bursting of the endosome [79]. Hence, the cargo is released to the cytosol and can take full effect. For instance, the MSN system utilized by Wu et al. released siRNA and doxorubicin into the cytoplasm after the coating with poly-($\beta$-aminoesters) induced endosome bursting [17]. In the same way, cationic polyethyleneimine (PEI) coating can trigger the proton sponge effect which was applied by Finlay and colleagues to deliver TWIST1 siRNA to xenograft tumors and reduce tumor burden [80].

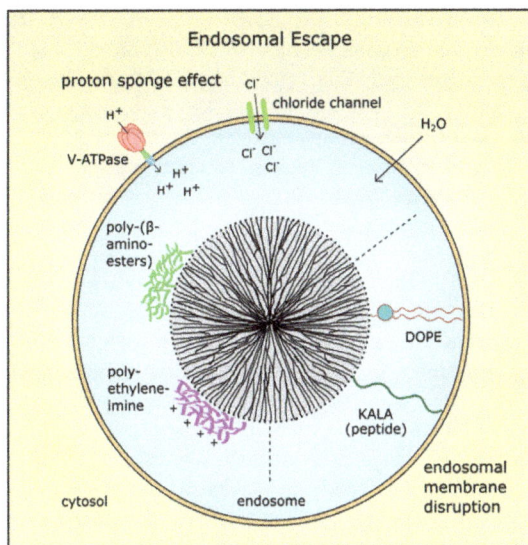

**Figure 3.** Endosomal escape mechanisms. After MSN were taken up by endocytosis, an endosomal escape is mandatory for drug efficacy. Coating with cationic polymers such as polyethyleneimine or poly-($\beta$-aminoesters) induces the proton sponge effect. The proton concentration increases during hydrolysis which leads to an increase in membrane potential and influx of counter-ions such as chloride ions. Finally, osmotic swelling by water inflow bursts the endosome and the MSN with its cargo is delivered into the cytosol. Also, fusogenic peptides such as KALA or zwitterionic co-lipids such as dioleoyl-phosphatidylethanolamine (DOPE) can destabilize the endosomal membrane resulting in MSN release.

Other methods use fusion lipids, cationic polymers or peptides to destabilize the endosomal membrane by proton absorption and acidification [81]. One example is the zwitterionic co-lipid dioleoylphosphatidyl-ethanolamine (DOPE) which was also utilized in combination with a polymer to coat MSN and improve drug release [82,83]. Moreover, Ashely et al. employed a fusogenic peptide to enhance endosomal escape of protocells in hepatocellular carcinoma [25]. Fusogenic peptides referred to as KALA were conjugated to PEI-coated MSN by Li and colleagues and were used to deliver (VEGF) targeting siRNA in a xenograft tumor model. The siRNA-loaded MSN with KALA peptides inhibited tumor proliferation significantly compared to control particles without siRNA or control siRNA, respectively [84]. However, many so far developed MSN systems relied on the proton sponge effect for endosomal escape. Aside from endocytosis, other mechanisms for nanoparticle uptake are possible, thus endosomal escape is not always necessary for drug efficacy.

Overall, surface modifications play an important role for efficient drug transport via MSN, MSN targeting and drug release. However, the "perfect" system does not exist and is unlikely to be invented due to the heterogeneity of cancer.

## 4. Biocompatibility of MSN

A major advantage of MSN is its high biocompatibility in vivo. Several studies examined biodistribution, toxicity and excretion of MSN. The FDA classified silica as "Generally Recognized as Safe" and silica is used as a food-additive and in cosmetics [11]. In general, silica particles are degraded into water-soluble orthosilicic acid ($Si(OH)_4$) which is also absorbed by humans to form silica as a trace element [85]. Many in vitro studies showed no toxicity for up to 100 $\mu$g mL$^{-1}$ MSN in cell culture [48,58,59,85]. Sometimes even higher concentrations were tested without significant toxicity [49,86]. It is generally recognized that crystalline silica nanoparticles can cause reactive oxygen species (ROS) formation which compromises cellular viability [87]. Yet, MSN seem to induce ROS formation only in high concentrations. For example, MSN concentrations of 1 mg mL$^{-1}$ and higher exhibited ROS in colon carcinoma cells while 200 $\mu$g mL$^{-1}$ did not induce ROS [88]. Furthermore, a relatively small MSN concentration did not promote ROS formation in hepatocellular carcinoma cells [89]. Elle and colleagues covalently coated MSN with antioxidants to reduce ROS formation and rutin decreased ROS formation dose-dependently in a keratinocyte cell line and dose-independently in colon carcinoma cells [20]. However, ROS formation after MSN application has been rarely examined due to the overall good biocompatibility.

One of the first in vivo studies was conducted by Park and colleagues who examined biodistribution of silica for four weeks. A relatively low dose of 20 mg kg$^{-1}$ MSN (126 nm diameter) was administered intravenously into mice and the body weight increased in the same manner as in the control group. The nanoparticles predominantly accumulated in MPS-related organs such as the liver and spleen. Yet, after one week MSN were mostly cleared from the analyzed organs (liver, spleen, heart, kidney, brain and lung) and almost completely vanished after four weeks. Moreover, histopathological analysis indicated no significant toxicity compared to controls, even though apparently macrophages in the liver (Kupffer cells) were swollen one day after MSN injection. The authors assumed that MSN were degraded and then excreted via the kidneys [85]. Furthermore, He and coworkers thoroughly studied nanoparticle excretion and biodistribution in vivo. On that account, MSN and PEGylated MSN of several sizes (80 nm, 120 nm, 200 nm and 360 nm) were analyzed. Fluorescently labeled MSN were evaluated in different organs with fluorescence intensity measurements of homogenized samples at several time points after injection of 20 mg kg$^{-1}$ nanoparticles. Most nanoparticles accumulated in spleen and liver, 30 min after injection also in the lungs, and low accumulation was detected in heart and kidneys. PEGylation reduced accumulation of larger particles in the lung and overall in the spleen 30 min after injection. However, after one month smaller particles were only observed in liver and spleen in low concentrations while 200 nm particles were also detected in even lower concentrations in heart, lung and kidneys. Regarding 360 nm MSN, the lowest concentrations were found after one month, whereas PEGylated MSN were still visible in all examined organs. Besides, nanoparticle concentration of larger particles in liver and spleen decreased over time. Blood clearance of MSN was slower for PEGylated particles and after eight hours particles were barely detectable, yet the smallest MSN had the longest blood circulation time. With regard to excretion, MSN and PEGylated MSN were mainly already excreted after 30 min and smaller particles mostly within five days. However, after one month larger particles were still detectable in urine. Histopathological evaluation showed no significant tissue toxicity and inflammation one month after injection for all particle sizes compared to controls [90]. In a study conducted by the group of Tamanoi biodistribution, biocompatibility and drug-delivery efficiency of MSN was analyzed in a xenograft tumor model. First, they determined a maximal tolerated dose of 50 mg kg$^{-1}$ spherical MSN (100–130 nm) after intravenous injection and monitoring for ten days. Then MSN were administered intraperitoneally with the same concentration in 18 doses over two months for long-term toxicity profiling. No unusual responses or behaviors compared to controls were observed and all measured hematologic factors were within normal ranges, proposing that the treatment did not induce an inflammatory response. However, all experiments were conducted in nude mice which lacked a thymus and therefore a possible T-cell response. Good biocompatibility could also be due to the fact that more than 90% of the administered silicon concentration was excreted via feces and urine within 4 days. Moreover, in a xenograft breast cancer tumor model MSN were mainly found in tumor, lung and kidneys 24 h after tail-vein injection, while 48 h after injection the spleen exhibited increased silicon concentration. Targeting with folic acid increased tumor accumulation of nanoparticles. Furthermore, camptothecin-loaded MSN reduced tumor size faster and greater after in total 18 intraperitoneal injections over nine weeks. Also, hematology profiling suggested reduced toxicity of camptothecin-loaded MSN compared to the free drug [40]. In a more recent study, Liu et al. evaluated 120 nm hollow MSN with a pH-dependent gatekeeper system in a xenograft hepatocellular carcinoma model. The untargeted but PEGylated MSN were loaded with doxorubicin and inhibited tumor proliferation over a time period of 21 days. At the same time, the mice weight increased while mice treated with the free drug lost weight. The same tendency was observed in survival analysis where mice treated with free doxorubicin all died shortly after the treatment stopped. However, half of the mice treated with PEGylated MSN survived more than one month after the last injection until the end of the experiment. In a biodistribution study after a single injection most of the particles accumulated in liver, spleen and lung whereas PEGylated MSN exhibited less accumulation. During the first week after injection, naked particle concentrations increased in liver and spleen while PEGylated MSN also increased in lung tissue. Only low concentrations of nanocarriers were detected in heart and kidney tissues. Yet, after one month MSN concentrations were decreased as expected [68]. Zhou and colleagues utilized a relatively high concentration of 100 mg kg$^{-1}$ rituximab-conjugated MSN for toxicity and distribution analysis in vivo. After seven MSN doses during three weeks the body weight increased correspondingly to control mice and histological analysis indicated no significant pathological lesions or damages in the major organs. Still, experiments were conducted in immunodeficient nude mice and therefor a lack of pathological damages is not surprising [59].

In brief, MSN exhibited remarkable good biocompatibility in many in vivo studies so far while tested particle concentrations increased over time. Still, accumulation of nanoparticles in MPS-related organs presents a challenge but this seemed to have no major impact on the animals' constitution and inflammatory responses remained mild. However, most studies were performed in immunodeficient mice decreasing the chances for a severe immune response. So, more studies in rodents with intact an immune system are necessary to fully evaluate the toxic profile of MSN before clinical trials. Nevertheless, the first early phase 1 clinical trial involving targeted silica nanoparticles for image-guided operative sentinel lymph node mapping is realized [10]. In conclusion, MSN are promising drug delivery vehicles for cancer therapy from a biocompatibility perspective.

## 5. Possible Challenges of MSN Application in Cancer Therapy

As mentioned above, MSN perform well in preclinical tests, yet only one clinical trial is currently performed. As for all new drugs and medicinal formulations, regulations by the FDA or the European Medicines Agency (EMA), respectively, present comprehensible and essential hurdles: From scale up of MSN synthesis to required dosage to acceptable pharmacokinetic and pharmacodynamic profiles.

In the case of MSN, synthesis of large amounts with consistent characteristics and quality might be challenging. Moreover, drug loading must be steady and not all drugs can be incorporated in MSN in a suitable concentration. The amount of loaded drug also influences the required nanoparticle dose. The maximal tolerated dose for unmodified MSN in a murine model was found to be 50 mg kg$^{-1}$ and an appropriate dose for human use needs to be evaluated in phase 1 clinical trials [40]. Yet, biocompatibility and efficiency are also dependent on the modifications such as targeting ligands and gatekeeper systems. For example, rituximab-conjugated MSN were even tolerated at 100 mg kg$^{-1}$ when intravenously injected in nude mice [59]. The nanoparticle distribution and excretion was also evaluated in murine models as mentioned above [90], but more data concerning immune response and possible side-effects, especially in an functional immunogenic environment, are needed. Also, particle accumulation in liver, spleen and other normal tissue poses a hurdle for clinical translation. Failure of the MSN system might lead to a burst drug release beyond the tumor tissue and could result in systemic toxicity, so the safety of the nanoparticle system needs to be critically evaluated. However, this is also dependent on the drug, for example siRNA would probably be degraded by nucleases and not pose a substantial threat while chemotherapeutic drugs could be harmful or even life threating. When the drug vehicles reached their site of action (tumor microenvironment or tumor cells), a controlled and efficient drug release has to be ensured. Assuming a MSN system for drug delivery fulfills the above mentioned criteria, it is most likely advantageous compared to liposomal or free drug formulations. It can be loaded with higher doses, can be targeted and drug release can be controlled. This would reduce the required dose and side-effects which are due to systemic delivery. Also, delicate drugs such as siRNA could be delivered. Upon degradation MSN are broken down into non-toxic silicic acid moieties which are easily excreted via the kidneys [85].

In conclusion, even though MSN provided good results in preclinical studies, clinical translation progresses slowly.

## 6. Summary

New and innovative approaches are needed to combat the heterogeneous disease cancer. Here, MSN were reviewed as drug delivery vehicles to improve efficacy and reduce side-effects. MSN ideally suit the criteria for nanoparticulate carriers since their structure allows high drug loading capacity and a plethora of surface modifications. MSN can be synthesized in different sizes with distinct pore sizes. Moreover, drug release can be finely tuned through various gatekeeper systems which are pH-sensitive or redox-sensitive, for example. PEGylation promotes escape from the MPS, so circulation time and availability are prolonged. Furthermore, MSN can either enter tumors by the EPR effect or can be actively targeted by various ligands. Another huge advantage of MSN is their biodegradability and high biocompatibility in vivo. However, clinical translation still remains a challenge and needs to be addressed. All in all, mesoporous silica nanoparticles are a promising tool for innovative cancer therapy.

**Acknowledgments:** A.W. was supported by Max Planck Graduate Center mit der Johannes Gutenberg-Universität Mainz.

**Author Contributions:** A.W. and J.B. wrote the paper.

**Conflicts of Interest:** The authors declare no conflict of interest.

## References

1. Siegel, R.L.; Miller, K.D.; Jemal, A. Cancer statistics, 2017. CA. *Cancer J. Clin.* **2017**, *67*, 7–30. [CrossRef] [PubMed]
2. Matsumura, Y.; Maeda, H. A New Concept for Macromolecular Therapeutics in Cancer Chemotherapy: Mechanism of Tumoritropic Accumulation of Proteins and the Antitumor Agent Smancs. *Cancer Res.* **1986**, *46*, 6387–6392. [CrossRef] [PubMed]
3. Bobo, D.; Robinson, K.J.; Islam, J.; Thurecht, K.J.; Corrie, S.R. Nanoparticle-Based Medicines: A Review of FDA-Approved Materials and Clinical Trials to Date. *Pharm. Res.* **2016**, *33*, 2373–2387. [CrossRef] [PubMed]
4. Allen, T.M.; Martin, F.J. Advantages of liposomal delivery systems for anthracyclines. *Semin. Oncol.* **2004**, *31*, 5–15. [CrossRef] [PubMed]
5. Susewind, M.; Schilmann, A.-M.; Heim, J.; Henkel, A.; Link, T.; Fischer, K.; Strand, D.; Kolb, U.; Tahir, M.N.; Brieger, J.; et al. Silica-coated Au@ZnO Janus particles and their stability in epithelial cells. *J. Mater. Chem. B* **2015**, *3*, 1813–1822. [CrossRef]
6. Wang, Y.X.J.; Hussain, S.M.; Krestin, G.P. Superparamagnetic iron oxide contrast agents: Physicochemical characteristics and applications in MR imaging. *Eur. Radiol.* **2001**, *11*, 2319–2331. [CrossRef] [PubMed]
7. Bianco, A.; Kostarelos, K.; Prato, M. Applications of carbon nanotubes in drug delivery. *Curr. Opin. Chem. Biol.* **2005**, *9*, 674–679. [CrossRef] [PubMed]
8. Bruchez, M., Jr.; Moronne, M.; Gin, P.; Weiss, S.; Alivisatos, A.P. Semiconductor Nanocrystals as Fluorescent Biological Labels. *Science* **1998**, *281*, 2013–2016. [CrossRef] [PubMed]
9. Lu, J.; Liong, M.; Zink, J.I.; Tamanoi, F. Mesoporous Silica Nanoparticles as a Delivery System for Hydrophobic Anticancer Drugs. *Small* **2007**, *3*, 1341–1346. [CrossRef] [PubMed]
10. Targeted Silica Nanoparticles for Image-Guided Intraoperative Sentinel Lymph Node Mapping in Head and Neck Melanoma, Breast and Gynecologic Malignancies. Available online: https://clinicaltrials.gov/ct2/show/NCT02106598?term=silica+nanoparticle&rank=1 (accessed on 6 April 2017).
11. US Food and Drug Administration GRAS Substances (SCOGS) Database—Select Committee on GRAS Substances (SCOGS) Opinion: Silicates. Available online: https://www.fda.gov/food/ingredientspackaginglabeling/gras/scogs/ucm260849.htm (accessed on 7 April 2017).
12. Liberman, A.; Mendez, N.; Trogler, W.C.; Kummel, A.C. Synthesis and surface functionalization of silica nanoparticles for nanomedicine. *Surf. Sci. Rep.* **2014**, *69*, 132–158. [CrossRef] [PubMed]
13. Kresge, C.T.; Leonowicz, M.E.; Roth, W.J.; Vartuli, J.C.; Beck, J.S. Ordered mesoporous molecular sieves synthesized by a liquid-crystal template mechanism. *Nature* **1992**, *359*, 710–712. [CrossRef]
14. Tang, F.; Li, L.; Chen, D. Mesoporous silica nanoparticles: Synthesis, biocompatibility and drug delivery. *Adv. Mater.* **2012**, *24*, 1504–1534. [CrossRef] [PubMed]
15. Zhao, D.; Feng, J.; Huo, Q.; Melosh, N.; Fredrickson, G.; Chmelka, B.; Stucky, G. Triblock copolymer syntheses of mesoporous silica with periodic 50 to 300 angstrom pores. *Science* **1998**, *279*, 548–552. [CrossRef] [PubMed]
16. Na, H.-K.; Kim, M.-H.; Park, K.; Ryoo, S.-R.; Lee, K.E.; Jeon, H.; Ryoo, R.; Hyeon, C.; Min, D.-H. Efficient functional delivery of siRNA using mesoporous silica nanoparticles with ultralarge pores. *Small* **2012**, *8*, 1752–1761. [CrossRef] [PubMed]
17. Wu, M.; Meng, Q.; Chen, Y.; Zhang, L.; Li, M.; Cai, X.; Li, Y.; Yu, P.; Zhang, L.; Shi, J. Large Pore-Sized Hollow Mesoporous Organosilica for Redox-Responsive Gene Delivery and Synergistic Cancer Chemotherapy. *Adv. Mater.* **2016**, *28*, 1963–1969. [CrossRef] [PubMed]
18. Roberts, C.M.; Shahin, S.A.; Wen, W.; Finlay, J.B.; Dong, J.; Wang, R.; Dellinger, T.H.; Zink, J.I.; Tamanoi, F.; Glackin, C.A. Nanoparticle delivery of siRNA against TWIST to reduce drug resistance and tumor growth in ovarian cancer models. *Nanomed. Nanotechnol. Biol. Med.* **2016**. [CrossRef] [PubMed]
19. Liu, X.; Yu, D.; Jin, C.; Song, X.; Cheng, J.; Zhao, X.; Qi, X.; Zhang, G. A dual responsive targeted drug delivery system based on smart polymer coated mesoporous silica for laryngeal carcinoma treatment. *New J. Chem.* **2014**, *38*, 4830–4836. [CrossRef]
20. Ebabe Elle, R.; Rahmani, S.; Lauret, C.; Morena, M.; Bidel, L.P.R.; Boulahtouf, A.; Balaguer, P.; Cristol, J.P.; Durand, J.O.; Charnay, C.; et al. Functionalized Mesoporous Silica Nanoparticle with Antioxidants as a New Carrier That Generates Lower Oxidative Stress Impact on Cells. *Mol. Pharm.* **2016**, *13*, 2647–2660. [CrossRef] [PubMed]

21. Albanese, A.; Tang, P.S.; Chan, W.C.W. The Effect of Nanoparticle Size, Shape, and Surface Chemistry on Biological Systems. *Annu. Rev. Biomed. Eng.* **2012**, *14*, 1–16. [CrossRef] [PubMed]

22. Xiao, Y.; Wang, T.; Cao, Y.; Wang, X.; Zhang, Y.; Liu, Y.; Huo, Q. Enzyme and voltage stimuli-responsive controlled release system based on β-cyclodextrin-capped mesoporous silica nanoparticles. *Dalton Trans.* **2015**, *44*, 4355–4361. [CrossRef] [PubMed]

23. Lin, D.; Cheng, Q.; Jiang, Q.; Huang, Y.; Yang, Z.; Han, S.; Zhao, Y.; Guo, S.; Liang, Z.; Dong, A. Intracellular cleavable poly(2-dimethylaminoethyl methacrylate) functionalized mesoporous silica nanoparticles for efficient siRNA delivery in vitro and in vivo. *Nanoscale* **2013**, *5*, 4291–4301. [CrossRef] [PubMed]

24. Möller, K.; Müller, K.; Engelke, H.; Bräuchle, C.; Wagner, E.; Bein, T. Highly efficient siRNA delivery from core-shell mesoporous silica nanoparticles with multifunctional polymer caps. *Nanoscale* **2016**, *8*, 4007–4019. [CrossRef] [PubMed]

25. Ashley, C.E.; Carnes, E.C.; Phillips, G.K.; Padilla, D.; Durfee, P.N.; Brown, P.A.; Hanna, T.N.; Liu, J.; Phillips, B.; Carter, M.B.; et al. The targeted delivery of multicomponent cargos to cancer cells by nanoporous particle-supported lipid bilayers. *Nat. Mater.* **2011**, *10*, 389–397. [CrossRef] [PubMed]

26. Kaszuba, M.; McKnight, D.; Connah, M.T.; McNeil-Watson, F.K.; Nobbmann, U. Measuring sub nanometre sizes using dynamic light scattering. *J. Nanopart. Res.* **2008**, *10*, 823–829. [CrossRef]

27. Fröhlich, E. The role of surface charge in cellular uptake and cytotoxicity of medical nanoparticles. *Int. J. Nanomed.* **2012**, *7*, 5577–5591. [CrossRef] [PubMed]

28. Slowing, I.; Trewyn, B.G.; Lin, V.S.-Y. Effect of surface functionalization of MCM-41-type mesoporous silica nanoparticles on the endocytosis by human cancer cells. *J. Am. Chem. Soc.* **2006**, *128*, 14792–14793. [CrossRef] [PubMed]

29. Nel, A.E.; Mädler, L.; Velegol, D.; Xia, T.; Hoek, E.M.V.; Somasundaran, P.; Klaessig, F.; Castranova, V.; Thompson, M. Understanding biophysicochemical interactions at the nano-bio interface. *Nat. Mater.* **2009**, *8*, 543–557. [CrossRef] [PubMed]

30. Gratton, S.E.A.; Ropp, P.A.; Pohlhaus, P.D.; Luft, J.C.; Madden, V.J.; Napier, M.E.; DeSimone, J.M. The effect of particle design on cellular internalization pathways. *Proc. Natl. Acad. Sci. USA* **2008**, *105*, 11613–11618. [CrossRef] [PubMed]

31. Lu, F.; Wu, S.H.; Hung, Y.; Mou, C.Y. Size effect on cell uptake in well-suspended, uniform mesoporous silica nanoparticles. *Small* **2009**, *5*, 1408–1413. [CrossRef] [PubMed]

32. Sorkin, A.; Von Zastrow, M. Signal transduction and endocytosis: Close encounters of many kinds. *Nat. Rev. Mol. Cell Biol.* **2002**, *3*, 600–614. [CrossRef] [PubMed]

33. He, Q.; Zhang, J.; Shi, J.; Zhu, Z.; Zhang, L.; Bu, W.; Guo, L.; Chen, Y. The effect of PEGylation of mesoporous silica nanoparticles on nonspecific binding of serum proteins and cellular responses. *Biomaterials* **2010**, *31*, 1085–1092. [CrossRef] [PubMed]

34. Yang, Y.; Yu, C. Advances in silica based nanoparticles for targeted cancer therapy. *Nanomed. Nanotechnol. Biol. Med.* **2016**, *12*, 317–332. [CrossRef] [PubMed]

35. Folkman, J.; Hanahan, D. Switch to the angiogenic phenotype during tumorigenesis. *Princess Takamatsu Symp.* **1991**, *22*, 339–347. [PubMed]

36. Wilhelm, S.; Tavares, A.J.; Dai, Q.; Ohta, S.; Audet, J.; Dvorak, H.F.; Chan, W.C.W. Analysis of nanoparticle delivery to tumours. *Nat. Rev. Mater.* **2016**, *1*, 1–12. [CrossRef]

37. Fang, J.; Nakamura, H.; Maeda, H. The EPR effect: Unique features of tumor blood vessels for drug delivery, factors involved, and limitations and augmentation of the effect. *Adv. Drug Deliv. Rev.* **2011**, *63*, 136–151. [CrossRef] [PubMed]

38. Van Furth, R.; Cohn, Z.A.; Hirsch, J.G.; Humphrey, J.H.; Spector, W.G.; Langevoort, H.L. The mononuclear phagocyte system: A new classification of macrophages, monocytes, and their precursor cells. *Bull. World Health Organ.* **1972**, *46*, 845–852. [PubMed]

39. Huang, X.; Li, L.; Liu, T.; Hao, N.; Liu, H.; Chen, D.; Tang, F. The shape effect of mesoporous silica nanoparticles on biodistribution, clearance, and biocompatibility in vivo. *ACS Nano* **2011**, *5*, 5390–5399. [CrossRef] [PubMed]

40. Lu, J.; Liong, M.; Li, Z.; Zink, J.I.; Tamanoi, F. Biocompatibility, biodistribution, and drug-delivery efficiency of mesoporous silica nanoparticles for cancer therapy in animals. *Small* **2010**, *6*, 1794–1805. [CrossRef] [PubMed]

41. Malugin, A.; Ghandehari, H. Cellular uptake and toxicity of gold nanoparticles in prostate cancer cells: A comparative study of rods and spheres. *J. Appl. Toxicol.* **2010**, *30*, 212–217. [CrossRef] [PubMed]

42. Mishra, S.; Webster, P.; Davis, M.E. PEGylation significantly affects cellular uptake and intracellular trafficking of non-viral gene delivery particles. *Eur. J. Cell Biol.* **2004**, *83*, 97–111. [CrossRef] [PubMed]

43. Zhu, Y.; Fang, Y.; Borchardt, L.; Kaskel, S. PEGylated hollow mesoporous silica nanoparticles as potential drug delivery vehicles. *Microporous Mesoporous Mater.* **2011**, *141*, 199–206. [CrossRef]

44. Heldin, C.-H.; Rubin, K.; Pietras, K.; Östman, A. High interstitial fluid pressure—An obstacle in cancer therapy. *Nat. Rev. Cancer* **2004**, *4*, 806–813. [CrossRef] [PubMed]

45. Parker, N.; Turk, M.J.; Westrick, E.; Lewis, J.D.; Low, P.S.; Leamon, C.P. Folate receptor expression in carcinomas and normal tissues determined by a quantitative radioligand binding assay. *Anal. Biochem.* **2005**, *338*, 284–293. [CrossRef] [PubMed]

46. Qi, X.; Yu, D.; Jia, B.; Jin, C.; Liu, X.; Zhao, X.; Zhang, G. Targeting CD133$^+$ laryngeal carcinoma cells with chemotherapeutic drugs and siRNA against ABCG2 mediated by thermo/pH-sensitive mesoporous silica nanoparticles. *Tumor Biol.* **2016**, *37*, 2209–2217. [CrossRef] [PubMed]

47. Zhang, H.; Zhang, W.; Zhou, Y.; Jiang, Y.; Li, S. Dual Functional Mesoporous Silicon Nanoparticles Enhance the Radiosensitivity of VPA in Glioblastoma. *Transl. Oncol.* **2017**, *10*, 229–240. [CrossRef] [PubMed]

48. Cheng, W.; Nie, J.; Xu, L.; Liang, C.; Peng, Y.; Liu, G.; Wang, T.; Mei, L.; Huang, L.; Zeng, X. A pH-sensitive delivery vehicle based on folic acid-conjugated polydopamine-modified mesoporous silica nanoparticles for targeted cancer therapy. *ACS Appl. Mater. Interfaces* **2017**. [CrossRef] [PubMed]

49. Chen, X.; Sun, H.; Hu, J.; Han, X.; Liu, H.; Hu, Y. Transferrin gated mesoporous silica nanoparticles for redox-responsive and targeted drug delivery. *Colloids Surf. B Biointerfaces* **2017**, *152*, 77–84. [CrossRef] [PubMed]

50. Chen, N.-T.; Souris, J.S.; Cheng, S.-H.; Chu, C.-H.; Wang, Y.-C.; Konda, V.; Dougherty, U.; Bissonnette, M.; Mou, C.-Y.; Chen, C.-T.; et al. Lectin-functionalized mesoporous silica nanoparticles for endoscopic detection of premalignant colonic lesions. *Nanomed. Nanotechnol. Biol. Med.* **2017**, *13*, 1941–1952. [CrossRef] [PubMed]

51. Sweeney, S.K.; Luo, Y.; O'Donnell, M.A.; Assouline, J.G. Peptide-Mediated Targeting Mesoporous Silica Nanoparticles: A Novel Tool for Fighting Bladder Cancer. *J. Biomed. Nanotechnol.* **2017**, *13*, 232–242. [CrossRef]

52. Hirano, Y.; Kando, Y.; Hayashi, T.; Goto, K.; Nakajima, A. Synthesis and cell attachment activity of bioactive oligopeptides: RGD, RGDS, RGDV, and RGDT. *J. Biomed. Mater. Res.* **1991**, *25*, 1523–1534. [CrossRef] [PubMed]

53. Wang, F.; Li, Y.; Shen, Y.; Wang, A.; Wang, S.; Xie, T. The functions and applications of RGD in tumor therapy and tissue engineering. *Int. J. Mol. Sci.* **2013**, *14*, 13447–13462. [CrossRef] [PubMed]

54. Pan, L.; Liu, J.; He, Q.; Shi, J. MSN-mediated sequential vascular-to-cell nuclear-targeted drug delivery for efficient tumor regression. *Adv. Mater.* **2014**, *26*, 6742–6748. [CrossRef] [PubMed]

55. Hicke, B.J.; Stephens, A.W.; Gould, T.; Chang, Y.-F.; Lynott, C.K.; Heil, J.; Borkowski, S.; Hilger, C.-S.; Cook, G.; Warren, S.; et al. Tumor targeting by an aptamer. *J. Nucl. Med.* **2006**, *47*, 668–678. [PubMed]

56. Babaei, M.; Abnous, K.; Taghdisi, S.M.; Amel Farzad, S.; Peivandi, M.T.; Ramezani, M.; Alibolandi, M. Synthesis of theranostic epithelial cell adhesion molecule targeted mesoporous silica nanoparticle with gold gatekeeper for hepatocellular carcinoma. *Nanomedicine* **2017**. [CrossRef] [PubMed]

57. Tang, Y.; Hu, H.; Zhang, M.G.; Song, J.; Nie, L.; Wang, S.; Niu, G.; Huang, P.; Lu, G.; Chen, X. An aptamer-targeting photoresponsive drug delivery system using "off-on" graphene oxide wrapped mesoporous silica nanoparticles. *Nanoscale* **2015**, *7*, 6304–6310. [CrossRef] [PubMed]

58. Durfee, P.N.; Lin, Y.S.; Dunphy, D.R.; Muñiz, A.J.; Butler, K.S.; Humphrey, K.R.; Lokke, A.J.; Agola, J.O.; Chou, S.S.; Chen, I.M.; et al. Mesoporous Silica Nanoparticle-Supported Lipid Bilayers (Protocells) for Active Targeting and Delivery to Individual Leukemia Cells. *ACS Nano* **2016**, *10*, 8325–8345. [CrossRef] [PubMed]

59. Zhou, S.; Wu, D.; Yin, X.; Jin, X.; Zhang, X.; Zheng, S.; Wang, C.; Liu, Y. Intracellular pH-responsive and rituximab-conjugated mesoporous silica nanoparticles for targeted drug delivery to lymphoma B cells. *J. Exp. Clin. Cancer Res.* **2017**, *36*, 24. [CrossRef] [PubMed]

60. Maloney, D.G.; Grillo-López, A.J.; White, C.A.; Bodkin, D.; Schilder, R.J.; Neidhart, J.A.; Janakiraman, N.; Foon, K.A.; Liles, T.-M.; Dallaire, B.K.; et al. IDEC-C2B8 (Rituximab) Anti-CD20 Monoclonal Antibody Therapy in Patients With Relapsed Low-Grade Non-Hodgkin's Lymphoma. *Blood* **1997**, *9*, 2188–2195.

61. Chen, F.; Hong, H.; Shi, S.; Goel, S.; Valdovinos, H.F.; Hernandez, R.; Theuer, C.P.; Barnhart, T.E.; Cai, W. Engineering of Hollow Mesoporous Silica Nanoparticles for Remarkably Enhanced Tumor Active Targeting Efficacy. *Sci. Rep.* **2014**, *4*, 5080. [CrossRef] [PubMed]
62. Chen, F.; Nayak, T.R.; Goel, S.; Valdovinos, H.F.; Hong, H.; Theuer, C.P.; Barnhart, T.E.; Cai, W. In vivo tumor vasculature targeted PET/NIRF Imaging with TRC105(Fab)-conjugated, dual-labeled mesoporous silica nanoparticles. *Mol. Pharm.* **2014**, *11*, 4007–4014. [CrossRef] [PubMed]
63. Mekaru, H.; Lu, J.; Tamanoi, F. Development of mesoporous silica-based nanoparticles with controlled release capability for cancer therapy. *Adv. Drug Deliv. Rev.* **2015**, *95*, 40–49. [CrossRef] [PubMed]
64. Gatenby, R.A.; Gillies, R.J. Why do cancers have high aerobic glycolysis? *Nat. Rev. Cancer* **2004**, *4*, 891–899. [CrossRef] [PubMed]
65. Meng, H.; Xue, M.; Xia, T.; Zhao, Y.L.; Tamanoi, F.; Stoddart, J.F.; Zink, J.I.; Nel, A.E. Autonomous in vitro anticancer drug release from mesoporous silica nanoparticles by pH-sensitive nanovalves. *J. Am. Chem. Soc.* **2010**, *132*, 12690–12697. [CrossRef] [PubMed]
66. Xiong, L.; Bi, J.; Tang, Y.; Qiao, S.Z. Magnetic Core-Shell Silica Nanoparticles with Large Radial Mesopores for siRNA Delivery. *Small* **2016**, *12*, 4735–4742. [CrossRef] [PubMed]
67. Popat, A.; Liu, J.; Lu, G.Q.M.; Qiao, S.Z. A pH-responsive drug delivery system based on chitosan coated mesoporous silica nanoparticles. *J. Mater. Chem.* **2012**, *22*, 11173–11178. [CrossRef]
68. Liu, J.; Luo, Z.; Zhang, J.; Luo, T.; Zhou, J.; Zhao, X.; Cai, K. Hollow mesoporous silica nanoparticles facilitated drug delivery via cascade pH stimuli in tumor microenvironment for tumor therapy. *Biomaterials* **2016**, *83*, 51–65. [CrossRef] [PubMed]
69. Estrela, J.M.; Ortega, A.; Obrador, E. Glutathione in Cancer Biology and Therapy. *Crit. Rev. Clin. Lab. Sci.* **2006**, *43*, 143–181. [CrossRef] [PubMed]
70. Kim, H.; Kim, S.; Park, C.; Lee, H.; Park, H.J.; Kim, C. Glutathione-induced intracellular release of guests from mesoporous silica nanocontainers with cyclodextrin gatekeepers. *Adv. Mater.* **2010**, *22*, 4280–4283. [CrossRef] [PubMed]
71. Liu, R.; Zhao, X.; Wu, T.; Feng, P. Tunable Redox-Responsive Hybrid Nanogated Ensembles. *J. Am. Chem. Soc.* **2008**, *130*, 14418–14419. [CrossRef] [PubMed]
72. Li, Y.; Hei, M.; Xu, Y.; Qian, X.; Zhu, W. Ammonium salt modified mesoporous silica nanoparticles for dual intracellular-responsive gene delivery. *Int. J. Pharm.* **2016**, *511*, 689–702. [CrossRef] [PubMed]
73. Mondragón, L.; Mas, N.; Ferragud, V.; de la Torre, C.; Agostini, A.; Martínez-Máñez, R.; Sancenón, F.; Amorós, P.; Pérez-Payá, E.; Orzáez, M. Enzyme-responsive intracellular-controlled release using silica mesoporous nanoparticles capped with ε-poly-L-lysine. *Chemistry* **2014**, *20*, 5271–5281. [CrossRef] [PubMed]
74. Bernardos, A.; Mondragón, L.; Aznar, E.; Marcos, M.D.; Martínez-Máñez, R.; Sancenón, F.; Soto, J.; Barat, J.M.; Pérez-Payá, E.; Guillem, C.; et al. Enzyme-responsive intracellular controlled release using nanometric silica mesoporous supports capped with "saccharides". *ACS Nano* **2010**, *4*, 6353–6368. [CrossRef] [PubMed]
75. Guardado-Alvarez, T.M.; Sudha Devi, L.; Russell, M.M.; Schwartz, B.J.; Zink, J.I. Activation of snap-top capped mesoporous silica nanocontainers using two near-infrared photons. *J. Am. Chem. Soc.* **2013**, *135*, 14000–14003. [CrossRef] [PubMed]
76. Croissant, J.; Maynadier, M.; Gallud, A.; Peindy N'Dongo, H.; Nyalosaso, J.L.; Derrien, G.; Charnay, C.; Durand, J.O.; Raehm, L.; Serein-Spirau, F.; et al. Two-photon-triggered drug delivery in cancer cells using nanoimpellers. *Angew. Chem. Int. Ed.* **2013**, *52*, 13813–13817. [CrossRef] [PubMed]
77. Chen, P.-J.; Hu, S.-H.; Hsiao, C.-S.; Chen, Y.-Y.; Liu, D.-M.; Chen, S.-Y. Multifunctional magnetically removable nanogated lids of $Fe_3O_4$–capped mesoporous silica nanoparticles for intracellular controlled release and MR imaging. *J. Mater. Chem.* **2011**, *21*, 2535. [CrossRef]
78. Thomas, C.R.; Ferris, D.P.; Lee, J.H.; Choi, E.; Cho, M.H.; Kim, E.S.; Stoddart, J.F.; Shin, J.S.; Cheon, J.; Zink, J.I. Noninvasive remote-controlled release of drug molecules in vitro using magnetic actuation of mechanized nanoparticles. *J. Am. Chem. Soc.* **2010**, *132*, 10623–10625. [CrossRef] [PubMed]
79. Freeman, E.C.; Weiland, L.M.; Meng, W.S. Modeling the proton sponge hypothesis: examining proton sponge effectiveness for enhancing intracellular gene delivery through multiscale modeling. *J. Biomater. Sci. Polym. Ed.* **2013**, *24*, 398–416. [CrossRef] [PubMed]
80. Finlay, J.; Roberts, C.M.; Dong, J.; Zink, J.I.; Tamanoi, F.; Glackin, C.A. Mesoporous silica nanoparticle delivery of chemically modified siRNA against TWIST1 leads to reduced tumor burden. *Nanomedicine* **2015**, *11*, 1657–1666. [CrossRef] [PubMed]

81. Ma, D. Enhancing endosomal escape for nanoparticle mediated siRNA delivery. *Nanoscale* **2014**, *6*, 6415–6425. [CrossRef] [PubMed]
82. Hoekstra, D.; Martin, O.C. Transbilayer redistribution of phosphatidylethanolamine during fusion of phospholipid vesicles. Dependence on fusion rate, lipid phase separation, and formation of nonbilayer structures. *Biochemistry* **1982**, *21*, 6097–6103. [CrossRef] [PubMed]
83. Zhang, X.; Li, F.; Guo, S.; Chen, X.; Wang, X.; Li, J.; Gan, Y. Biofunctionalized polymer-lipid supported mesoporous silica nanoparticles for release of chemotherapeutics in multidrug resistant cancer cells. *Biomaterials* **2014**, *35*, 3650–3665. [CrossRef] [PubMed]
84. Li, X.; Chen, Y.; Wang, M.; Ma, Y.; Xia, W.; Gu, H. A mesoporous silica nanoparticle—PEI—Fusogenic peptide system for siRNA delivery in cancer therapy. *Biomaterials* **2013**, *34*, 1391–1401. [CrossRef] [PubMed]
85. Park, J.-H.; Gu, L.; von Maltzahn, G.; Ruoslahti, E.; Bhatia, S.N.; Sailor, M.J. Biodegradable luminescent porous silicon nanoparticles for in vivo applications. *Nat. Mater.* **2009**, *8*, 331–336. [CrossRef] [PubMed]
86. Wu, M.; Meng, Q.; Chen, Y.; Du, Y.; Zhang, L.; Li, Y.; Zhang, L.; Shi, J. Large-pore ultrasmall mesoporous organosilica nanoparticles: Micelle/precursor co-templating assembly and nuclear-targeted gene delivery. *Adv. Mater.* **2015**, *27*, 215–222. [CrossRef] [PubMed]
87. Duan, J.; Yu, Y.; Li, Y.; Yu, Y.; Li, Y.; Zhou, X.; Huang, P.; Sun, Z. Toxic effect of silica nanoparticles on endothelial cells through DNA damage response via Chk1-dependent G2/M checkpoint. *PLoS ONE* **2013**, *8*, e62087. [CrossRef] [PubMed]
88. Heikkilä, T.; Santos, H.A.; Kumar, N.; Murzin, D.Y.; Salonen, J.; Laaksonen, T.; Peltonen, L.; Hirvonen, J.; Lehto, V.P. Cytotoxicity study of ordered mesoporous silica MCM-41 and SBA-15 microparticles on Caco-2 cells. *Eur. J. Pharm. Biopharm.* **2010**, *74*, 483–494. [CrossRef] [PubMed]
89. Tarn, D.; Ashley, C.E.; Xue, M.; Carnes, E.C.; Zink, J.I.; Brinker, C.J. Mesoporous silica nanoparticle nanocarriers: Biofunctionality and biocompatibility. *Acc. Chem. Res.* **2013**, *46*, 792–801. [CrossRef] [PubMed]
90. He, Q.; Zhang, Z.; Gao, F.; Li, Y.; Shi, J. In vivo biodistribution and urinary excretion of mesoporous silica nanoparticles: Effects of particle size and PEGylation. *Small* **2011**, *7*, 271–280. [CrossRef] [PubMed]

© 2017 by the authors. Licensee MDPI, Basel, Switzerland. This article is an open access article distributed under the terms and conditions of the Creative Commons Attribution (CC BY) license (http://creativecommons.org/licenses/by/4.0/).

*nanomaterials*

MDPI

*Article*

# Cellular Response to Titanium Dioxide Nanoparticles in Intestinal Epithelial Caco-2 Cells is Dependent on Endocytosis-Associated Structures and Mediated by EGFR

**Kristin Krüger, Katrin Schrader and Martin Klempt \***

Max Rubner-Institut (MRI), Federal Research Institute for Nutrition and Food, Department of Safety and Quality of Milk and Fish Products, Hermann-Weigmann-Straße 1, 24103 Kiel, Germany; kruegerkristin@gmx.net (K.K.); katrin.schrader@mri.bund.de (K.S.)
\* Correspondence: martin.klempt@mri.bund.de; Tel.: +49-431-609-2256; Fax: +49-431-609-2300

Academic Editors: Dong-Wook Han and Wojciech Chrzanowski
Received: 3 March 2017; Accepted: 5 April 2017; Published: 7 April 2017

**Abstract:** Titanium dioxide ($TiO_2$) is one of the most applied nanomaterials and widely used in food and non-food industries as an additive or coating material (E171). It has been shown that E171 contains up to 37% particles which are smaller than 100 nm and that $TiO_2$ nanoparticles (NPs) induce cytotoxicity and inflammation. Using a nuclear factor Kappa-light-chain enhancer of activated B cells (NF-κB) reporter cell line (Caco-2[nfkb-RE]), Real time polymerase chain reaction (PCR), and inhibition of dynamin and clathrin, it was shown that cellular responses induced by 5 nm and 10 nm $TiO_2$ NPs (nominal size) depends on endocytic processes. As endocytosis is often dependent on the epithelial growth factor receptor (EGFR), further investigations focused on the involvement of EGFR in the uptake of $TiO_2$ NPs: (1) inhibition of EGFR reduced inflammatory markers of the cell (i.e., nuclear factor (NF)-κB activity, mRNA of IL8, CCL20, and CXCL10); and (2) exposure of Caco-2 cells to $TiO_2$ NPs activated the intracellular EGFR cascade beginning with EGFR-mediated extracellular signal-regulated kinases (ERK)1/2, and including transcription factor ELK1. This was followed by the expression of ERK1/2 target genes CCL2 and CXCL3. We concluded that $TiO_2$ NPs enter the cell via EGFR-associated endocytosis, followed by activation of the EGFR/ERK/ELK signaling pathway, which finally induces NF-κB. No changes in inflammatory response are observed in Caco-2 cells exposed to 32 nm and 490 nm $TiO_2$ particles.

**Keywords:** titanium dioxide nanoparticles; intestinal epithelial cells; inflammation; endocytosis; EGFR; ERK1/2

## 1. Introduction

Titanium dioxide ($TiO_2$) is one of the most applied nanomaterials and is widely used in the food and non-food industry as additive or coating material (for review see [1]). Food-grade $TiO_2$ is coded E171 and contains up to 37% nanoparticles (NP), resulting in an estimated exposure to $TiO_2$ NPs of 1 mg kg$^{-1}$·day$^{-1}$ in adults and up to 2–3 mg kg$^{-1}$·day$^{-1}$ in children [2].

Studies using Caco-2 cell lines have shown that $TiO_2$ NPs induced oxidative damage [3], influenced metabolic activity and cytotoxicity [4,5], produced reactive oxygen species (ROS) [6,7], and induced the expression of interleukin 8 (IL8) [8] through the activation of nuclear factor (NF)-κB and p38 mitogen activated protein kinase (MAPK) pathways [9]. Any influence of $TiO_2$ NPs on Caco-2 cells is caused by primary contact, which is followed by an interaction which is still unclear. Several studies discuss a possible activation of toll-like receptors (TLR) [10,11] or suggest an endocytic uptake [12–15]. In vitro studies on Caco-2 cells demonstrated that $TiO_2$ NPs are taken up by the cell:

TiO$_2$ NPs have been detected intracellularly [16], surrounded by cytoplasmic vesicles [17] without disruption of junctional complexes [18]. Endocytosis, as the initial cellular reaction and beginning of the inflammatory response of TiO$_2$ NPs, has not been investigated so far.

Several mechanisms responsible for cellular uptake of materials are named endocytosis: phagocytosis, which is reserved by specialized cell types and pinocytosis, which can be divided in micropinocytosis, clathrin-mediated endocytosis, caveolae-mediated endocytosis, and endocytosis independently of clathrin and caveolae [19]. Figure 1 provides an overview of the different endocytosis pathways and the inhibitors used in this study.

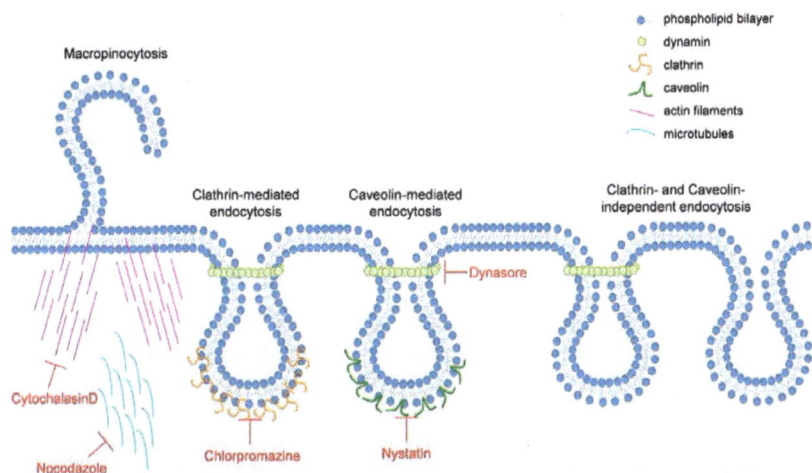

**Figure 1.** Illustration of different manners in which endocytosis occurs. Shown are the different methods of endocytosis and the mechanisms of action for the inhibitors used in this study. The illustration has been adapted from [19].

In the present study, we exposed Caco-2 cells to TiO$_2$ particles with nominal particle diameters of 5 nm, 10 nm, 32 nm, and 490 nm. We could show that TiO$_2$ NPs of 5 nm and 10 nm express their inflammatory potential mainly via an EGFR-mediated endocytosis process. Actin filaments and tubulin microtubules, as well as dynamin and, partly, clathrin, are involved in this process. Further investigations revealed that TiO$_2$ NPs induced an EGFR dependent activation of downstream ERK1/2 and of the ERK-related transcription factor ELK1 (ETS (E26 transformation-specific) domain-containing protein) and consequently an increased expression of chemokine (C–C motif) ligand 2 (CCL2) and chemokine (C–X–C motif) ligand 3 (CXCL3) genes.

## 2. Results

### 2.1. Characterization of TiO$_2$ Particles in Cell Medium

TEM of the particles in cell culture medium revealed single particle sizes of 10.7 nm (mean), 10.0 nm (median), and 9.0 nm (mode) for NP1, 12.1 nm (mean), 11.0 nm (median), and 10.0 (mode) for NP2, 23.1 nm (mean), 17.0 nm (median), 16.0 nm (mode) for NP3, and 138.9 nm (mean), 134.0 nm (median), and 78.0 nm (mode) for NP4 (Figure 2a–d). Agglomerates were observed in NP1, NP2, and NP4, while no agglomeration could be detected in NP3. By static light scattering SLS analysis (Figure 2e–h) we could demonstrate that the agglomerates of NP4 dispersed, while NP1 and NP2 showed a wide size distribution of the particles in a range from 40 nm to 57 μm with peaks at 300 nm and 4.5 μm for NP1 and peaks at 400 nm, 2.5 μm and 30 μM for NP2. NP3 and NP4 showed a unimodal size distribution with a mean agglomeration size of 96 nm for NP3 and 390 nm for NP4.

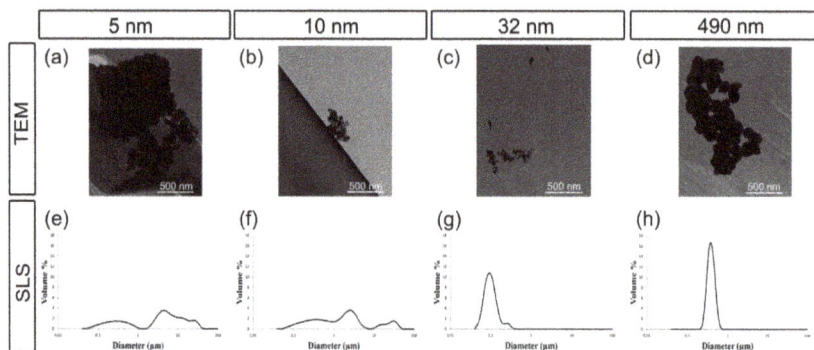

**Figure 2.** Characterization of TiO$_2$ particles used in this study. (**a–d**) TEM images of TiO$_2$ particles in medium of nominal (**a**) 5 nm (NP1); (**b**) 10 nm (NP2); (**c**) 32 nm (NP3); and (**d**) 490 nm (NP4) size. The bar corresponds to 500 nm. (**e–h**) Size distribution measured by static light scattering SLS of (**e**) np1; (**f**) np2; (**g**) np3; and (**h**) NP4. See text for details.

## 2.2. TiO$_2$ Particles Are Located within the Cell

Based on our previous study [9], and others [12–15], we reasoned that TiO$_2$ particles are endocytosed by Caco-2 cells. To investigate this possibility, we used laser scanning confocal microscopy identification of 10 nm TiO$_2$ NPs within the Caco-2 cells (Figure 3a–c). We chose 10 nm particles as these caused an inflammatory response in Caco-2 cells, as shown in our previous study [9]. Three sections along the z-axis ((a) apical side; (b) within; (c) basal side) of the cells provide information about TiO$_2$ NPs localization. Confocal images show nuclei (blue) surrounded by actin filaments (red). Aggregates of 10 nm TiO$_2$ NPs (green) are located at the apical side of the cells (Figure 3a), in the cell membrane, and within the cell (Figure 3b,c).

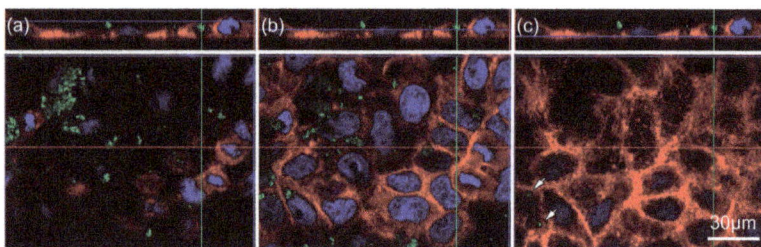

**Figure 3.** Confocal laser scanning microscopy of Caco-2 cells treated with NP2. (**a–c**). Monolayers of Caco-2 cells were treated with 10 nm TiO$_2$ particles (NP2, concentration of 40 μg cm$^{-2}$ of cell growth surface) and incubated for 6 h. Cells were stained for actin with Alexa Fluor 633 Phalloidin (red), nuclei were stained with DAPI (blue) and TiO$_2$ NPs fluorescence (green). The upper panel shows a cross-section of Caco-2 cells. Shown are three images along the z-axis from the apical side (**a**) to the basal side (**c**) of the epithelium. The blue line in the upper image indicates the position of the image below, whereas the red line marks the position of the cross-section. Arrows demonstrate TiO$_2$ particles within the cell membrane and intracellular. The bar corresponds to 30 μm.

## 2.3. TiO$_2$ NP-Induced Inflammatory Response Is Reduced by Inhibitors of Endocytosis

To further prove that NPs are endocytosed by Caco-2 cells, we used inhibitors of molecules which are involved in endocytosis (see Figure 1) and measured the NF-κB response of Caco-2 cells after administration of TiO$_2$ NPs. NP1 and NP2 induced NF-κB activity in Caco-2$^{nfkb-RE}$ cells by a factor of $7.0 \pm 0.6$ and $7.7 \pm 0.7$, respectively, NP3 and NP4 showed no effect on NF-κB activity ($1.0 \pm 0.0$ and $1.4 \pm 0.1$, respectively) (Figure 4a). Nystatin (50 µg mL$^{-1}$), a caveolae-mediated endocytosis inhibitor, had no effect on NP-induced NF-κB activation. Treatment with 10 µM chlorpromazine (a clathrin-mediated endocytosis inhibitor) reduced NP-induced luciferase activity by 74% and 75% in NP1 and NP2, respectively. Dynasore (100 µM) (a dynamin inhibitor) reduced activation of NF-κB by 68% and 69% in NP1 and NP2, respectively. As actin filaments and microtubules play an essential role in all types of endocytosis, we investigated the effects of cytochalasin D (0.1 µg mL$^{-1}$) and nocodazole (2 µM) on NP-induced NF-κB response. Treatment with cytochalasin D reduced NF-κB activity by 64% in both, NP1 and NP2 treated cells. Nocodazole reduced NP-induced NF-κB activity by 45% (NP1) and 42% (NP2). Since NP2 showed pronounced effects, we chose this NP for all further experiments. Expression of IL8 mRNA, chemokine (C–C motif) ligand 20 (CCL20) mRNA and C–X–C motif chemokine 10 (CXCL10) mRNA as indicators for inflammatory response of the cell was increased by $6.7 \pm 1.1$-fold, $9.6 \pm 1.4$-fold, and $9.2 \pm 2.9$-fold 3 h after NP2 exposure compared to untreated control cells (Figure 4b). Co-treatment with dynasore reduced the response by 76.2%, 66.7%, and 84.3% (IL8, CCL20, CXCL10), cytochalasin D altered the expression by 77.6%, 76.3%, and 87.1% and nocodazol lowered the reduced the response cells by 59.2%, 77.3%, and 89.5%. Treatment with nystatin did not change the NP-induced increase in mRNA expression of IL8, CCL20, or CXCL20, while chlorpromazine only reduced CCL20 expression by 47.3%.

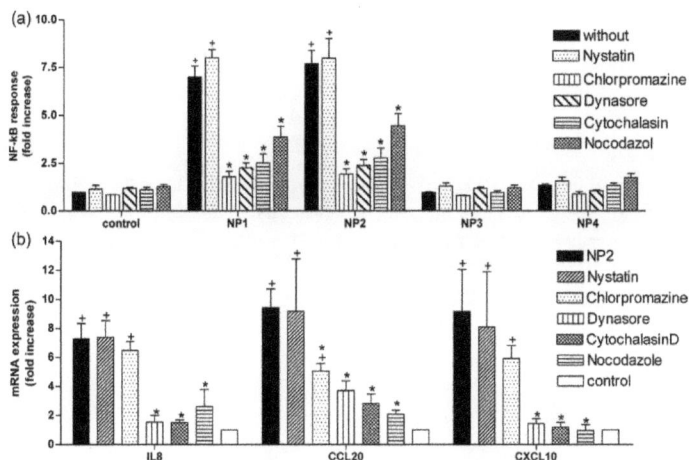

**Figure 4.** Effect of endocytic inhibitors on luciferase activity of Caco-2$^{\text{nfkb-RE}}$ cells and expression of IL8 mRNA, CCL20 mRNA, and CXCL10 mRNA. (**a**) Treatment of nystatin (50 µg mL$^{-1}$), chlorpromazine (10 µM), dynasore (100 µM), cytochalasin D (0.1 µg mL$^{-1}$), and nocodazole (2 µM) on luciferase activity of Caco-2$^{\text{nfkb-RE}}$ cells 6 h after treatment with NP1, NP2, NP3, or NP4. Before NPs treatment (concentration of 40 µg cm$^{-2}$ of the cell growth surface) confluent monolayers of Caco-2$^{\text{nfkb-RE}}$ cells were incubated with inhibitors. Luminescence of treated cells are presented as the fold change to untreated controls; (**b**) mRNA expression of IL8, CCL20, and CXCL10 3 h after NP2 exposure pretreated with the indicated inhibitors. mRNA expression of the treated cells are presented as the fold change to untreated controls. Mean $\pm$ SEM ($n \geq 3$). Non-parametric analysis of variance was followed by Bonferroni post-tests ($p < 0.05$). Asterisks (*) represent significant differences to NPs treated cells, plus (+) represent significant differences to untreated controls.

## 2.4. Activation of EGFR Is Essential for the Cellular Response to TiO$_2$ NPs

As receptor mediated endocytosis is often initiated by activating the EGFR receptor, we aimed to suppress the NP-induced inflammatory response by pretreatment of Caco-2 cells with EGFR inhibitors

BIBX 1382 and CL-387785. This treatment resulted in a reduced NF-κB activity in NP2 treated cells by 51.3% and 44.6%, respectively (Figure 5a), 6h after NPs exposure. Further, 3 h after NPs exposure IL8 mRNA, CCL20 mRNA, and CXCL10 mRNA expression was reduced by 81.6%, 75.3%, and 73.7%, respectively, in BIBX 1382 treated cells and by 89.1%, 77.4%, and 73.2%, respectively, in CL-387785 treated cells (Figure 5b).

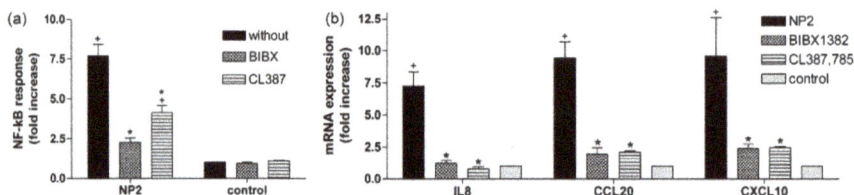

**Figure 5.** Effect of EGFR inhibitors on luciferase activity of Caco-2$^{nfkb-RE}$ cells and expression of IL8 mRNA, CCL20 mRNA, and CXCL10 mRNA. (**a**) Effect of BIBX 1382 and CL-387785 on luciferase activity of Caco-2$^{nfkb-RE}$ cells 6 h after NP2 (concentration of 40 μg cm$^{-2}$ of the cell growth surface) treatment. Luminescence of treated cells are presented as the fold change to untreated control; (**b**) mRNA expression of IL8, CCL20, and CXCL10 3 h after NP2 exposure pretreated with inhibitors BIBX 1382 and CL-387785. mRNA expression of the treated cells are presented as the fold change to untreated controls. Mean ± SEM ($n \geq 3$). Non-parametric analysis of variance was followed by Bonferroni post-tests ($p < 0.05$). Asterisks (*) represent significant differences to NP2 treated cells, plus (+) represent significant differences to untreated control cells.

### 2.5. NPs Are Colocalized with Lysosomes

To gain evidence that TiO$_2$ NP2 after being taken up by endocytosis are transported though the cell within lysosomes, we used live cell imaging. Caco-2 cells treated with NP2 showed a fluorescence signal of the NPs (green, Figure 6a), whereas untreated control cells showed no signal (Figure 6d). Lysosomes labeled by LysoTracker® Red DND-99 are present in NP-treated cells (Figure 6b) and in untreated control cells (Figure 6e). Overlay of both images indicate NPs localized to lysosomes 3 h after exposure (marked by the arrows, Figure 6c,f).

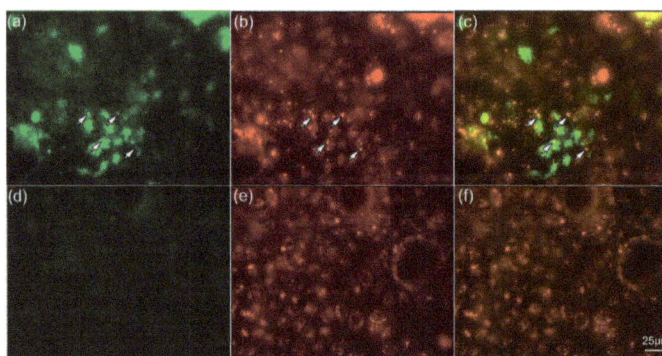

**Figure 6.** NPs are co-localized with lysosomes. Confluent monolayers of Caco-2 cells (**a–c**) treated with NP2 for 3 h and (**d–f**) untreated control cells, co-incubated with 50 nm LysoTracker® Red DND-99. Fluorescence imaging of (**a,d**) NPs (green), visualized by using FITC filter set; (**b,e**) lysosomes labeled by LysoTracker® Red DND-99 (red); and (**c,f**) overlay demonstrated co-localized NPs with lysosomes. Arrows demonstrate the same position of NPs and lysosomes. The bar corresponds to 25 μm.

## 2.6. TiO$_2$ NPs Activate EGFR/ERK/ELK Signaling Pathway

To show that after endocytosis and transport via lysosomes gene expression is mediated via the known EGFR pathway, we analyzed the downstream elements of the pathway using Western blot and EMSA analyses. NP2 treatment of Caco-2 cells induced a transient activation of ERK1/2 compared to untreated control cells since levels of p-ERK1/2 increased in a time dependent manner until 60 min after NPs exposure. Activation of ERK1/2 faded 90 min after NPs exposure (Figure 7a). To analyze the involvement of EGFR in the NP-induced ERK activation we used EGFR inhibitor BIBX 1382. As shown by Western blot analysis p-EKR1/2 was induced 30 min after NPs and EGF stimulation by a factor of $2.5 \pm 0.4$ and $4.2 \pm 0.9$ compared to untreated control cells, and was completely abolished using BIBX 1382 (Figure 7b).

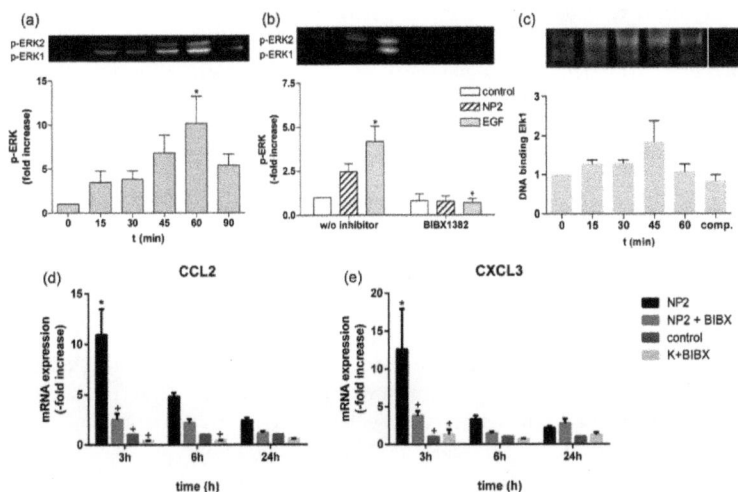

**Figure 7.** TiO$_2$ NPs induced thee EGFR/ERK/ELK signaling pathway. (**a**) TiO$_2$ NPs induced activation of p-ERK1/2. Caco-2 cells were stimulated with NP2 (concentration of 40 µg cm$^{-2}$ of the cell growth surface) for 15, 30, 45, 60, and 90 min. Phosphorylated ERK1/2 was evaluated by Western blot analyses as shown in the upper row and presented in the bar diagram as the fold change to unstimulated samples = 0 min. Expression of p-ERK1/2 was normalized to total protein expression measured by densitometry; (**b**) the effect of EGFR inhibitor BIBX 1382 on NP- and EGF-induced ERK1/2 phosphorylation. Caco-2 cells were pre-treated with the inhibitor and exposed to NP2 or 100 ng mL$^{-1}$ EGF for 30 min or remained untreated (w/o inhibitor). Treated samples are presented as the fold change to the untreated control sample; (**c**) TiO$_2$ NPs induced binding capacity of ELK1. Caco-2 cells were stimulated with NP2 for 15, 30, 45, and 60 min. ELK1 binding was measured by EMSA and shown as the binding complex of ELK1 and ELK1-specific biotin-labeled oligonucleotides. For competition reaction (comp.) a 1000-fold molar excess of unlabeled oligonucleotides were used. Binding capacity was quantified by densitometry and treated samples are presented as the fold change to the untreated control. Presented are representative blots of $n \geq 3$ independent experiments; (**d,e**) mRNA expression analyses of CCL2 (**d**) and CXCL3 (**e**) in Caco-2 cells. Caco-2 cells were pre-treated with 50 µM BIBX 1382 and exposed to NP2 or remained untreated for indicated time. The expression of CCL2 mRNA and the expression of CXCL3 mRNA was determined and values of the treated samples are presented as the fold change to the untreated control sample at the same time point. All data are presented the mean $\pm$ SEM of $n \geq 3$ independent experiments. One- and two-way ANOVA, followed by Tukey's multiple comparisons test ($p < 0.05$) were conducted. Asterisks (*) represent significant differences to untreated control w/o inhibitor, plus (+) represent significant differences to NP2 or EGF-treated cells without inhibitor.

The best-studied nuclear target of phosphorylated ERK1/2 is the transcription factor ELK1 [20]. Stimulation with NP2 increased the binding of ELK1 to its DNA recognition site (Figure 7c) and declined 60 min after NPs treatment. Densitometric analyses revealed a $1.8 \pm 0.5$-fold increase of the binding complex compared to untreated control. Expression analysis revealed an increase by factor $11.0 \pm 2.6$ and $12.6 \pm 5.3$ 3 h after NPs exposure of CCL2 and CXCL3 mRNA, respectively (Figure 7d). 6 h after NPs treatment mRNA expression of CCL2 and CXCL3 was induced by $4.8 \pm 0.4$-fold and by $3.3 \pm 0.5$-fold, respectively. Pre-incubation with EGFR inhibitor BIBX 1382 decreased the NP-induced mRNA expression of CCL2 and CXCL3 after 3 h by 77% and by 70%, and after 6 h by 54% and 58%.

## 3. Discussion

Although it has been shown that intestinal epithelial Caco-2 cells exposed to TiO$_2$ NPs induced a transient inflammatory response, the interaction mechanism of TiO$_2$ NPs with the cell surface remains unclear. By CLSM we were able to show that TiO$_2$ NPs are internalized in Caco2-cells (Figure 3) and confirmed previously-published data using TEM [16–18,21]. The mechanism of internalization is still unknown. Some studies suggest that TiO$_2$ NPs are internalized by endocytic mechanisms [12] or an interaction with the Toll-like receptor 4 [10,11]. To gain deeper insight into potential uptake mechanisms, we focused on endocytic pathways and EGFR, which is influenced by NPs [22,23].

As shown in Figure 1, four different mechanisms (micropinocytosis, clathrin-mediated endocytosis, caveolae-mediated endocytosis, and endocytosis independent of clathrin and caveolae [19]) are known in epithelial cells. Actin and microtubule dynamics are involved in all types of endocytosis [24,25]. Our results showed that TiO$_2$ NP-induced NF-κB response was reduced by cytochalasin D, which inhibits actin polymerization [26,27] and in part by nocodazole, which depolymerizes microtubules [28,29] (Figure 4a). Expression analyses of typical inflammatory genes [30] confirm these observations: TiO$_2$ NPs of 5 nm and 10 nm induced expression of different inflammatory markers, as shown in our previous study [9], and were reduced by treatment with cytochalasin D and nocodazole (Figure 4b). It can be concluded that an intact cytoskeleton is essential for internalization of TiO$_2$ NPs in Caco2 cells as has been demonstrated in human keratinocytes [31], alveolar macrophages, [32] and fibroblasts [33].

Transport within cells requires forming of subcellular structures. In Caco-2 cells these vesicles can be formed caveolae-mediated [15], clathrin-mediated [34], dynamin-dependent [35], and clathrin- and caveolae-independent [36]. To investigate the mechanism involved, we used specific inhibitors of the key molecules: First, we used nystatin, a sterol-binding drug, which prevents the formation of lipid rafts [37]. Although there have been reports that nystatin inhibits cellular uptake of TiO$_2$ NPs in Caco-2 cells [12] as well as in intestinal rainbow trout cells [38], we could not detect an altered cellular response to TiO$_2$ NPs after treatment with nystatin (Figure 4) at concentrations which do not alter cell viability. Next, we treated Caco-2 cells with chlorpromazine. This cationic amphiphilic drug prevents coated pit assembly at the cell surface in clathrin-mediated endocytosis (CME) [39]. CME is characterized by the formation of a clathrin lattice around the invaginated membrane. This premature vesicle is pinched off by the GTPase dynamin to form clathrin-coated vesicles (CCV) [40]. We observed a reduction in the NP-induced NF-κB activation in Caco-2 cells (Figure 4a) but we could not detect an influence in the expression of IL8 mRNA and CXCL10 mRNA (Figure 4b). Therefore, we suggest the inflammatory response to TiO$_2$ NPs in Caco-2 cells is only partly mediated by clathrin. This confirms a previous study, which has shown an involvement of CME in the uptake of TiO$_2$ NPs in Caco-2 cells [12].

Clathrin-dependent endocytosis is mostly associated with receptor-mediated endocytosis [22], and NPs induced an EGFR mediated signaling pathway via extracellular signal-regulated kinase (ERK) in lung epithelial cells [41,42]. We, thus, examined the effect of EGFR kinase inhibitors BIBX1382 [23] and CL-387785 [43] to the cellular response of TiO$_2$ NPs. Both block phosphorylation of tyrosine kinase and prevent activation of downstream signal transduction pathways. The NP-induced activation of NF-κB and mRNA expression of IL8, CCL20, and CXCL10 was decreased after treatment with EGFR inhibitors (Figure 5a,b). The results demonstrate that activation of EGFR is essential for initiating cellular response to $TiO_2$ NPs. We assume $TiO_2$ NPs activate EGFR and will be internalized into vesicles. As receptor-ligand interactions are very specific, it is most likely that recognition is not based on the metallic $TiO_2$, but on some of the proteins present in the protein corona [44–46].

As activation of EGFR is required for cellular response to $TiO_2$ NPs, we further analyzed downstream effectors of EGFR. ERK1/2 is part of the Ras/Raf/MEK/ERK signal transduction cascade [47]. Our results show that NPs induced phosphorylation of ERK1/2 (Figure 7). This activation was followed by an increased binding capacity of the transcription factor ELK1 and an increased mRNA expression of ERK1/2 target genes CCL2 and CXCL3 [48]. The NP-induced activation was reduced after EGFR inhibition. Various studies using different types of NPs have shown an activation of ERK signaling pathway [49,50] with a previous activation of EGFR in epithelial cells [51–55]. Further, it was shown that the expression of CCL2 and CXCL3 mRNA was increased after different inflammatory stimuli and was part of an inflammatory response, mediated by EGFR [56,57].

In conclusion, we show that in Caco-2 cells $TiO_2$ NPs, probably due to their protein corona, are recognized by EGFR, which is internalized via clathrin-dependent and clathrin-independent mechanisms, but not by caveolin-mediated endocytosis. The involvement of cytoskeletal, as well as dynamin, in this endocytosis process has been documented. $TiO_2$ NPs activate the complete EGFR/ERK/ELK signaling pathway, including the expression of the effector mRNA CCL2 and CXCL3.

## 4. Material and Methods

### 4.1. Preparation of TiO₂ NPs

Titanium (IV) dioxide (anatase) particles were obtained from Alfa Aesar (Alfa Aesar GmbH and CoKG, Karlsruhe, Germany). Four differently-sized particles were used: 1. 5 nm (NP1; Stock Number 44689, Lot F11T023, specific surface area (SSA) 210 m$^2$ g$^{-1}$), 2. 10 nm (NP2; Stock Number 44690, Lot B19T020; SSA 120 m$^2$ g$^{-1}$), 3. 32 nm (NP3; Stock Number 39953, Lot F23T043, SSA 51 m$^2$ g$^{-1}$) and 4. 490 nm (NP4; Stock Number 36199, Lot G02S013; SSA not specified). Stock dispersions of all particles were prepared with deionized water to a final concentration of 2 mg mL$^{-1}$. The dispersion was sonicated at 23 KHZ and 150 W (MSE Ltd., London, UK) for 2 min and finally autoclaved. Immediately before treatment of the cells the dispersion was diluted in culture medium as indicated below.

### 4.2. Nanoparticle Characterization

Characterizations of the particles were performed by transmission-electron microscopy (TEM) using freeze-fracture preparation technique and by static light scattering (SLS). Sonicated and autoclaved stock solutions of NPs were diluted 1:10 in cell culture medium and cryoprotected by immersion in glycerol solution. Samples were cryofixed into melting Freon 22 and liquid nitrogen. Freeze fracturing took place at −120 °C with a BAF 400 (Bal-Tec, Balzers, Liechtenstein). Freeze-fractured specimens were replicated by application of Pt/C and C by electron-gun evaporation. The replicas were cleaned in concentrated sodium hypochlorite and in acetone and examined with a Tecnai 10 (FEI Company, Hillsboro, OR, USA) transmission-electron microscope operated at 80 kV. Size measurement of single particles and particle aggregations using the TEM images was conducted with the program "Bild-Vermessen 1.0" (CAD-KAS Kassler Computersoftware GbR, Markranstädt, Germany). At least 30 single particles and 20 aggregates of the different NPs were measured. Particle size distribution was measured by SLS using a LS 230 (Beckmann Coulter, Krefeld, Germany). The particle dispersion was dosed into the instrument without special sample preparation. The volume fraction-length mean diameter was measured.

### 4.3. Cell Culture

Human colon adenocarcinoma cell line Caco-2 (Toni Lindl, Munich, Germany) between passage 24 and 50 were cultured in Caco-2 medium (45% Dulbecco's Modified Eagle Medium (DMEM), low glucose, 45% Ham's F12, 9% fetal calf serum (FCS), 0.9% non-essential amino acids) (all PAA

Laboratories GmbH, Austria), and insulin (10 µg mL$^{-1}$) (Biochrome AG, Berlin, Germany) at 37 °C and 5% $CO_2$. For all experiments cells were seeded at a density of $1 \times 10^5$–$2 \times 10^5$ cells cm$^{-2}$. TiO$_2$ particles were added to the cells at a final concentration of 40 µg cm$^{-2}$ cell growth surface as this concentration is in accordance with the observed exposure in humans [2] and in the range of other comparable references [3,5,18,58]. We determined the time we exposed the cells to NPs prior to the described experiments by a number time course analyses.

*4.4. mRNA Expression Analysis*

For mRNA expression analysis cells were seeded in 24-well or six-well culture plates (Sarstedt AG and Co., Nümbrecht, Germany). After reaching confluency, TiO$_2$ particles of indicated sizes or water (control) were added at a final concentration of 40 µg cm$^{-2}$ cell growth surface. At indicated time points cells were washed twice with PBS and proceeded to RNA extraction. RNA was extracted using the GeneJET RNA Purification Kit (Thermo Fisher Scientific GmbH, Dreieich, Germany) and first-strand cDNA synthesis was prepared using a RevertAid™ H Minus First Strand cDNA Synthesis Kit (Thermo Fisher Scientific GmbH, Dreieich, Germany) as described by the manufacturer. Real-time PCR was performed on a 7500 Real-Time PCR Systems (Life Technologies Inc., Carlsbad, CA, USA) using HOT FIREPol® EvaGreen® qPCR Mix Plus ROX (Solis BioDyne, Tartu, Estonia). Sequences of primers (used at 0.2 µM) were as follows: B2M (Beta-2 microglobulin) forward: GCAAGGACTGGTCTTTCTATCT, reverse: TAACTATCTTGGGCTGTG-ACAAA; CCL2 (chemokine (C–C motif) ligand 2) forward: CCCA AAGAAGCTGTGATCTTCA; reverse: TCTGGGGAAAGCTAGGGGAA, CCL20 (chemokine (C–C motif) ligand 20) forward: CGAATCAGAAGCAGCAAGCAA, reverse: TTGCGCACACAGACAACTTT; CXCL3 (chemokine (C–X–C motif) ligand 3) forward: CCCAAACCGAAGTCATAGCCA, reverse: ACCC TGCAGGAAG-TGTCAA; CXCL10 (chemokine (C–X–C motif) ligand 10) forward: GCCATTCT GATTTGCTGCCTT, reverse: GCTCCCCTCTGGTTTTAAGGA; GAPDH (glyceraldehyde 3-phosphate dehydrogenase) forward: AGAGCACAAGAGGAAGAGAGAG, reverse: GGTTGAGCACA GGGTAC TTTATT; IL8 (interleukin 8) forward: CACCGGAAGGAACCATCTCA, reverse: TGGCAAAACTG CACCTTC-ACA. Parameters for qPCR were as follows: 95 °C for 15 min, 40 cycles of 10 s at 95 °C, 30 s at 60 °C, and 30 s at 72 °C. After cycling, melting curve analysis was performed. Expressions of the different genes was normalized to the expressions of GAPDH and B2M and compared to control.

*4.5. NF-κB Reporter Gene Assay*

Caco-2$^{nfkb-RE}$ [9] cells were seeded in 96-well plates (Corning Incorporated, New York, NY, USA). At confluency, inhibitors of endocytosis were added to the cells and incubated as indicated below. Subsequently, cells were treated with TiO$_2$ NPs or water (control) at a final concentration of 40 µg cm$^{-2}$ cell growth surface. After 6 h cells were harvested and luciferase activity was measured for 5 s in a CHAMELEON™ V plate reader (Hidex, Finland) using Beetle-Juice substrate (PJK GmbH, Kleinblittersdorf, Germany) according to the manufacturer's instructions. Luciferase activity is expressed in relation to untreated controls.

*4.6. Fluorescence and Confocal Laser Scanning Microscopy (CLSM)*

Detection of lysosomes was performed by fluorescence microscopy. Caco-2 cells were seeded in eight-well on cover glass II (Sarstedt AG and Co., Nümbrecht, Germany). At confluency, prepared medium with TiO$_2$ NPs or water (control) was added. Simultaneously LysoTracker® Red DND-99 (Molecular Probes, Eugene, OR, USA) was added to medium at a final concentration of 50 nM. After incubation time of 3 h cells were observed by fluorescence microscopy. Co-localization of lysosomes and NPs was detected by different filter sets. NPs were visualized by using FITC filter set and LysoTracker® Red DND-99 by using filter set appropriate to Texas Red® dye. To investigate intracellular distribution of the NPs, confocal laser scanning microscopy was employed. Caco-2 cells were seeded in a removable 12-well chamber (ibidi GmbH, Munich, Germany). At confluency, prepared medium with TiO$_2$ NPs or water (control) was added. After incubation time of 6 h cells were harvested and stained for actin with Alexa Fluor 633 Phalloidin (Molecular Probes, Eugene, OR, USA) as described by the manufacturer. Nuclei were stained with Dapi-Fluoromount-G™ clear mounting media (SouthernBiotech™, Birmingham, AL, USA). NPs were excited by a laser at 488 nm and the emission was measured from 490 nm to 501 nm. Excitation of phalloidin took place by a laser at 633 nm and emission was measured from 642 nm to 655 nm to allow a clear separation of both signals. Confocal and fluorescence images were acquired with a confocal laser scanning microscope TCS SP8 (Leica Microsystems GmbH, Wetzlar, Germany) using a 40× objective and 63× glycerol immersion objective.

### 4.7. Western Blot Analyses

Caco-2 cells were seeded in six-well plates (8.87 $cm^{-2}$ per well) (Sarstedt, Germany) at a density of 1–2 × $10^5$ cells $cm^{-2}$. At confluency, prepared medium with 10 nm TiO2 NPs was added to the cells at a final concentration of 40 µg $cm^{-2}$ cell growth surface or remained untreated. EGF at a concentration of 100 ng $mL^{-1}$ was used as a positive control in the EGFR/ERK signaling pathway as described [59]. Cells were washed twice with cold PBS and lysed 20 min on ice in RIPA buffer (Sigma-Aldrich, St. Louis, MO, USA) added with inhibitors (1 mM PMSF, 1 mM $Na_3VO_4$, 1× complete protease inhibitor cocktail (Santa Cruz Biotechnology, Santa Cruz, CA, USA), 1× phosphatase inhibitor cocktail 3 (Sigma-Aldrich, St. Louis, MO, USA). Lysates were clarified by centrifugation (12,000× $g$ for 6 min) and protein concentration was measured using bicinchoninic acid assay (BCA) reagent. 40 µg protein from each preparation were separated by sodium dodecyl sulfate-polyacrylamide gel electrophoresis (SDS-PAGE) and transferred to a nitrocellulose membrane (Carl Roth GmbH + Co.KG, Karlsruhe, Germany). Membranes were blocked in 5% ($w/v$) BSA/TBST (20 mM Tris–HCl (pH 7.6), 150 mM NaCl, and 0.1% ($v/v$) Tween 20) for 1 h at room temperature. Primary antibody incubation p-ERK1/2 1:200 (sc-7383, Santa Cruz Biotechnology, Santa Cruz, CA, USA) in 5% BSA/TBST took place at 4 °C overnight, followed by secondary antibody incubation 1:500 in TBST (anti-mouse HRP (PAB10782, Abnova, Taiwan)) for 2 h at room temperature. The antibody was detected with enhanced chemiluminescence (ECL) reagent (MBL International Corporation, Woburn, MA, USA) and visualized with Fusion Solo S (Vilber Lourmat Deutschland GmbH, Eberhardzell, Germany). Densitometry of the bands was analyzed using FusionCapt Advance Solo 4s software and the ratio of p-ERK to total protein, detected by Ponceau staining, was determined. Expressions of the treated samples were set in relation to the untreated control sample. Experiments were performed at least three times.

### 4.8. Electrophoretic Mobility Shift Assay (EMSA)

Binding of ELK1 was investigated by non-radioactive EMSA using a LightShift EMSA Optimization and Control Kit (Thermo Fisher Scientific Inc., Waltham, MA, USA) according to manufacturer's protocol. In detail, Caco-2 cells were seeded in six-well plates (Sarstedt AG and Co, Nümbrecht, Germany) at a density of 1–2 × $10^5$ cells $cm^{-2}$. At confluency, cells were stimulated with 10 nm $TiO_2$ NPs at a final concentration of 40 µg $cm^{-2}$ cell growth surface or remained unstimulated. Cells were washed twice with ice cold PBS and lysed 20 min on ice in lysis buffer (10 mM HEPES, 1.5 mM $MgCl_2$, 10 mM KCl, 1 mM DTT, 1 mM PMSF, 1× complete protease inhibitor cocktail (Santa Cruz Biotechnology, Santa Cruz, CA, USA)) adapted from [60]. Cells were scraped, placed in a microliter tube, and centrifuged 20 min at 11,000× $g$ at 4 °C to separate cytosolic and nuclear fractions. Supernatant was decanted and pellet was suspended in extraction buffer to lyse nuclei (20 mM HEPES, 1.5 mM $MgCl_2$, 0.42 M NaCl, 0.2 mM EDTA, 25% ($v/v$) glycerol, 1 mM DTT, 1 mM PMSF, 1× complete protease inhibitor cocktail, [60]). 2 µg of nucleus extraction were used for binding reaction with biotin-labeled DNA for ELK1: 5′-TTTGCAAAATGCAGGAATTGTTTTCACAGT-3′ [61]. Samples were separated by 6% native polyacrylamide gel and transferred to a nylon membrane (GE Healthcare, Little Chalfont, UK). Biotin was detected with peroxidase-coupled streptavidin and ECL (Thermo Fisher Scientific Inc., Waltham, MA, USA) and visualized and quantified with Fusion Solo S (Vilber Lourmat Deutschland GmbH, Eberhardzell, Germany). Densitometry of the binding complex was analyzed using FusionCapt

Advance Solo 4s software and expressions of the treated samples were set in relation to the untreated control sample. Experiments were performed at least three times.

### 4.9. Cell Treatment

To investigate the mechanisms involved in the $TiO_2$ NPs-induced inflammatory response in Caco-2 cells, we used various inhibitors of endocytic processes (Figure 1, Table 1). Confluent monolayers of Caco-2 cells were pre-incubated with inhibitors (cytochalasin D (AppliChem, Darmstadt, Germany), nocodazole (AppliChem, Darmstadt, Germany), nystatin (AppliChem, Darmstadt, Germany), dynasore (Santa Cruz Biotechnology, Inc., Santa Cruz, CA, USA), chlorpromazine (Sigma-Aldrich, St. Louis, MS, USA), BIBX 1382 (Santa Cruz Biotechnology, Inc., Santa Cruz, CA, USA), and CL-387785 (Santa Cruz Biotechnology, Inc., Santa Cruz, CA, USA)). The final concentrations that were used in medium, incubation time, and effect of inhibitors are summarized in Table 1. After pre-incubation, the medium with $TiO_2$ NPs or water (control) was added at a final concentration of 40 $\mu$g cm$^{-2}$ cell growth surface and incubated for 3 h or 6 h. Cytotoxic activity was tested by the ToxiLight™ bioassay kit (Lonza Group AG, Walkersville, MD, USA). No cytotoxic effect was observed for inhibitors at the used concentration (data not shown). After particle treatment, cells were washed twice with PBS and further treated with the stated methods.

**Table 1.** Inhibitors used in this study. Presented are the effects of the inhibitors, the corresponding reference for the used concentration, and pre-incubation time.

| Inhibitor | Effect | Reference | Concentration | Preincubated Time |
|---|---|---|---|---|
| Cytochalasin D | Inhibits actin polymerization | [26,27] | 0.1 $\mu$g mL$^{-1}$ | 30 min |
| Nocodazole | Depolymerizes microtubules | [28,29] | 2 $\mu$M | 30 min |
| Nystatin | Inhibits Caveolae-dependent endocytosis | [37] | 50 $\mu$g mL$^{-1}$ | 30 min |
| Dynasore | Inhibits dynamin | [62,63] | 100 $\mu$M | 1 h |
| Chlorpromazine | Inhibitor of clathrin-mediated endocytosis | [35,64] | 100 $\mu$M | 1 h |
| BIBX 1382 | EGFR kinase inhibitor | [23] | 46 $\mu$M | 1 h |
| CL-387785 | EGFR kinase inhibitor | [43] | 20 $\mu$M | 3 h |

Epithelial growth factor (EGF) at a concentration of 100 ng mL$^{-1}$ was used as a positive control in the EGFR/ERK signaling pathway, as described [61]. A confluent monolayer of Caco-2 cells was incubated for 1 h with 50 $\mu$M BIBX 1382 and subsequently stimulated with NPs, EGF, or left unstimulated. After treatment, cells were washed twice with cold PBS and processed as described above. Results of the treated cells are presented as the fold change to untreated cells.

### 4.10. Data Analysis

We performed every experiment with, at least, replicate samples for each treatment with the appropriate controls carried out on three different days (i.e., at least technical duplicates with three biological replicates). Data of RT-qPCR were analyzed using the delta-delta $C_t$ method [65]. Results are presented as means $\pm$ standard error (SEM) and expressed as the fold change to the untreated control. Statistical analysis was carried out using GraphPad Prism 4 software (GraphPad Software, Inc., La Jolla, CA, USA). After testing for Gausian distribution (Kolmogorov-Smirnoff test) and homogeneity of variance (Levene test) comparisons of the groups were either carried out by non-parametric analysis, followed by Bonferroni post-tests or by ANOVA, followed by Tukey's multiple comparison test. Differences were considered to be statistically significant at $p < 0.05$.

**Acknowledgments:** We kindly thank Silvia Kaschner, Kevin Pohl, and Kerstin Hansen for expert technical assistance. This work was supported by the German Federal Ministry of Food and Agriculture.

**Author Contributions:** Kristin Krüger and and Martin Klempt conceived and designed the experiments; Kristin Krüger performed the experiments and analyzed the data; Kristin Krüger and Martin Klempt wrote the paper; Katrin Schrader contributed analysis tools for characterization of the NPs.

**Conflicts of Interest:** The authors declare no conflict of interest.

## References

1.  Schrand, A.M.; Rahman, M.F.; Hussain, S.M.; Schlager, J.J.; Smith, D.A.; Syed, A.F. Metal based nanoparticles and their toxicity assessment. *Wiley Interdiscip. Rev. Nanomed. Nanobiotechnol.* **2010**, *2*, 544–568. [CrossRef] [PubMed]
2.  Weir, A.; Westerhoff, P.; Fabricius, L.; Hristovski, K.; von Goetz, N. Titanium dioxide nanoparticles in food and personal care products. *Environ. Sci. Technol.* **2012**, *46*, 132242–132250. [CrossRef] [PubMed]
3.  Barone, F.; de Berardis, B.; Bizzarri, L.; Degan, P.; Andreoli, C.; Zijno, A.; de Angelis, I. Physico-chemical characteristics and cyto-genotoxic potential of ZnO and TiO$_2$ nanoparticles on human colon carcinoma cells. *J. Phys. Conf. Ser.* **2011**, *304*, 012047. [CrossRef]
4.  Gerloff, K.; Fenoglio, I.; Carella, E.; Kolling, J.; Albrecht, C.; Boots, A.W.; Förster, I.; Schins, R.P.F. Distinctive toxicity of TiO$_2$ rutile/anatase mixed phase nanoparticles on Caco-2 cells. *Chem. Res. Toxicol.* **2012**, *25*, 646–655. [CrossRef] [PubMed]
5.  Gerloff, K.; Albrecht, C.; Boots, A.W.; Förster, I.; Schins, R.P.F. Cytotoxicity and oxidative DNA damage by nanoparticles in human intestinal Caco-2 cells. *Nanotoxicology* **2009**, *3*, 355–364. [CrossRef]
6.  Ivask, A.; Titma, T.; Visnapuu, M.; Vija, H.; Käkinen, A.; Sihtmäe, M.; Pokhrel, S.; Mädler, L.; Heinlaan, M.; Kisand, V.; et al. Toxicity of 11 Metal Oxide Nanoparticles to Three Mammalian Cell Types In 9 itro. *Curr. Top. Med. Chem.* **2015**, *15*, 1914–1929. [CrossRef] [PubMed]
7.  Fisichella, M.; Berenguer, F.; Steinmetz, G.; Auffan, M.; Rose, J.; Prat, O. Intestinal toxicity evaluation of TiO$_2$ degraded surface-treated nanoparticles: A combined physico-chemical and toxicogenomics approach in caco-2 cells. *Part. Fibre Toxicol.* **2012**, *9*, 18. [CrossRef] [PubMed]
8.  Chalew, T.E.A.; Schwab, K.J. Toxicity of commercially available engineered nanoparticles to Caco-2 and SW480 human intestinal epithelial cells. *Cell Biol. Toxicol.* **2013**, *29*, 101–116. [CrossRef] [PubMed]
9.  Krüger, K.; Cossais, F.; Neve, H.; Klempt, M. Titanium dioxide nanoparticles activate IL8-related inflammatory pathways in human colonic epithelial Caco-2 cells. *J. Nanopart. Res.* **2014**, *16*, 2402. [CrossRef]
10. Mano, S.S.; Kanehira, K.; Taniguchi, A. Comparison of Cellular Uptake and Inflammatory Response via Toll-Like Receptor 4 to Lipopolysaccharide and Titanium Dioxide Nanoparticles. *Int. J. Mol. Sci.* **2013**, *14*, 13154–13170. [CrossRef] [PubMed]
11. Chen, P.; Migita, S.; Kanehira, K.; Sonezaki, S.; Taniguchi, A. Development of sensor cells using NF-κB pathway activation for detection of nanoparticle-induced inflammation. *Sensors* **2010**, *11*, 7219–7230. [CrossRef] [PubMed]
12. Gitrowski, C.; Al-Jubory, A.R.; Handy, R.D. Uptake of different crystal structures of TiO$_2$ nanoparticles by Caco-2 intestinal cells. *Toxicol. Lett.* **2014**, *226*, 264–276. [CrossRef] [PubMed]
13. Geiser, M.; Rothen-Rutishauser, B.; Kapp, N.; Schürch, S.; Kreyling, W.; Schulz, H.; Semmler, M.; Hof, V.I.; Heyder, J.; Gehr, P. Ultrafine Particles Cross Cellular Membranes by Nonphagocytic Mechanisms in Lungs and in Cultured Cells. *Environ. Health Perspect.* **2005**, *113*, 1555–1560. [CrossRef] [PubMed]
14. He, B.; Lin, P.; Jia, Z.; Du, W.; Qu, W.; Yuan, L.; Dai, W.; Zhang, H.; Wang, X.; Wang, J.; et al. The transport mechanisms of polymer nanoparticles in Caco-2 epithelial cells. *Biomaterials* **2013**, *34*, 6082–6098. [CrossRef] [PubMed]
15. Bannunah, A.M.; Vllasaliu, D.; Lord, J.; Stolnik, S. Mechanisms of Nanoparticle Internalization and Transport across an Intestinal Epithelial Cell Model: Effect of Size and Surface Charge. *Mol. Pharm.* **2014**, *11*, 4363–4373. [CrossRef] [PubMed]
16. Janer, G.; del Molino, E.M.; Fernández-Rosas, E.; Fernández, A.; Vázquez-Campos, S. Cell uptake and oral absorption of titanium dioxide nanoparticles. *Toxicol. Lett.* **2014**, *228*, 103–110. [CrossRef] [PubMed]
17. Brun, E.; Barreau, F.; Veronesi, G.; Fayard, B.; Sorieul, S.; Chanéac, C.; Carapito, C.; Rabilloud, T.; Mabondzo, A.; Herlin-Boime, N.; et al. Titanium dioxide nanoparticle impact and translocation through ex vivo, in vivo and in vitro gut epithelia. *Part. Fibre Toxicol.* **2014**, *11*, 13. [CrossRef] [PubMed]
18. Koeneman, B.A.; Zhang, Y.; Westerhoff, P.; Chen, Y.; Crittenden, J.C.; Capco, D.G. Toxicity and cellular responses of intestinal cells exposed to titanium dioxide. *Cell Biol. Toxicol.* **2009**, *26*, 225–238. [CrossRef] [PubMed]
19. Conner, S.D.; Schmid, S.L. Regulated portals of entry into the cell. *Nature* **2003**, *422*, 37–44. [CrossRef] [PubMed]

20. Yoon, S.; Seger, R. The extracellular signal-regulated kinase: Multiple substrates regulate diverse cellular functions. *Growth Factors* **2006**, *24*, 21–44. [CrossRef] [PubMed]
21. Faust, J.J.; Doudrick, K.; Yang, Y.; Westerhoff, P.; Capco, D.G. Food grade titanium dioxide disrupts intestinal brush border microvilli in vitro independent of sedimentation. *Cell Biol. Toxicol.* **2014**, *30*, 169–188. [CrossRef] [PubMed]
22. Apodaca, G. Endocytic Traffic in Polarized Epithelial Cells: Role of the Actin and Microtubule Cytoskeleton. *Traffic* **2001**, *2*, 149–159. [CrossRef] [PubMed]
23. Solca, F.F.; Baum, A.; Langkopf, E.; Dahmann, G.; Heider, K.; Himmelsbach, F.; van Meel, J.C.A. Inhibition of Epidermal Growth Factor Receptor Activity by Two Pyrimidopyrimidine Derivatives. *J. Pharmacol. Exp. Ther.* **2004**, *311*, 502–509. [CrossRef] [PubMed]
24. Qualmann, B.; Kessels, M.M.; Kelly, R.B. Molecular Links between Endocytosis and the Actin Cytoskeleton. *J. Cell Biol.* **2000**, *150*, 111–116. [CrossRef]
25. Schafer, D.A. Coupling actin dynamics and membrane dynamics during endocytosis. *Curr. Opin. Cell Biol.* **2002**, *14*, 76–81. [CrossRef]
26. Goddettes, D.W.; Frieden, C. Actin Polymerization. The mechanism of action of cytochalasin D. *J. Biol. Chem.* **1986**, *261*, 15974–15980.
27. Baker, N.T.; Graham, L.L. Campylobacter fetus translocation across Caco-2 cell monolayers. *Microb. Pathog.* **2010**, *49*, 260–272. [CrossRef] [PubMed]
28. Ben-Ze'ev, A.; Farmer, S.R.; Penman, S. Mechanisms of Regulating Tubulin Synthesis in Cultured Mammalian Cells. *Cell* **1979**, *17*, 319–325. [CrossRef]
29. Vasquez, R.J.; Howell, B.; Yvon, A.M.; Wadsworth, P.; Cassimeris, L. Nanomolar concentrations of nocodazole alter microtubule dynamic instability in vivo and in vitro. *Mol. Biol. Cell* **1997**, *8*, 973–985. [CrossRef] [PubMed]
30. Wang, D.; Dubois, R.N.; Richmond, A. The role of chemokines in intestinal inflammation and cancer. *Curr. Opin. Pharmacol.* **2009**, *9*, 688–696. [CrossRef] [PubMed]
31. Busch, W.; Bastian, S.; Trahorsch, U.; Iwe, M.; Kühnel, D.; Meißner, T.; Springer, A.; Gelinsky, M.; Richter, V.; Ikonomidou, C.; et al. Internalisation of engineered nanoparticles into mammalian cells in vitro: Influence of cell type and particle properties. *J. Nanopart. Res.* **2010**, *13*, 293–310. [CrossRef]
32. Scherbart, A.M.; Langer, J.; Bushmelev, A.; van Berlo, D.; Haberzettl, P.; van Schooten, F.-J.; Schmidt, A.M.; Rose, C.R.; Schins, R.P.; Albrecht, C. Contrasting macrophage activation by fine and ultrafine titanium dioxide particles is associated with different uptake mechanisms. *Part. Fibre Toxicol.* **2011**, *8*, 31. [CrossRef] [PubMed]
33. Allouni, Z.E.; Høl, P.J.; Cauqui, M.A.; Gjerdet, N.R.; Cimpan, M.R. Role of physicochemical characteristics in the uptake of $TiO_2$ nanoparticles by fibroblasts. *Toxicol. In Vitro* **2012**, *26*, 469–479. [CrossRef] [PubMed]
34. Torgersen, M.L.; Skretting, G.; van Deurs, B.; Sandvig, K. Internalization of cholera toxin by different endocytic mechanisms. *J. Cell Sci.* **2001**, *114*, 3737–3747. [PubMed]
35. Krieger, S.E.; Kim, C.; Zhang, L.; Marjomaki, V.; Bergelson, J.M. Echovirus 1 entry into polarized Caco-2 cells depends on dynamin, cholesterol, and cellular factors associated with macropinocytosis. *J. Virol.* **2013**, *87*, 8884–8895. [CrossRef] [PubMed]
36. Sato, K.; Nagai, J.; Mitsui, N.; Yumoto, R.; Takano, M. Effects of endocytosis inhibitors on internalization of human IgG by Caco-2 human intestinal epithelial cells. *Life Sci.* **2009**, *85*, 800–807. [CrossRef] [PubMed]
37. Rothberg, K.G.; Heuser, J.E.; Donzell, W.C.; Ying, Y.-S.; Glenney, J.R.; Anderson, R.G.W. Caveolin, a protein component of caveolae membrane coats. *Cell* **1992**, *68*, 673–682. [CrossRef]
38. Al-Jubory, A.R.; Handy, R.D. Uptake of titanium from $TiO_2$ nanoparticle exposure in the isolated perfused intestine of rainbow trout: Nystatin, vanadate and novel $CO_2$—Sensitive components. *Nanotoxicology* **2013**, *7*, 1282–1301. [CrossRef] [PubMed]
39. Wang, L.; Rothberg, K.G.; Anderson, R.G.W. Mis-Assembly of Clathrin Lattices on Endosomes Reveals a Regulatory Switch for Coated Pit Formation Materials and Methods. *J. Cell Biol.* **1993**, *123*, 1107–1117. [CrossRef] [PubMed]
40. Hinshaw, J.E. Dynamin and its role in Membrane Fission. *Annu. Rev. Cell Dev. Biol.* **2000**, *16*, 483–519. [CrossRef] [PubMed]

41. Tamaoki, J.; Isono, K.; Takeyama, K.; Tagaya, E.; Nakata, J.; Nagai, A. Ultrafine carbon black particles stimulate proliferation of human airway epithelium via EGF receptor-mediated signaling pathway. *Am. J. Physiol. Lung Cell. Mol. Physiol.* **2004**, *287*, 1127–1133. [CrossRef] [PubMed]

42. Unfried, K.; Sydlik, U.; Bierhals, K.; Weissenberg, A.; Abel, J. Carbon nanoparticle-induced lung epithelial cell proliferation is mediated by receptor-dependent Akt activation. *Am. J. Physiol. Lung Cell. Mol. Physiol.* **2008**, *294*, L358–L367. [CrossRef] [PubMed]

43. Discafani, C.M.; Carroll, M.L.; Floyd, M.B.; Hollander, I.J.; Husain, Z.; Johnson, B.D.; Kitchen, D.; May, M.K.; Malo, M.S.; Minnick, A.A.; et al. Irreversible Inhibition of Epidermal Growth Factor Receptor Tyrosine Kinase with In Vivo Activity by N-[4-[(3-Bromophenyl) Amino]-6-Quinazolinyl]-2-Butynamide (CL-387,785). *Biochem. Pharmacol.* **1999**, *57*, 917–925. [CrossRef]

44. Giudice, M.C.L.; Herda, L.M.; Polo, E.; Dawson, K.A. In situ characterization of nanoparticle biomolecular interactions in complex biological media by flow cytometry. *Nat. Commun.* **2016**, *7*, 13475. [CrossRef] [PubMed]

45. Strojan, K.; Leonardi, A.; Bregar, V.B.; Križaj, I.; Svete, J.; Pavlin, M. Dispersion of Nanoparticles in Different Media Importantly Determines the Composition of Their Protein Corona. *PLoS ONE* **2017**, *12*, e0169552. [CrossRef] [PubMed]

46. Marucco, A.; Gazzano, E.; Ghigo, D.; Enrico, E.; Fenoglio, I. Fibrinogen enhances the inflammatory response of alveolar macrophages to $TiO_2$, $SiO_2$ and carbon nanomaterials. *Nanotoxicology* **2014**, *10*, 1–9. [CrossRef] [PubMed]

47. Roskoski, R. ERK1/2 MAP kinases: Structure, function, and regulation. *Pharmacol. Res.* **2012**, *66*, 105–143. [CrossRef] [PubMed]

48. Schweppe, R.E.; Tom, H.C.; Ahn, N.G. Global gene expression analysis of ERK5 and ERK1/2 signaling reveals a role for HIF-1 in ERK5-mediated responses. *J. Biol. Chem.* **2006**, *281*, 20993–21003. [CrossRef] [PubMed]

49. Totlandsdal, A.I.; Refsnes, M.; Låg, M. Mechanisms involved in ultrafine carbon black- induced release of IL-6 from primary rat epithelial lung cells. *Toxicol. In Vitro* **2010**, *24*, 10–20. [CrossRef] [PubMed]

50. Park, E.-J.; Yi, J.; Chung, K.-H.; Ryu, D.-Y.; Choi, J.; Park, K. Oxidative stress and apoptosis induced by titanium dioxide nanoparticles in cultured BEAS-2B cells. *Toxicol. Lett.* **2008**, *180*, 222–229. [CrossRef] [PubMed]

51. Comfort, K.K.; Maurer, E.I.; Braydich-Stolle, L.K.; Hussain, S.M. Interference of Silver, Gold, and Iron Oxide Nanoparticles on Epidermal Growth Factor Signal Transduction in Epithelial Cells. *ACS Nano* **2011**, *5*, 10000–10008. [CrossRef] [PubMed]

52. Weissenberg, A.; Sydlik, U.; Peuschel, H.; Schroeder, P.; Schneider, M.; Schins, R.P.F.; Abel, J.; Unfried, K. Reactive oxygen species as mediators of membrane-dependent signaling induced by ultrafine particles. *Free Radic. Biol. Med.* **2010**, *49*, 597–605. [CrossRef] [PubMed]

53. Peuschel, H.; Sydlik, U.; Haendeler, J.; Büchner, N.; Stöckmann, D.; Kroker, M.; Wirth, R.; Brock, W.; Unfried, K. C-Src-mediated activation of Erk1/2 is a reaction of epithelial cells to carbon nanoparticle treatment and may be a target for a molecular preventive strategy. *Biol. Chem.* **2010**, *391*, 1327–1332. [CrossRef] [PubMed]

54. Sydlik, U.; Bierhals, K.; Soufi, M.; Abel, J.; Schins, R.P.F.; Unfried, K. Ultrafine carbon particles induce apoptosis and proliferation in rat lung epithelial cells via specific signaling pathways both using EGF-R. *Am. J. Physiol. Lung Cell. Mol. Physiol.* **2006**, *291*, L725–L733. [CrossRef] [PubMed]

55. Skuland, T.; Øvrevik, J.; Låg, M.; Schwarze, P.; Refsnes, M. Silica nanoparticles induce cytokine responses in lung epithelial cells through activation of a p38/TACE/TGF-$\alpha$/EGFR- pathway and NF-$\kappa$B signaling. *Toxicol. Appl. Pharmacol.* **2014**, *279*, 76–86. [CrossRef] [PubMed]

56. Pastore, S.; Mascia, F.; Mariotti, F.; Dattilo, C.; Mariani, V.; Pastore, S.; Mascia, F.; Mariotti, F.; Dattilo, C.; Mariani, V.; et al. ERK1/2 Regulates Epidermal Chemokine Expression and Skin Inflammation. *J. Immunol.* **2005**, *174*, 5047–5056. [CrossRef] [PubMed]

57. Rada, B.; Gardina, P.; Myers, T.G.; Leto, T.L. Reactive oxygen species mediate inflammatory cytokine release and EGFR-dependent mucin secretion in airway epithelial cells exposed to Pseudomonas pyocyanin. *Mucosal Immunol.* **2011**, *4*, 158–171. [CrossRef] [PubMed]

58. Xiong, S.; Tang, Y.; Ng, H.S.; Zhao, X.; Jiang, Z.; Chen, Z.; Ng, K.W.; Loo, S.C.J. Specific surface area of titanium dioxide ($TiO_2$) particles influences cyto- and photo-toxicity. *Toxicology* **2013**, *304*, 132–140. [CrossRef] [PubMed]

59. Kaulfuss, S.; Burfeind, P.; Gaedcke, J.; Scharf, J.-G. Dual silencing of insulin-like growth factor-I receptor and epidermal growth factor receptor in colorectal cancer cells is associated with decreased proliferation and enhanced apoptosis. *Mol. Cancer Ther.* **2009**, *8*, 821–833. [CrossRef] [PubMed]

60. Zhang, L.; Li, N.; Caicedo, R.; Neu, J. Alive and dead Lactobacillus rhamnosus GG decrease tumor necrosis factor-alpha-induced interleukin-8 production in Caco-2 cells. *J. Nutr.* **2005**, *135*, 1752–1756. [PubMed]

61. Hennenberg, M.; Strittmatter, F.; Beckmann, C.; Rutz, B.; Füllhase, C.; Waidelich, R.; Montorsi, F.; Hedlund, P.; Andersson, K.E.; Stief, C.G.; et al. Silodosin Inhibits Noradrenaline-Activated Transcription Factors Elk1 and SRF in Human Prostate Smooth 7 Muscle. *PLoS ONE* **2012**, *7*, e50904. [CrossRef] [PubMed]

62. Macia, E.; Ehrlich, M.; Massol, R.; Boucrot, E.; Brunner, C.; Kirchhausen, T. Dynasore, a cell-permeable inhibitor of dynamin. *Dev. Cell* **2006**, *10*, 839–850. [CrossRef] [PubMed]

63. Kim, C.; Bergelson, M.J. Echovirus 7 Entry into Polarized Intestinal Epithelial Cells Requires. *Am. Soc. Microbiol.* **2012**, *3*, 1–10. [CrossRef] [PubMed]

64. Broeck, D.V.; de Wolf, M. Selective blocking of clathrin-mediated endocytosis by RNA interference: Epsin as target protein. *Biotechniques* **2006**, *41*, 475–484. [CrossRef]

65. Pfaffl, M.W.; Horgan, G.W.; Dempfle, L. Relative expression software tool (REST) for group-wise comparison and statistical analysis of relative expression results in real-time PCR. *Nucleic Acids Res.* **2002**, *30*, e36. [CrossRef] [PubMed]

© 2017 by the authors. Licensee MDPI, Basel, Switzerland. This article is an open access article distributed under the terms and conditions of the Creative Commons Attribution (CC BY) license (http://creativecommons.org/licenses/by/4.0/).

*nanomaterials*

MDPI

*Article*

# The Effects of Silica Nanoparticles on Apoptosis and Autophagy of Glioblastoma Cell Lines

Rafał Krętowski [1,*], Magdalena Kusaczuk [1], Monika Naumowicz [2], Joanna Kotyńska [2], Beata Szynaka [3] and Marzanna Cechowska-Pasko [1]

[1]   Department of Pharmaceutical Biochemistry, Medical University of Białystok, Mickiewicza 2A, 15-222 Białystok, Poland; mkusaczuk@wp.pl (M.K.); mapasko@gmail.com (M.C.-P.)
[2]   Institute of Chemistry, University of Bialystok, K. Ciołkowskiego 1K, 15-245 Białystok, Poland; monikan@uwb.edu.pl (M.N.); joannak@uwb.edu.pl (J.K.)
[3]   Department of Histology and Embryology, Medical University of Białystok, Waszyngtona 13, 15-269 Białystok, Poland; beataszynaka@gmail.com
*   Correspondence: r.kretowski@umb.edu.pl; Tel.: +48-85-748-56-39; Fax: +48-85-748-56-91

Received: 15 July 2017; Accepted: 11 August 2017; Published: 21 August 2017

**Abstract:** Silica nanoparticles (SiNPs) are one of the most commonly used nanomaterials in various medical applications. However, possible mechanisms of the toxicity caused by SiNPs remain unclear. The study presented here provides novel information on molecular and cellular effects of SiNPs in glioblastoma LBC3 and LN-18 cells. It has been demonstrated that SiNPs of 7 nm, 5–15 nm and 10–20 nm induce time- and dose-dependent cytotoxicity in LBC3 and LN-18 cell lines. In contrast to glioblastoma cells, we observed only weak reduction in viability of normal skin fibroblasts treated with SiNPs. Furthermore, in LBC3 cells treated with 5–15 nm SiNPs we noticed induction of apoptosis and necrosis, while in LN-18 cells only necrosis. The 5–15 nm SiNPs were also found to cause oxidative stress, a loss in mitochondrial membrane potential, and changes in the ultrastructure of the mitochondria in LBC3 cells. Quantitative real-time PCR results showed that in LBC3 cells the mRNA levels of pro-apoptotic genes *Bim*, *Bax*, *Puma*, and *Noxa* were significantly upregulated. An increase in activity of caspase-9 in these cells was also observed. Moreover, the activation of SiNP-induced autophagy was demonstrated in LBC3 cells as shown by an increase in LC3-II/LC3-I ratio, the upregulation of *Atg5* gene and an increase in AVOs-positive cells. In conclusion, this research provides novel information concerning molecular mechanisms of apoptosis and autophagy in LBC3 cells.

**Keywords:** autophagy; apoptosis; glioblastoma multiforme; nanotoxicity; silica nanoparticles

## 1. Introduction

*Glioblastoma multiforme* (GBM) is the most frequently diagnosed and highly aggressive form of primary brain tumor [1]. The median survival time of GBM patients is less than 15 months [2]. Although multidisciplinary approaches of treatment, including maximal tumor resection and the combination of irradiation and conventional chemotherapy are applied, GBM is still associated with poor prognosis and remains incurable [3]. It is believed that two factors make GBM treatment extremely difficult. Firstly, the brain itself has limited capacity of regeneration, and secondly, GBM is extremely invasive and therapy-resistant [2,4]. Therefore, extensive efforts to develop new therapeutic strategies relying on selective destruction of cancer cells are currently being explored. One of the latest solutions in cancer treatment is the application of nanoparticle-based technologies.

Recent development of nanotechnology raised the need of intensive investigation of the cytotoxic effects of nanomaterials [5]. To date, the cytotoxicity of different nanoparticles (NPs) has been demonstrated in various in vivo and in vitro studies [6]. This cell-damaging property of nanoparticles has prompted a widespread quest of nanomaterials with possible application in cancer research. Given this, nanoparticles have already been used in controllable drug delivery [7,8], and theranostics [9].

Silica nanoparticles (SiNPs) are one of the most commonly used nanomaterials in biomedical research due to their certain benefits e.g.,: biocompatibility, large surface area for biomacromolecules loading, relative stability, and low production costs [10,11]. SiNPs have widely been explored as biosensors, biomarkers, cancer therapeutics, DNA or drug delivery systems, and additives for food and cosmetics [12]. However, their cytotoxic effects have also been reported [13].

To date, the mechanisms by which SiNPs induce cytotoxicity are not completely clear. Heterogeneity of physicochemical parameters of SiNPs, for example: size, shape, structure, and elemental constituents allow them to display multidirectional mechanisms of action in cancer cells [14]. The key mechanisms that seem to be connected with silica nanotoxicity include production of the reactive oxygen species (ROS), DNA destruction or aberrant aggregation of nucleoplasmic proteins [12,13,15,16]. These cellular disturbances caused by SiNPs lead primarily to the apoptotic death of damaged cells. Apoptosis plays a pivotal role in the control of tumor growth [17]. It has been demonstrated that SiNPs can trigger apoptosis through the activation of various apoptotic pathways [18,19]. The death receptor-mediated apoptosis of SiNPs-treated cells has been confirmed in vitro in A549 cell line [18]. Other reports emphasize the role of the mitochondrial pathway initiated after exposure to SiNPs [20,21]. It has been shown that treatment with SiNPs resulted in generation of oxidative stress and ROS production, which in turn led to apoptosis by intrinsic apoptotic pathway [21]. The dose-dependent upregulation of *caspase-9* and -3 genes in A431 and A549 cell lines has been noticed [21]. Ahmad et al. have proven that proapoptotic *Bax* and *caspase-3* genes were upregulated, while the anti-apoptotic *Bcl-2* gene was downregulated in human liver HepG2 cell line [20]. In addition to apoptosis, in much research SiNPs-mediated necrotic cell death has also been reported [22–24]. Exposition of human umbilical vein endothelial cells (HUVECs) to SiNPs with diameters of 304 and 310 nm resulted in enhanced necrosis, while treatment of alveolar macrophages with SiNPs resulted in 80% of apoptosis and 20% of necrosis in these cells [22]. Additionally, Corbalan et al. showed that low [NO]/[ONOO−] ratio advisable increased nitroxidative or oxidative stress and is closely correlated with endothelial inflammation and necrosis [23].

Recently, autophagy has been identified as a novel mechanism induced in cells after exposition to nanoparticles. Autophagy can be triggered by a variety of microorganisms (bacteria, viruses) or parasites. Since NPs may present similar sizes to the microorganisms, they can possibly be perceived as a foreign bodies and cause autophagy activation [25–27]. Autophagy can be defined as a cytoprotective mechanism aiming at lysosomal degradation and recycling of proteins and damaged organelles, to maintain cellular homeostasis [27]. However, prolonged and uncontrolled autophagy may cause harmful cellular dysfunction and results in cell death [28]. A growing body of evidence suggests that deregulation of autophagy may also contribute to the toxicity evoked by SiNPs [26,27]. Duan et al. have demonstrated that SiNPs-induced autophagy and endotelial dysfunction in HUVECs cell line occur through the PI3K/Akt/mTOR signaling pathway [26]. The same research group has found that in HepG2 cells, the SiNPs-induced autophagy and autophagic cell death were triggered by ROS generation, suggesting that SiNPs can be a potential factor disrupting cellular homeostasis [27].

In light of the available data, cellular and molecular effects of SiNPs treatment seem to vary in size- and cell type-dependent manner. Although a considerable amount of data concerning SiNPs nanotoxicity exists, still little is known about the mechanism of SiNPs cytotoxicity in aggressive type of tumors such as GBM. In our study, we focused on highlighting the cellular and molecular effects responsible for the cytotoxicity of 5–15 nm SiNPs in human glioblastoma LBC3 and LN-18 cell lines. The mechanisms of apoptosis and autophagy, as two crucial processes of damaged cells elimination, were investigated.

## 2. Materials and Methods

### 2.1. Reagents

The Dulbecco's modified Eagle's medium (DMEM), containing glucose at 4.5 mg/mL with GlutaMax™, trypsin-EDTA, penicillin, streptomycin and fetal bovine serum Gold (FBS Gold) were provided by Gibco (San Diego, CA, USA). Passive lysis buffer, ReliaPrep RNA Cell Miniprep System and luminescent Caspase-Glo 9 Assay were provided by Promega (Madison, WI, USA), and BCA Protein Assay Kit by Thermo Scientific (Rockford, IL, USA). Annexin V Apoptosis Detection Kit I, JC-1 MitoScreen Kit, APO-Direct Kit were product of BD Pharmingen™ (San Diego, CA, USA). Sigma-Fast BCIP/NBT reagent, acridine orange, fumed silica dioxide amorphous powder 7 nm, silica dioxide spherical, porous nanopowder 5–15 nm and silica dioxide nanopowder 10–20 nm, 3-(4,5-dimethylthiazol-2-yl)-2,5-diphenyltetrazolium bromide and dichlorodihydrofluorescein diacetate were provided by Sigma (St. Louis, MO, USA). Polyclonal (rabbit) anti-human LC3 antibody and alkaline phosphatase-labeled anti-rabbit immunoglobulin G were provided by Cell Signaling Technology (Boston, MA, USA). High Capacity RNA-to-cDNA Kit was purchased from Applied Biosystems (Foster City, CA, USA).

### 2.2. Cell Cultures and Exposure to Silica Nanoparticles

The LBC3 cell line was developed from *glioblastoma multiforme* tissue taken from 56-year-old female patient subjected to surgical tumor resection, and was kindly given to us by Prof. Cezary Marcinkiewicz (Department of Neuroscience, Temple University, Philadelphia, PA, USA) [29]. The LN-18 and human skin fibroblasts (CRL1474) cell lines were provided by American Type Culture Collection (ATCC). The LBC3, LN-18 cells and fibroblasts were cultured in DMEM, supplemented with heat-inactivated, 10% (FBS Gold), streptomycin (100 µg/mL) and penicillin (100 U/mL). The cells were cultured in Falcon flasks (BD Pharmingen™, San Diego, CA, USA) at 37 °C, 5% $CO_2$ and 95% air in an incubator Galaxy S+ (RS Biotech, Irvine, UK). At approximately 70% confluence, cells were detached with 0.05% trypsin, 0.02% EDTA and counted in a Scepter cell counter (Millipore, Billerice, MA, USA). Next, $2.0 \times 10^5$ cells were seeded in 2 mL of DMEM in six-well plates. In order to minimize the aggregation of SiNPs, prior to the experiments nanoparticles were dispersed in deionized water by a sonicator (Sonopuls, Bandelin, Berlin, Germany), on ice for 10 min (160 W, 20 kHz,). After 24 h incubation, DMEM was removed and replaced with DMEM containing SiNPs suspensions at three different sizes: 7 nm, 5–15 nm and 10–20 nm, at concentrations ranging from 12.5 to 1000 µg/mL. The LBC3 and LN-18 cells not treated with SiNPs served as the negative controls. Next, the cells were incubated for 24 and 48 h and retained for further analyses.

### 2.3. Cell Viability

Cell viability was measured according to the manner of Carmichael et al. using 3-(4,5-dimethylthiazol-2-yl)-2,5-diphenyltetrazolium bromide (MTT) [30]. The LBC3, LN-18 cells and fibroblasts, at a density of $2.0 \times 10^5$ per well, were seeded in 6-well plates. After 24 h DMEM was removed and replaced with DMEM containing SiNPs suspensions at three different sizes: 7, 5–15 and 10–20 nm, at the concentrations ranging from 12.5 to 1000 µg/mL. The untreated LBC3 and LN-18 cells served as the negative controls. Then, the both cell lines were incubated with SiNPs for 24 and 48 h. The LBC3 and LN-18 cells were washed three times with phosphate buffer saline (PBS) and then incubated with 1 mL of MTT solution (0.25 mg/mL in PBS) in 5% $CO_2$ incubator, at 37 °C, for 4 h. The DMEM was removed and 1 mL of 0.1 mol/L HCl in absolute isopropanol was added. The absorbance of converted dye in living cells was measured at the wavelength of 570 nm on an Infinite M200 microplate reader (Tecan, Salzburg, Austria). The viability of LBC3 and LN-18 cells as well as fibroblasts cultured with SiNPs was calculated as the percentage of the untreated cells. All the experiments were done in duplicates in at least three cultures.

## 2.4. Characterization of Silica Nanoparticles

The characterization of SiNPs in deionized water or DMEM at different time points was carried out using the electrophoretic light scattering technique on Zetasizer Nano ZS analyzer equipped with a 4 mW He-Ne laser (Malvern Instruments, Malvern, UK). The voltage selection and measurement duration were performed using default settings of the apparatus. The size of the SiNPs was measured by the Dynamic Light Scattering (DLS) technique. Zeta potentials were determined on the basis of electrophoresis experiment (Electrophoretic Light Scattering, ELS) of the sample and by measurements of the velocity of the nanoparticles using Laser Doppler Velocimetry (LDV) method. In order to minimize particles aggregation, before addition to deionized water or DMEM, the stock suspensions of SiNPs (0.1 mg/mL in deionized water) were homogenized through a sonicator (Sonopuls, Bandelin, Berlin, Germany), 160 W, 20 kHz, for 10 min. The final concentration of nanoparticles in individual media was 100 µg/mL.

## 2.5. Detection of Apoptosis and Necrosis

Apoptosis and necrosis of LBC3 and LN-18 cell lines were evaluated by flow cytometry on FACSCanto II cytometer (BD, San Diego, CA, USA). The cells ($2.0 \times 10^5$) were seeded in 2 mL of DMEM in six-well plates. After 24 h, the DMEM was removed, replaced with the 5–15 nm SiNPs suspension in DMEM, at 50 or 100 µg/mL concentrations. Both cell lines were incubated for 24 and 48 h. The cells were detached, resuspended in DMEM and then in binding buffer. Subsequently, the cells were stained with FITC Annexin V and PI (FITC Annnexin V apoptosis detection Kit I, (BD Pharmingen™, San Diego, CA, USA) at room temperature, in the dark, for 15 min. Data were analyzed using FACSDiva software (BD Pharmingen™, San Diego, CA, USA).

## 2.6. Intracellular ROS Detection

The level of intracellular ROS was determined using dichlorodihydrofluorescein diacetate (DCFH-DA) assay, (Sigma, St. Louis, MO, USA). After diffusion through the cell membrane, DCFH-DA is deacetylated by cellular esterases to a non-fluorescent compound, which is later oxidized by intracellular ROS into a fluorescent 2′,7′-dichlorofluorescein (DCF). The LBC3 cells ($10 \times 10^5$) were seeded in 200 µL of DMEM in 96-well black plates. After 24 h, DMEM was removed and the cells were stained with 10 µM of DCFH-DA in PBS at 37 °C, 5% $CO_2$ incubator, for 45 min. Then, the dye was removed and replaced with the 5–15 nm SiNPs suspensions in DMEM, at 50 or 100 µg/mL concentrations and incubated for 24 and 48 h. The DCF fluorescence intensity was measured by Infinite M200 microplate reader (Tecan, Salzburg, Austria), at the excitation wavelength of 485 nm and the emission wavelength of 535 nm. The intracellular ROS generation in SiNPs-stimulated LBC3 cells was shown as the intensity of fluorescence of the DCF.

## 2.7. Mitochondrial Membrane Potential (ΔΨm) Analysis

The LBC3 cells ($2.0 \times 10^5$) were incubated in 2 mL of DMEM in six-well plates. After 24 h, DMEM was removed and replaced with the 5–15 nm SiNP suspensions in DMEM at the concentrations of 50 or 100 µg/mL and incubated for further 24 and 48 h. Then, the cells were detached and at $1 \times 10^6$ cells per mL were suspended in PBS. Subsequently, disruption of the mitochondrial membrane potential in LBC3 cells was assessed using MitoScreen kit (BD, San Diego, CA, USA), following the manufacturer's instructions. Briefly, supravital cells were washed and resuspended in PBS supplemented with 10 mg/mL the lipophilic cationic probe 5,5,6,6-tetrachloro-1,1,3,3-tetraethylbenzimidazolcarbocyanine iodide (JC-1). Then, the LBC3 cells were incubated at 37 °C for 15 min, washed and resuspended in PBS, and analyzed using flow cytometry (BD FACSCanto II, San Diego, CA, USA). The percentage of cells with disrupted mitochondrial membrane potential (MMP) was calculated using the FACSDiva software (BD, San Diego, CA, USA).

*2.8. Caspase-9 Activity Analysis*

The activity of caspase 9 was measured using luminescent Caspase-Glo 9 Assay (Promega, Madison, WI, USA) according to the manufacturer's protocol. Briefly, the LBC3 cells were seeded in 200 µL of DMEM in 96-well white-walled culture plates ($1 \times 10^5$ cells per well) and incubated for 24 h. Next, the medium was removed and replaced with the 5–15 nm SiNP suspensions in DMEM, at the concentrations of 50 or 100 µg/mL and incubated for 24 and 48 h. Next, a 100 µL of the Caspase-Glo 9 reagent was added. The resultant luminescence was measured in a microplate reader (Tecan, Salzburg, Austria) and presented as relative light units (RLU).

*2.9. RNA Isolation*

Total RNA was purified using ReliaPrep RNA Cell Miniprep System (Promega, Madison, WI, USA) with DNase I treatment according to the manufacturer's protocol. In order to assess the quantity and quality of the extracted RNA, spectrophotometric analysis was performed (NanoPhotometer, Implen, Munich, Germany). The concentration of RNA, as well as A260/280 and A260/230 ratios were measured.

*2.10. Gene Expression Analysis*

cDNA synthesis was performed using High Capacity RNA-to-cDNA Kit (Gibco, San Diego, CA, USA) following the producer's indications. Briefly, 1 µg of purified total RNA was reversely transcribed in a reaction mixture containing oligo dT-16 primers, random octamers, dNTPs and MuLV reverse transcriptase (RT) in total volume of 20 µL. As a template for real-time qPCR reactions 2 µL of cDNA was used. Amplification of the product was done using 2xHS-PCR Master Mix SYBR A (A&A Biotechnology, Gdynia, Poland). Primer sequences for the following genes: *Bim*, *Puma*, *Noxa*, *Bax* and housekeeping *RPL13a* were as described in our previous papers [31,32]. Primer sequences for the *Atg5* gene were as described by Alirezaei et al. [33]. Additional evaluation of primer accuracy was performed using the Primer-BLAST software. The applied reaction parameters were as follows: initial denaturation at 95 °C for 3 min, followed by 40 cycles of 95 °C for 1 min, 58 to 62 °C for 30 s, and 72 °C for 45 s. To perform real-time qPCR assay the CFX Connect Real-Time PCR System (Bio-Rad, Hercules, CA, USA) was used. Reactions were run in triplicates and the quantification of gene expression was analyzed using the relative quantification method with modification of Pfaffl [34].

*2.11. Transmission Electron Microscopy*

The morphological changes in human LBC3 cells were evaluated by the transmission electron microscopy (TEM). The LBC3 cells ($2.5 \times 10^5$) were seeded in 2 mL of DMEM in six-well plates. After 24 h, the medium was removed and replaced by the 5–15 nm SiNPs suspensions in DMEM at the concentration of 50 µg/mL. Subsequently, the LBC3 cells were incubated for 24 and 48 h. After incubation, the cells were centrifuged ($1000 \times g$, 5 min), fixed in a mixture of 2.5% glutaraldehyde and 2% paraformaldehyde in 0.1 M cacodylate buffer (CB) at pH 7.0, at 4 °C, for 1 h, and taken up into the agar blocks. Then, samples were washed in CB at 4 °C, for 1 h, post-fixed in 1% osmium tetroxide in CB at 4 °C, for 1 h and next dehydrated through a graded series of ethanol and embedded in glycid ether 100 (Epon 812). Ultrathin sections were contrasted with uranyl acetate and lead citrate, mounted on nickel grids and evaluated in a transmission electron microscope OPTON 900 (Zeiss, Oberkochen, Germany).

*2.12. Western Analysis*

Cells were washed with cold PBS and solubilized in 100 µL per well of passive lysis buffer. The lysates from each well were centrifuged at $10,000 \times g$, at 4 °C, for 10 min. Samples of lysates containing 20 µg of protein were subjected to SDS-PAGE, as described by Laemmli [35]. The 12% polyacrylamide gel and constant current (25 mA) were used. The proteins were transferred to nitrocellulose membranes and subsequently pre-treated with Tris-buffered saline (TBS) containing 0.05% Tween 20 (TBS-T) and 5% non-fat dry milk at room temperature, for 2 h. Membranes were probed with the primary polyclonal (rabbit) anti-human LC3 I/II antibody (1:1000) in 5% dry milk in TBS-T at 4 °C, for 16 h. Subsequently, the alkaline phosphatase-conjugated secondary antibody against rabbit IgG (whole molecule) at 1:2500 dilution was added in TBS-T with slow shaking for 1 h. The membranes were washed with TBS-T and exposed to Sigma-Fast BCIP/NBT reagent (Sigma, St. Louis, MO, USA).

*2.13. Protein Assay*

Protein concentration in cell lysates was measured by the method of Smith et al. using BCA Protein Assay Kit (Thermo Scientific, Rockford, IL, USA). Bovine serum albumin was used as a standard [36].

*2.14. Fluorescent Microscopy Assay*

The acidic vesicular organelles (AVOs) formation were visualized by the acridine orange (AO). An acidotropic dye is used as a marker and suggests the occurrence of autophagy in the analyzed cells. AO is a fluorescent dye that moves freely across biological membranes and stains the DNA and the cytoplasm bright green. In lysosomes and acidic organelles, acridine orange is protonated, forms aggregates and display bright red fluorescence [37].

The LBC3 cells ($2.5 \times 10^5$) were seeded in 2 mL of medium in six-well plates. After 24 h DMEM was removed and replaced the 5–15 nm SiNPs suspensions in DMEM, at concentrations of 50 or 100 µg/mL. Next, the cells were incubated for 24 and 48 h. After incubation, the cells were washed twice with PBS and stained with 1mL of the dye, 10 µM acridine orange, at 37 °C, for 10 min. Subsequently the cells were analyzed by fluorescence microscopy (Olympus CXK41, U-RLFT50, Tokyo, Japan) equipped with a tungsten 50 W lamp, a 490 nm band-pass blue excitation filter, a 500 nm diachronic mirror, and a 515 nm long pass barrier filter. Next, the medium with staining solution was removed, and the cell layer was washed with PBS and analyzed under a fluorescent microscope, at 200-fold magnification. The hundred cells per sample were examined by fluorescence microscopy, according to the following criteria: the AVOs-positive cells were determined by counting the number of cells with red signal from AVOs staining in cytoplasm in comparison to the number of cells without red signal from AVOs.

*2.15. Statistical Analysis*

Mean values from three independent experiments ± standard deviations (SD) were calculated. The data were statistically analyzed using one way-ANOVA followed by Tukey's post hoc *t*-test analysis. The significant differences of means were determined at the level of * $p < 0.05$ or ** $p < 0.001$.

## 3. Results

*3.1. The Effect of Silica Nanoparticles on Cell Viability*

The antiproliferative effect of 7 nm, 5–15 nm and 10–20 nm SiNPs on LBC3 (Figure 1A,C,E), LN-18 cells (Figure 1B,D,F) and human skin fibroblasts (Figure 1G) was determined using the MTT assay. The cells were incubated with increasing concentrations of SiNPs (ranging from 12.5 to 1000 µg/mL), for 24 and 48 h. It has been shown that SiNPs of all sizes caused time-dependent and dose-dependent reduction of cell viability in both, LBC3 and LN-18 cell lines. The reduction of cell viability was dependent on the sizes of SiNPs. The decrease in cell viability of LBC3 and LN-18 cells was observed after 24 h of incubation in all used SiNPs sizes. In cells treated with higher concentrations of SiNPs, the effect on cell viability was markedly more pronounced in case of the LBC3 cells; while in LN-18 cells the viability was much greater. Prolongation of incubation time up to 48 h, in cells incubated with SiNPs, resulted in strong decrease in number of viable LBC3 cells in comparison to viable LN-18 cells. The highest cytotoxicity was observed in case of cells treated with medium size SiNPs (5–15 nm) in both, LBC3 and LN-18, cell lines. The strong cytotoxic effect was observed after both time points of incubation with 5–15 nm SiNPs in LBC3 cells. In contrast to LN-18 and LBC3 cell lines we observed only weak, dose-dependent reduction in viability of normal skin fibroblasts using SiNPs 5–15 nm in the concentration from 12.5 to 1000 µg/mL (Figure 1G). Given this, based on the MTT results, 5–15 nm SiNPs in two concentrations: 50 and 100 µg/mL were selected for further studies.

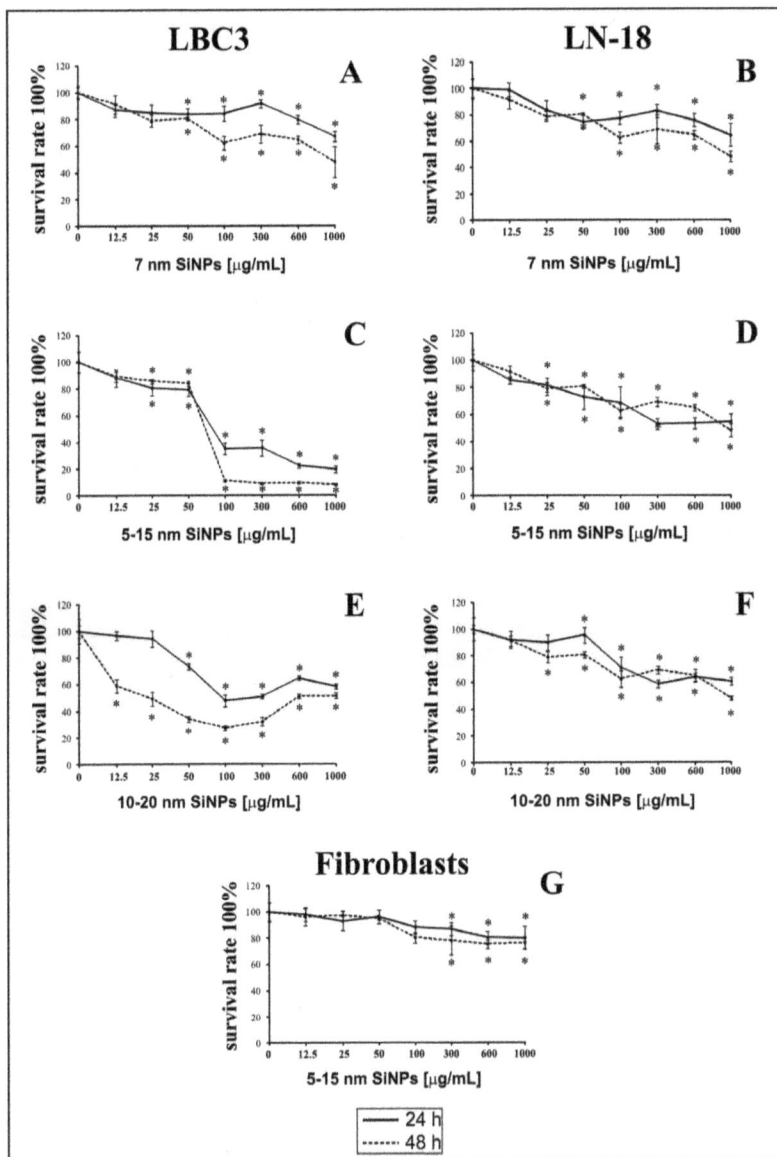

**Figure 1.** The viability of LBC3 (**A,C,E**) and LN-18 (**B,D,F**) cells treated with different concentrations (12.5 to 1000 µg/mL) of SiNPs in three different sizes: 7 nm (**A,B**), 5–15 nm (**C,D**) and 10–20 nm (**E,F**), for 24 and 48 h. Graph (**G**) presents the viability of human skin fibroblasts incubated with different concentrations, from 12.5 to 1000 µg/mL, of 5–15 nm SiNPs, for 24 and 48 h. Mean values from three independent experiments ± SD are presented. Significant alterations are expressed relative to controls and marked with asterisks. Statistical significance was considered if * $p < 0.05$.

*3.2. The Effect of the Dispersion Media on Zeta Potentials*

The effect of the dispersion medium on zeta potentials and size distributions of SiNPs was measured in both deionized water and DMEM containing 10% FBS, at 37 °C and different time points. The values of the determined parameters have been collected in Table 1. Size distribution of SiNPs, dispersed in deionized water (A) and DMEM (10% of FBS) (B), has been presented on Figure 2. Zeta potential is the pivotal parameter that controls electrostatic interactions in nanoparticle dispersions [38]. The magnitude of the zeta potential is predictive of the colloidal stability. It is well known that higher absolute value of zeta potential reflects higher stable state of colloidal systems. The values of $\zeta$-potentials higher than +30 mV or lower than −30 mV permits a basically stable suspensions, in which particles are not prone to aggregation or precipitation [39]. The $\zeta$-potentials of SiNPs dispersed in deionized water show more negative values ranging from −32.6 to −35.1 mV compared to particles dispersed in culture medium (−8.11 to −8.96 mV). Silica nanoparticle sizes determined instantly after dispersion in water exhibited a bimodal size distribution profile, with one population (representing approximately 99.1% of all nanoparticles) with a size of 60.4 nm, and the other (representing about 0.9% of the nanoparticles) with a size of 329.5 nm (Table 1). These values changed slightly in time, showing higher percentage of ~400 nm aggregates after 12, 24 and 48 h (Table 1, Figure 2). The sizes of SiNPs measured instantly after dispersion in DMEM with 10% of FBS, exhibited a mono-disperse pattern with a peak at 133.1 nm. After 24 h, the particles showed a bimodal pattern with a peak at 129.9 nm (representing approximately 94% of all particles), and a peak at 673 nm (representing approximately 6% of all particles). Similar pattern of size distribution was observed after 48 h (Table 1, Figure 2). For all the particles, peaks in the large size area might represent agglomeration of particles. The sizes of SiNPs measured by DLS technique were larger than their original size, in both deionized water and DMEM with 10% FBS, which might be due to the van der Waals force and hydrophobic interaction with surrounding media.

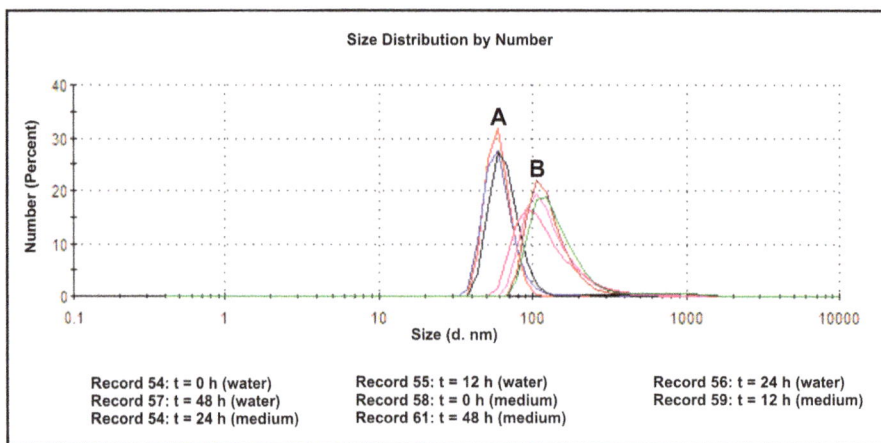

**Size Distribution by Number**

Record 54: t = 0 h (water)      Record 55: t = 12 h (water)      Record 56: t = 24 h (water)
Record 57: t = 48 h (water)      Record 58: t = 0 h (medium)      Record 59: t = 12 h (medium)
Record 54: t = 24 h (medium)    Record 61: t = 48 h (medium)

**Figure 2.** Size distribution of 5–15 nm SiNPs determined by DLS measurements: dispersed in deionized water (**A**), and dispersed in DMEM with 10% of FBS (**B**). Representative images generated by Zetasizer Nano ZS are presented.

**Table 1.** Zeta potential and size of SiNPs in deionized water (**A**) or DMEM with 10% of FBS (**B**) as dispersion medium at different time points (at 37 °C).

| Times (h) | Deionized Water (A) | | | Medium (DMEM with 10% FBS) (B) | | |
|---|---|---|---|---|---|---|
| | Zeta Potential (mV) | Diameter (nm) | | Zeta Potential (mV) | Diameter (nm) | |
| | | Fraction I | Fraction II | | Fraction I | Fraction II |
| 0 | −32.6 | 60.14 (99.1%) | 329.5 (0.9%) | −8.30 | 133.1 (100%) | - |
| 12 | −33.3 | 61.01 (97.5%) | 269.2 (2.5%) | −8.24 | 140.0 (100%) | - |
| 24 | −35.1 | 62.05 (98.2%) | 270.1 (1.8%) | −8.11 | 129.9 (94.0%) | 673.0 (6.0%) |
| 48 | −35.0 | 65.93 (98.4%) | 321.6 (1.6%) | −8.96 | 146.0 (95.9%) | 893.0 (4.1%) |

### 3.3. The Effect of Silica Nanoparticles on Apoptosis and Necrosis

The apoptosis and necrosis of LBC3 and LN-18 cells were evaluated by flow cytometry on FACSCanto II (BD, San Diego, CA, USA). The percent of apoptotic and necrotic LBC3 cells incubated with 50 and 100 µg/mL of 5–15 nm SiNPs for 24 and 48 h is showed in Figure 3A–C. As depicted in Figure 3B, after 24 h of incubation in the cells treated with 50 and 100 µg/mL of 5–15 nm SiNPs, the percent of apoptotic cells was significantly higher in comparison to the control cells. We found a time- and dose-dependent increase in apoptosis of LBC3 cells. The percent of necrotic LBC3 cells in cultures incubated with 50 and 100 µg/mL of 5–15 nm SiNPs for 24 and 48 h, is reflected in Figure 3C. The dose-dependent but not time-dependent increase in necrosis of LBC3 cells was observed. The percent of apoptotic and necrotic LN-18 cells in cultures incubated with 50 and 100 µg/mL of 5–15 nm SiNPs for 24 and 48 h is showed in Figure 3D–F. In contrast to LBC3 cells, we did not observe any effect of SiNPs on apoptosis of LN-18 cells. The percent of apoptotic LN-18 cells did not change independently on the incubation time and doses of the SiNPs. In contrast to apoptosis, we observed the strong time- and dose-dependent effect of SiNPs on necrosis of the LN-18 cells (Figure 4D,F). In case of cultures incubated in DMEM with 50 µg/mL of 5–15 nm SiNPs for 24 h we did not observe changes in the percent of necrotic cells. In the cells incubated with 100 µg/mL SiNPs we observed a marked increase in necrosis in comparison to control cells. After 48 h of incubation of LN-18 cells with both concentrations of 5–15 nm SiNPs we noticed subsequent rise of necrosis in comparison to the control.

In LN-18 cells the primary mechanism initiated after treatment with SiNPs was necrosis, which is basically known as uncontrolled and passive process. Accordingly, to further focus on the mechanisms of SiNPs-mediated cell death, we have decided to continue our research on LBC3 cell line.

### 3.4. The Effect of Silica Nanoparticles on Intracellular ROS Generation

Figure 4A shows the fluorescence intensity of 2′,7′-dichlorofluorescein (DCF) in LBC3 cells incubated with 50 or 100 µg/mL of 5–15 nm SiNPs for 24 and 48 h. The fluorescence of DCF was intensified with an increase in the intracellular ROS production and was dependent on time of incubation and concentration of SiNPs. After 24 h-incubation of LBC3 cells with 50 or 100 µg/mL of 5–15 nm SiNPs, the intracellular ROS production was approximately 2-fold higher in comparison to the untreated cells. After 48 h, resulted in about 2-fold higher ROS production in LBC3 cells incubated with 50 µg/mL of 5–15 nm SiNPs, and 3-fold higher in case of cells treated with 100 µg/mL, in comparison to the untreated controls (Figure 4A).

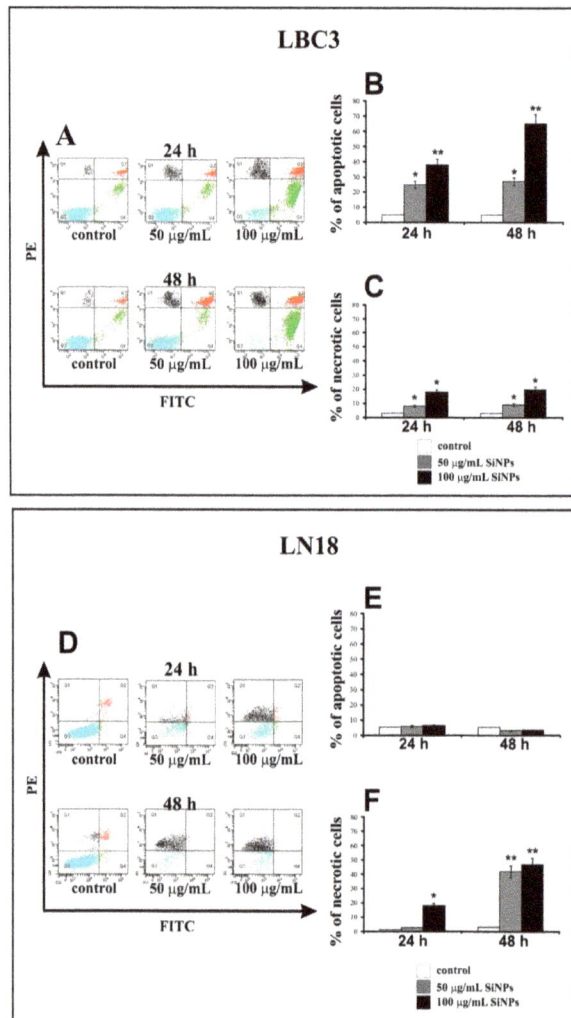

**Figure 3.** The effect of SiNPs on apoptosis (**A,B,D,E**) and necrosis (**A,C,D,F**) of LBC3 (**A–C**) and LN-18 (**D–F**) cell lines evaluated by annexin V assay. The cells were incubated for 24 and 48 h in DMEM with 50 and 100 µg/mL of 5–15 nm SiNPs. The cells were double-stained with FITC-Annexin V and PI. Representative dot plots for Annexin V-FITC/propidium iodide (PI) staining are shown (**A,D**). Following acquisition of sample, the cells were gated through the forward scatter FSC and side scatter SSC and analyzed for fluorescence intensity of FITC-Annexin V and PI. The cells were divided into four subpopulations: live cells—Q3 (annexin V-FITC−/PI−), early apoptotic cells—Q4 (annexin V FITC+/PI−), late apoptotic cells—Q2 (annexin V-FITC+/PI+), and necrotic cells—Q1 (annexin V FITC−/PI+). Percentage of apoptotic cells was the sum of percentage early apoptotic (Q4) and late apoptotic cells (Q2). Mean values of the percentage of apoptotic and necrotic cells, from three independent experiments ± SD are presented. Significant alterations are expressed relative to controls and marked with asterisks. Statistical significance was considered if * $p < 0.05$ or ** $p < 0.001$.

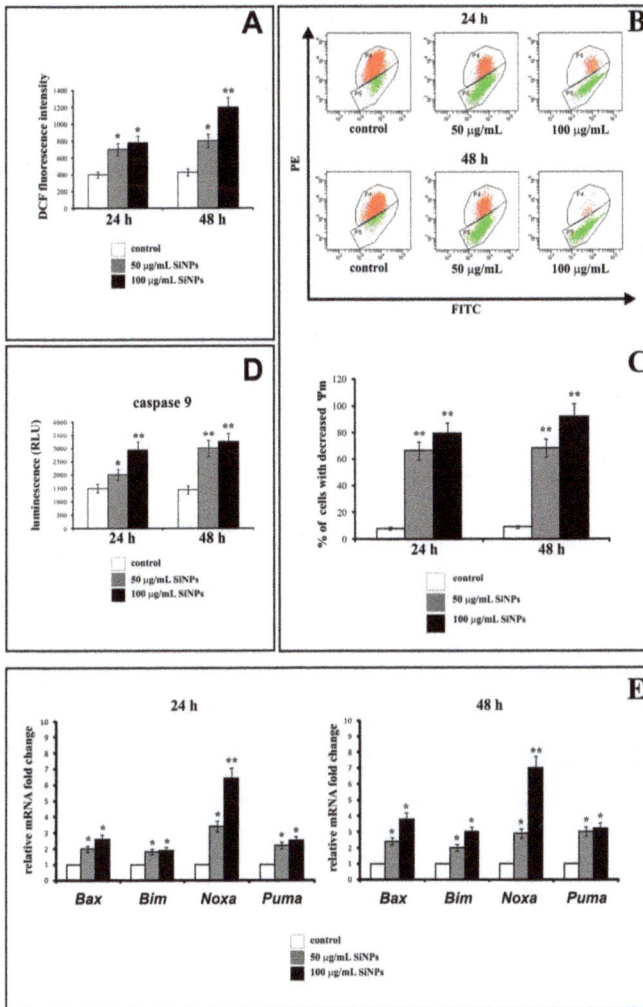

**Figure 4.** Oxidative stress and biochemical changes induced by 5–15nm SiNPs in LBC3 cells. The cells were treated with 50 and 100 µg/mL of 5–15 nm SiNPs for 24 and 48 h. Intracellular reactive oxygen species (ROS) production in LBC3 cells is presented on panel **A**. Panel **B** shows the flow cytometry analysis of $\Delta\Psi_m$ in LBC3 cells. The top panel **B** shows representative dot plots of LBC3 cells analyzed by JC-1 staining. X-axis and Y-axis are green and red fluorescence, respectively. The gate P4—populations of cells with normal $\Delta\Psi$m and gate P5—population of cells with decreased $\Delta\Psi$m. The lower panel **C** shows the percentage of LBC3 cells with decreased $\Delta\Psi$m. Panel **D** shows activity of caspase-9 in LBC3 cells. Relative quantification of proapoptotic genes expression in LBC3 cells is presented in panel **E**. Results are shown as a relative fold change in mRNA expression in comparison to untreated controls, where expression level was set as 1. Significant alterations are expressed relative to controls and marked with asterisks. Statistical significance was considered if * $p < 0.05$ or ** $p < 0.001$.

### 3.5. The Effect of Silica Nanoparticles on the Change of Mitochondrial Membrane Potential

In order to evaluate the influence of SiNPs treatment on permeabilization of mitochondrial membrane, the assessment of the changes in mitochondrial membrane potential ($\Delta\Psi$m) has been performed.

Figure 4B and C show the effect of 5–15 nm SiNPs on mitochondrial membrane potential ($\Delta\Psi_m$) in LBC3 cells. Figure 4B shows representative dot plot of LBC3 cells stained with JC-1, and Figure 4C shows the percentage of LBC3 cells with decreased $\Delta\Psi_m$. We observed the dose-dependent reduction in mitochondrial membrane potential in LBC3 cells. After 24 h as well as 48 h of incubation, a significant loss in $\Delta\Psi$m was observed. The cells treated with 50 μg/mL of SiNPs showed approximately 10-fold decrease in $\Delta\Psi$m, while cells incubated with 100 μg/mL of SiNPs showed nearly 11-fold reduction of $\Delta\Psi$m in comparison to the control cells.

### 3.6. The Effect of Silica Nanoparticles on Caspase-9 Activity

Figure 4D shows the activity of caspase-9 in LBC3 cells exposed to 5–15 nm SiNPs at concentrations of 50 and 100 μg/mL, for 24 and 48 h. The activity of caspase-9 was increased in dose- and time-dependent manner. We observed 2-fold rising of caspase-9 activity after 48 h for both concentrations of SiNPs in comparison to the untreated cells.

### 3.7. The Effect of Silica Nanoparticles on Proapoptotic Genes Expression

The ROS generation can lead to simultaneous activation of apoptotic and autophagic cell death. The gene expression of *Bax*, *Bim*, *Noxa* and *Puma* in LBC3 cells exposed to 5–15 nm SiNPs at concentrations of 50 and 100 μg/mL for 24 and 48 h, was measured and results presented on Figure 4E. Our results showed that mRNA levels of *Bax*, *Bim*, *Noxa*, and *Puma* genes were upregulated in dose- and time-dependent manner in cells incubated with 5–15 nm SiNPs.

### 3.8. The Effect of Silica Nanoparticles on Morphological Changes of LBC3 Cells

Figure 5 shows the morphological changes of LBC3 cells exposed to 5–15 nm SiNPs (50 μg/mL). After 24 and 48 h of incubation, control cells had oval or rod-shaped mitochondria with medium or high electron density matrix (Figure 5A,D, green arrows). In LBC3 cells exposed to 5–15 nm SiNPs for 24 h we noticed destruction of mitochondrial structure such as focal brightening in the matrix, mitochondrial membranes deformation, mitochondrial swelling and cristae rupturing (Figure 5C, red arrows). Except that, as presented on Figure 5B, part of the cell is composed of a small electron dense material (SiNPs) either free or as membrane-bound aggregates in the cytoplasm (Figure 5B, yellow arrows). The changes in LBC3 cells exposed to 5–15 nm SiNPs for 48 h were similar to those observed in cells incubated with SiNPs for 24 h. Some alterations were more pronounced after 48 h. Changes in the mitochondria consist of altered size, shape and the focal edema with the damage to the mitochondrial membranes, mitochondrial swelling and cristae rupturing (Figure 5F, red arrows). Moreover, we observed a number of SiNPs dispersed in cytosol (Figure 5E, yellow arrows).

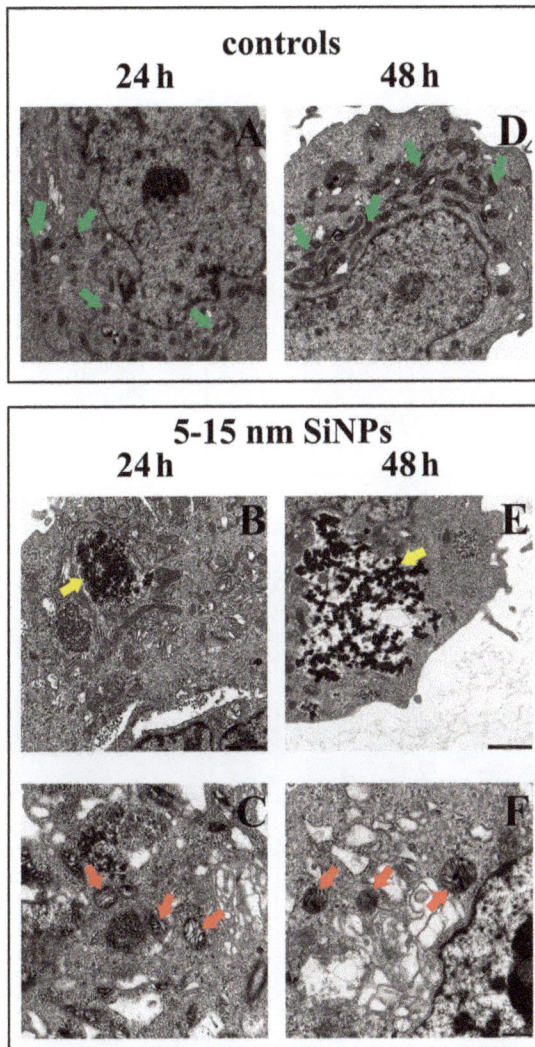

**Figure 5.** Morphological changes in LBC3 cells incubated with 50 μg/mL of 5–15 nm SiNPs for 24 (**A–C**) and 48 h (**D–F**) exposure. Control cells (**A**—left panel, **D**—right panel): numerous oval or rod-shaped mitochondria with clear intermembrane spaces were clearly visible (green arrows), (magnification 3000×). **B**—left panel and **E**—right panel: part of cell with electron dense regions located close to the membrane (yellow arrows), (magnification 7000×). **C**—left panel and **F**—right panel: a fragment of the cell with visible multiform mitochondria, mitochondrial swelling with firmly compacted mitochondrial matrix, bright spaces intermembrane (cristae), swelling and cristae rupturing (red arrows), are visible (magnification 12,000×).

### 3.9. The Effect of Silica Nanoparticles on Expression of Autophagy Markers

Figure 6A,B shows Western blot analysis of autophagy marker Light Chain 3 (LC3-I and LC3-II) expression in LBC3 cells incubated with 50 and 100 μg/mL SiNPs for 24 (lanes: 1, 2, 3) and 48 h (lanes: 4, 5, 6). The cells incubated for 24 and 48 h with 50 and 100 μg/mL of SiNPs demonstrate the expression of LC3-II form, and only week expression of LC3-I. Moreover, we observed that the expression of membrane-bound LC3-II, (the phosphatidylethanolamine-conjugated form) was increased in cells incubated with 50 µg/mL (lanes: 2, 5) and 100 µg/mL (lanes: 3, 6) of SiNPs after 24 and 48 h in comparison to the control cells (lanes: 1, 4). Next, we conducted densitometry analysis and calculated LC3-II/LC3-I ratio (Figure 6B). We noticed time- and dose-dependent increase of LC3-II/LC3-I ratio in cells incubated with both concentrations of SiNPs.

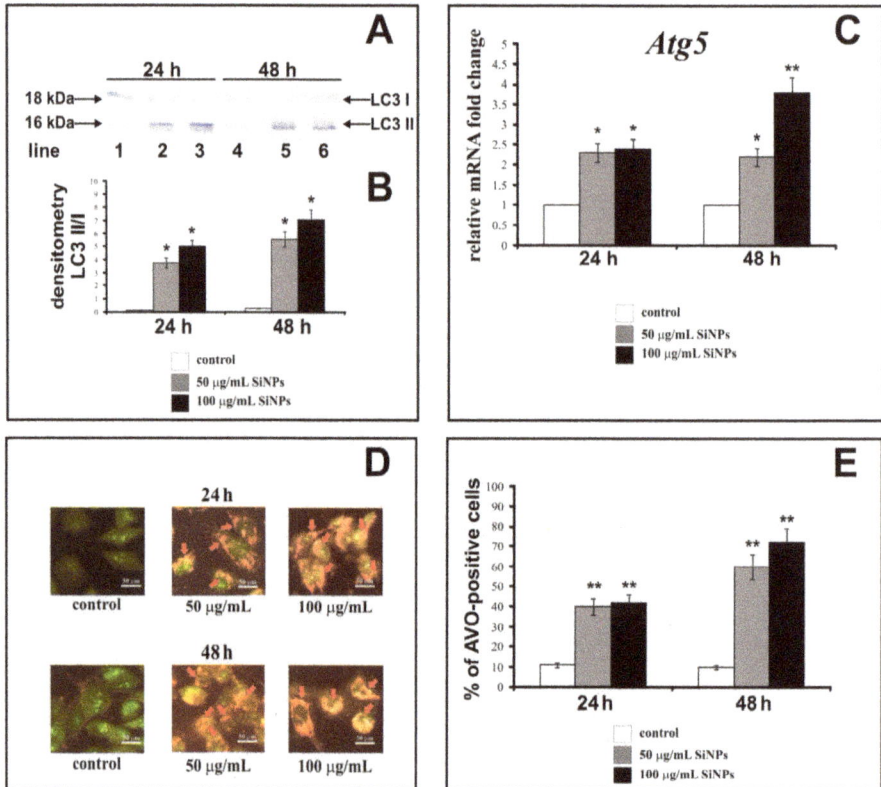

**Figure 6.** Western blot analysis (**A**) and densitometric analysis (**B**) of LC3-I and LC3-II ratio in LBC3 cells incubated in medium with 50 and 100 µg/mL of 5–15 nm SiNPs for 24 and 48 h. Samples containing 20 µg of protein were submitted to electrophoresis and immunoblotting. A representative Western blot is presented on panel A and densitometric analysis—on panel B. Relative quantification of *Atg5* gene expression in LBC3 cells (**C**). Results are shown as a relative fold change in mRNA expression in comparison to untreated controls, where expression level was set as 1. The effect of SiNPs on formation of AVOs in the LBC3 cell lines (**D,E**) evaluated by fluorescence microscope assay. The volume of the cellular acidic compartment was visualized by acridine orange staining. The cells were incubated in DMEM with 50 or 100 µg/mL of 5–15 nm SiNPs for 24 or 48 h and the cells were photographed under a fluorescence microscope at 200-fold magnification, (*scale bar* 50 µm) (**D**), or counted was the percentage of AVOs-positive cells (**E**). Representative images of AVOs positive cells (red arrows), from one of three independent experiments are shown. Significant alterations are expressed relative to controls and marked with asterisks. Mean values from three independent experiments ± SD are presented; * $p < 0.05$ or ** $p < 0.001$.

Figure 6C shows the level of *Atg5* gene expression in LBC3 cells exposed to 5–15 nm SiNPs at concentrations: 50 and 100 µg/mL, for 24 and 48 h. The transcript of *Atg5* was significantly upregulated in time- and dose-dependent manner in LBC3 cells (Figure 6C). Thus, we noticed the coexistence of increased LC3-II/LC3-I ratio with the upregulation of *Atg5* gene expression, confirming the occurrence of autophagy in cells subjected to SiNPs treatment.

*3.10. The Effect of Silica Nanoparticles on Acidic Vesicular Organelles Formation*

Autophagy can be characterized by the formation of AVOs (Acidic Vesicular Organelles), which can be detected by staining with acridine orange. This dye crosses the biological membranes and accumulates in acidic compartments, where it is seen as bright red fluorescence. Figure 6D shows AVOs formation in the cytoplasm, while Figure 6E shows the percent of AVOs-positive LBC3 cells after treatment with 50 or 100 µg/mL of 5–15 nm SiNPs, for 24 and 48 h. We observed a time- and dose-dependent (only after 48 h incubation) increase in AVOs formation in the cytoplasm of LBC3 cells. After 24 h of incubation, the percentages of AVOs-positive cells were about 4-fold higher in LBC3 cells treated with 50 µg/mL and 100 µg/mL of SiNPs, in comparison to control cells. The prolongation of the incubation time up to 48 h resulted in 6-fold higher percentage of AVOs-positive LBC3 cells when treated with 50 µg/mL of SiNPs, and 7-fold higher percentage of AVOs-positive cells, when treated with 100 µg/mL of SiNPs, in comparison to the control cells.

## 4. Discussion

A constant increase in number of currently diagnosed malignant diseases is calling for new treatment strategies involving the implementation of novel therapeutic approaches. Rapid progress in nanotechnology offers new solutions for development of new methods of cancer diagnosis and therapy. Biomedical research carried out in the last decade, showed that the chemical and physical properties of nanomaterials such as size or shape play an important role in determination for their cytotoxicity [40]. Furthermore, it is known that SiNPs are the most commonly used nanomaterials for medical applications because of the chemical stability, good biodistribution, cellular internalization and tumor penetration [21,41].

A great deal of research has demonstrated that SiNPs are cytotoxic and can cause ROS generation, DNA damage, aberrant nucleoplasmic protein aggregation, apoptosis and autophagy. These properties of SiNPs have prompted a widespread quest for the possible methods of their utilization as cancer therapeutics [42].

Moreover, preliminary reports suggest that SiNPs might be able to cross the blood-brain barrier, which makes them an interesting agent to use in glioblastoma research [43].

The charge, size and functionalization of NPs are important parameters determining the level of cellular damage of cancer cells. Our study has shown that SiNPs exerted cytotoxic effect on glioblastoma LBC3 and LN-18 cell lines but not in human skin fibroblasts. This cytotoxic effect in both glioblastoma cell lines was dependent on the size and dose of SiNPs as well as the time of incubation, which is in line with previous reports [44,45]. Kim et al. indicated, that the small-size SiNPs (20 nm) were more cytotoxic for U373MG human glioblastoma cells than the large-size ones (100 nm) [45]. Napierska et al. reported that in general, cells exposed to 5–15 nm SiNPs showed higher cytotoxic effect, than those exposed to 7 nm SiNPs, which is in agreement with our findings [46]. Similar results has been presented by Lin et al. who demonstrated that amorphous 5–15 nm SiNPs induced cytotoxic effect on human lung cancer cells [47]. Furthermore, Chang et al. demonstrated SiNPs-mediated cytotoxicity in lung, gastric and colon cancer cells as well as in normal fibroblasts [48]. Other authors concluded that the cytotoxicity of SiNPs was dependent on the cell type and population doubling time. Additionally, it has been proven that, intracellular distribution of SiNPs has a considerable influence on protein aggregation, gene expression and cell cytotoxicity [49]. Thus, SiNPs may penetrate cell membranes, lodge in the mitochondria, and lead to damage of cancer cells [49].

Physicochemical properties of SiNPs are important factors in nanotoxicity research [41]. We demonstrated that there were differences in $\zeta$-potential values and nanoparticles sizes dependently on the dispersion medium. These differences are due to the adsorption of the proteins on the nanoparticle surface. It has been known, that the contact of NPs with biological fluids such as serum or ions, lipids, and other macromolecules is followed by instant adsorption of proteins on the nanoparticle surface in a form of a layer known as the protein corona [50]. Since DMEM contains 10% of FBS, this layer is likely to adsorb on the SiNPs surface. Protein coating may then alter the behavior of the nanoparticles, potentially modifying the aggregation state and cellular response [51]. Similar observations were also reported by other authors, indicating a constant need of empirical evaluation of physicochemical parameters of commercially available nanoparticles in particular experimental conditions [51,52].

The mechanisms of silica nanoparticles cytotoxicity are not fully understood and depend on their physicochemical properties such as: size [52]. Wittmaack [53] suggested that SiNPs can interfere with the membrane-mediated processes by gravitational settling of high concentrations of nanoparticles on top of the cells in the culture. On the other hand, it has been proven that SiNPs can induce oxidative stress and generate intracellular ROS formation. Mitochondria are involved in the generation of ROS through one-electron carriers in the respiratory chain. There are three major types of ROS, among others: superoxide anion ($O_2^-$), hydroxyl radical (HO·) and hydrogen peroxide ($H_2O_2$), which play pivotal role in cell metabolism, signaling, and homeostasis. Accumulation of ROS in the cells increases damage of the proteins, lipids, and nucleic acids [54]. Therefore, we decided to study the effect of SiNPs on ROS generation in LBC3 cell line. Our research showed that SiNPs induce intracellular ROS generation in LBC3 cells in dose- and time-dependent manner. Another in vitro studies have shown that nanoparticles caused oxidative stress, which impaired the balance between cellular ROS production and the mechanisms of ROS detoxification [41]. The increase in ROS has been suggested to be involved in apoptosis of cancer cells [20,21,55].

Apoptosis and necrosis are two characteristic types of cell death, which are discussed in many reports [17,31,32]. The two main mechanisms of the induction of apoptosis, the mitochondrial (intrinsic) pathway and receptor-mediated (extrinsic) pathway, are commonly known [56]. In our study, we evaluated the apoptosis and necrosis occurring in LN-18 and LBC3 cells after stimulation with SiNPs. Our research has shown that 5–15 nm SiNPs induces apoptosis and necrosis in LBC3 cells, whereas in LN-18 cells only necrosis occurs. The mechanism of necrosis is already well understood. Necrosis occurs as a result of a strong damage to the cell membrane that leads to rapid reduction of intracellular ATP levels and loss of osmotic balance of the cells [57]. Apoptosis however, can be induced by oxidative stress, and thus mitochondria play a pivotal role in this process [58]. We speculate that different survival rates between both cell lines exposed to the same nanoparticles sizes may be caused by different rate of endocytosis of SiNPs. We believe that, necrosis could be caused by a very fast up-take of SiNPs by LN-18 cells. There have been LN-18 cells overloaded with nanoparticles, which might results in the necrotic break down of these cells in comparison to LBC3, where the SiNPs overload was not that pronounced.

Mitochondria play pivotal role in activation of the intrinsic pathway of apoptosis in mammalian cells [59,60]. In the present study, we observed loss of the mitochondrial membrane potential and destruction of mitochondrial ultrastructure. In LBC3 cells in the presence of 5–15 nm SiNPs, we noticed mitochondrial swelling, cristae rupturing, deformation of mitochondrial membrane and over all mitochondrial damage. Due to the low MMP and changes in mitochondria ultrastructure, we suggest that 5–15 nm SiNPs-induced apoptosis in LBC3 cells occurs through the mitochondrial pathway. Ahamed showed that SiNPs generated oxidative stress and induced apoptosis by opening of the mitochondrial permeability transition pores (MPTP) and subsequent decrease of the MMP [58]. Moreover, during the opening of the MPTP, oxidative damage of mitochondrial membrane may occur due to the lower rate of hydroperoxide removal [58,59]. Sun et al. demonstrated that mitochondrial pathway of apoptosis mediated by oxidative stress was a potential mechanism of cytotoxicity induced by 43 nm SiNPs in HepG2 cell line [57]. Accordingly, we suggest that the loss of the MMP through the oxidative stress induced by 5–15 nm SiNPs, results in increased generation of ROS and further cytotoxic effect in LBC3 cells.

Cytochrome c is a protein, which upon extrusion from intermembrane space of mitochondria into the cytosol, forms a complex with apoptotic protease-activating factor 1 (APAF-1) and procaspase-9 and leads to the assembly of the apoptosome. These results in the activation of the caspase cascade, which evokes a series of biochemical and morphological alterations characteristic to apoptosis [59]. Therefore, we examined the activity of caspase-9 in LBC3 cells exposed to 5–15 nm SiNPs. Our results showed that capsase-9 activity increased in dose- and time-dependent manner. Nowak et al. demonstrated that SiNPs induced the expression of caspases-3 and -9 in cells [61], while Tokgun et al. found that SiNPs activated extrinsic pathway of apoptosis through the activation of caspase-8 in A549 cell line [18]. Apart from that, Ahmed showed that exposure of A431 and A549 cells to SiNPs, evoked upregulation of *caspase-3* and *caspase-9* genes in dose-dependent manner [21]. Moreover, the enzymatic activities of caspase-3 and caspase-9 were also enhanced in comparison to the control cells.

Next, we decided to investigate the main proapoptotic factors coupled with mitochondrial pathway of apoptosis at the molecular level. After 24 and 48 h treatment of LBC3 cells with 50 and 100 µg/mL of 5–15 nm SiNPs showed the upregulation of *Bim*, *Bax*, *Puma* and *Noxa* mRNA. Induction of apoptosis by SiNPs can be connected with transcriptional induction of BH3-only proteins such as PUMA, BIM, NOXA, BIK. The BH3-only proteins antagonize with anti-apoptotic BCL-2 family proteins e.g., BCL-2, BCL-xL, Mcl-1 and can activate the proapoptotic proteins BAX and BAK [23]. Activated BAX and BAK lead to the formation of pores in the outer mitochondrial membrane [17]. This suggests that the deregulation of transcription of genes involved in mitochondria-mediated apoptosis might play a pivotal role in the regulation of SiNPs-dependent death of LBC3 cells.

Most of the previous studies concerning SiNPs were focused on resolving the mechanisms of apoptotic and necrotic cell death. To date little is known about the toxicological consequences of SiNP-induced autophagy in cancer cells. Autophagy, also termed as type II cell death, plays an important role in cell survival during stress conditions. Paradoxically, this process can protect cancer cells, but may also contribute to cell death when the stress is prolonged and insuperable [27]. Autophagy involves the sequestration and degradation of cellular components, organelles or proteins by the lysosomal pathway [27]. It is controlled by a group of evolutionarily conserved genes (*Atg* genes) and it can lead to the induction, activation and nucleation of autophagic vesicles. During autophagy, double-membrane autophagosomes assemble in order to engulf intracellular components. The cytosolic form of microtubule-associated protein 1A/1B-light chain 3 (LC3-I) is conjugated with phosphatidylethanolamine during the process of autophagosomal membrane formation and subsequently generates LC3-II [62]. Conversion of LC3-I into LC3-II is commonly used to monitor autophagy and is known as the biomarker of this process. Furthermore, autophagy can be characterized by AVOs formation, ultrastructural analysis of autophagosomes [27]. Interestingly, SiNPs have been accepted as a new class of autophagy activators, since they could be recognized by cells as the endosomal pathogens or proteins, which are commonly degraded by autophagy pathway [27].

In our studies we observed an increase in LC3-II/LC3-I ratio, an upregulation of *Atg5* gene transcript, and an increase in AVOs-positive cells. Furthermore, we noticed the coexistence of autophagy and apoptosis in LBC3 cells exposed to 5–15 nm SiNPs. Recently, a few studies demonstrated that autophagy increased significantly after exposure to SiNPs [27]. Yu et al. showed that SiNPs could induce autophagy and autophagic cell death via the ROS generation in HepG2 cell line [27]. Apart from that, it has also been known that autophagy can induce apoptosis in ATG5-dependent pathway. ATG5 protein is one of the components of the basic autophagic machinery [63]. The key point in this mechanism is the cleavage of the ATG5 by calpain to create a cut form of the protein, that translocates into the mitochondria membrane, and cause the opening of the MPTP and activation of the intrinsic apoptosis pathway [63,64]. Additionally, another possible mechanism for autophagy-dependent cancer cell death is the selective recruitment of cell survival factors (including growth factors), for degradation in autophagosomes. Moreover, mammalian cells have been shown to selective by recruit cytoplasmic catalase to autophagosomes, which further leads to the accumulation of intracellular ROS and mitochondrial-dependent apoptosis [65].

## 5. Conclusions

In conclusion, our results suggest that SiNPs can induce cytotoxicity in glioblastoma LBC3 and LN-18 cell lines, but not in human skin fibroblasts. Interestingly, we observed the coexistence of apoptosis and autophagy in LBC3 cells, while in LN-18 we noticed only necrosis. The SiNPs treatment resulted in oxidative stress and the loss of the MMP in LBC3 cells. Moreover, the upregulation of the proapoptotic genes: *Bim, Bax, Puma, Noxa* and increased activity of caspase-9 were observed in LBC3 glioma cells. These results may indicate that mitochondrial-dependent pathway of apoptosis is involved in SiNPs-mediated LBC3 cells death (Figure 7).

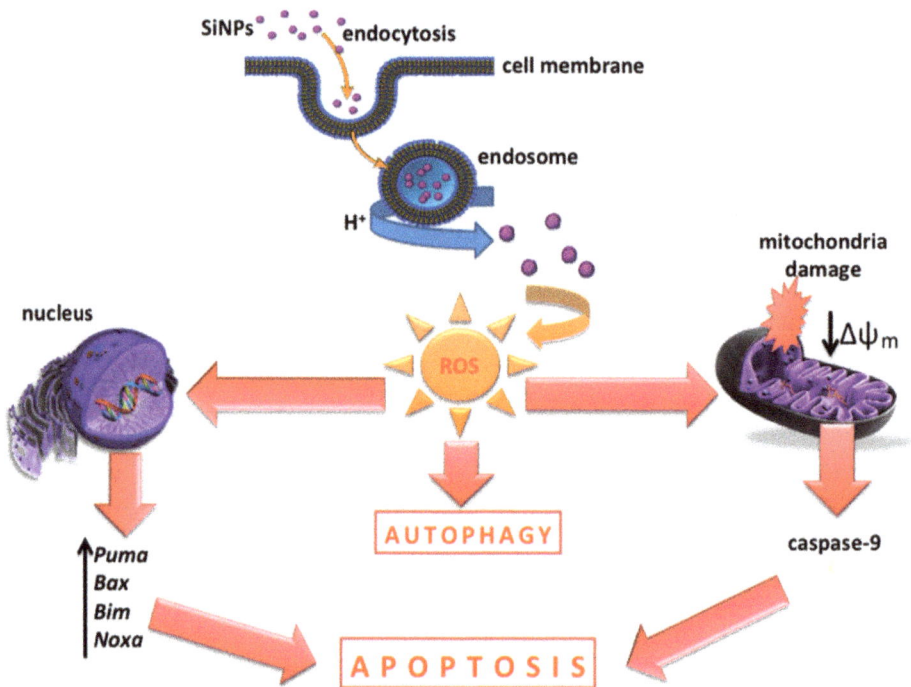

**Figure 7.** The effect of SiNPs on apoptosis, autophagy, and their suggested mechanism in glioblastoma LBC3 cells.

Although a great deal of information about SiNPs has already been available, it is still a subject of intensive investigations. Our findings demonstrate that SiNPs can act in a cell type-specific way and can initiate variable and complex mechanisms in response to their exposure. Therefore, it is worthwhile to comprehensively elucidate molecular mechanisms activated by SiNPs treatment to use it successfully as a potential therapeutic agent for glioblastoma multiforme therapy.

**Acknowledgments:** This study was conducted with the use of equipment purchased by Medical University of Białystok as part of the OP DEP 2007-2013, Priority Axis I.3, contract No. POPW.01.03.00-20-001/12. The ZetaSizer Nano ZS apparatus was funded by the European Funds for Regional Development and the National Founds of Ministry of Science and Higher Education, as part of the Operational Program Development of Eastern Poland 2007-3013, project: POPW.01.03.00-20-034/09-00.

**Author Contributions:** Rafał Krętowski—corresponding author, wrote the paper, planned experiments; performed the experiments; data analysis; Magdalena Kusaczuk—performed the genes expression experiments; Monika Naumowicz and Joanna Kotyńska—performed the characterization of silica nanoparticles; Beata Szynaka—performed the TEM experiments; Marzanna Cechowska-Pasko—analyzed the data.

**Conflicts of Interest:** The authors report no conflicts of interest in this work.

**Compliance with Ethical Standards:** The manuscript does not contain clinical studies or patient data.

## Abbreviations

| | |
|---|---|
| AVOs | acidic vacuolar organelles |
| DCFH-DA | dichlorodihydrofluorescein diacetate |
| GBM | glioblastoma multiforme |
| HUVECs | human umbilical vein endothelial cells |
| JC-1 | 5,5,6,6-tetrachloro-1,1,3,3-tetraethylbenzimidazolcarbocyanine iodide |
| LC3 | microtubule-associated protein 1A/1B-light chain |
| MMP | mitochondrial membrane potential |
| MPTP | mitochondrial permeability transition pores |
| MTT | 3-(4,5-dimethylthiazol-2-yl)-2,5-diphenyltetrazolium bromide |
| APAF-1 | apoptotic protease activating factor 1 |
| ROS | reactive oxygen species |
| SiNPs | silica nanoparticles |

## References

1. Thakkar, J.P.; Dolecek, T.A.; Horbinski, C.; Ostrom, Q.T.; Lightner, D.D.; Barnholtz-Sloan, J.S.; Villano, J.L. Epidemiologic and molecular prognostic review of glioblastoma. *Cancer Epidemiol. Biomark. Prev.* **2014**, *23*, 1985–1996. [CrossRef] [PubMed]
2. Caldarella, A.; Barchielli, A. Glioblastoma in the Canton of Zurich Switzerland revisited: 2005 to 2009. *Cancer* **2016**, *122*, 37–40. [CrossRef] [PubMed]
3. Verma, J.; Lal, S.; Van Noorden, C.J. Nanoparticles for hyperthermic therapy: Synthesis strategies and applications in glioblastoma. *Int. J. Nanomed.* **2014**, *10*, 2863–2877.
4. Yao, K.C.; Komata, T.; Kondo, Y.; Kanzawa, T.; Kondo, S.; Germano, I.M. Molecular response of human glioblastoma multiforme cells to ionizing radiation: Cell cycle arrest modulation of the expression of cyclin-dependent kinase inhibitors and autophagy. *J. Neurosurg.* **2003**, *98*, 378–384. [CrossRef] [PubMed]
5. Mamaeva, V.; Rosenholm, J.M.; Bate-Eya, L.T.; Bergman, L.; Peuhu, E.; Duchanoy, A.; Fortelius, L.E.; Landor, S.; Toivola, D.M.; Lindén, M.; et al. Mesoporous silica nanoparticles as drug delivery systems for targeted inhibition of Notch signaling in cancer. *Mol. Ther.* **2011**, *19*, 1538–1546. [CrossRef] [PubMed]
6. Petrache Voicu, S.N.; Dinu, D.; Sima, C.; Hermenea, A.; Ardelean, A.; Codrici, E.; Stan, M.S.; Zarnescu, O.; Dinischiotu, A. Silica Nanoparticles Induce Oxidative Stress and Autophagy but Not Apoptosis in the MRC-5 Cell Line. *Int. J. Mol. Sci.* **2015**, *16*, 29398–29416. [CrossRef] [PubMed]
7. Nakamura, T.; Sugihara, F.; Matsushita, H.; Yoshioka, Y.; Mizukami, S.; Kikuchi, K. Mesoporous silica nanoparticles for 19F magnetic resonance imaging fluorescence imaging and drug delivery. *Chem. Sci.* **2015**, *6*, 1986–1990. [CrossRef] [PubMed]
8. Wu, X.; Min, M.S.; Zao, J.X. Recent development of silica nanoparticles as delivery vectors from cancer imaging and therapy. *Nanomedicine* **2014**, *10*, 297–312. [CrossRef] [PubMed]
9. Kempen, P.J.; Greasley, S.; Parker, K.A.; Campbell, J.L.; Chang, H.Y.; Jones, J.R.; Sinclair, R.; Gambhir, S.S.; Jokerst, J.V. Theranostic mesoporous silica nanoparticles biodegrade after pro-survival drug delivery and ultrasound/magnetic resonance imaging of stem cells. *Theranostics* **2015**, *5*, 631–642. [CrossRef] [PubMed]
10. Kim, M.H.; Na, H.K.; Kim, Y.K.; Ryoo, S.R.; Cho, H.S.; Lee, K.E.; Jeon, H.; Ryoo, R.; Min, D.H. Facile synthesis of monodispersed mesoporous silica nanoparticles with ultra large pores and their application in gene delivery. *ACS Nano* **2011**, *5*, 3568–3576. [CrossRef] [PubMed]
11. Arap, W.; Pasqualini, S.; Montalti, M.; Petrizza, L.; Prodi, L.; Rampazzo, E.; Zaccheroni, N.; Marchio, S. Luminescent silica nanoparticles for cancer diagnosis. *Curr. Med. Chem.* **2013**, *20*, 2195–2211. [CrossRef] [PubMed]

12. Chen, M.; von Mikecz, A. Formation of nucleoplasmic protein aggregates impairs nuclear function in response to SiO₂ nanoparticles. *Exp. Cell Res.* **2005**, *305*, 51–62. [CrossRef] [PubMed]

13. Nabeshi, H.; Yoshikawa, T.; Matsuyama, K.; Nakazato, Y.; Tochigi, S.; Kondoh, S.; Hirai, T.; Akase, T.; Nagano, K.; Abe, Y.; et al. Amorphous nanosilica induce endocytosis-dependent ROS generation and DNA damage in human keratinocytes. *Part. Fibre Toxicol.* **2011**, *15*, 1. [CrossRef] [PubMed]

14. Kim, I.Y.; Joachim, E.; Choi, H.; Kim, K. Toxicity of silica nanoparticles depends on size dose and cell type. *Nanomedicine* **2015**, *11*, 1407–1416. [CrossRef] [PubMed]

15. Chen, M.; Singer, L.; Scharf, A.; von Mikecz, A. Nuclear polyglutamine-containing protein aggregates as active proteolytic centers. *J. Cell Biol.* **2008**, *180*, 697–704. [CrossRef] [PubMed]

16. Guo, C.; Xia, Y.; Niu, P.; Jiang, L.; Duan, J.; Yu, Y.; Zhou, X.; Li, Y.; Sun, Z. Silica nanoparticles induce oxidative stress, inflammation, and endothelial dysfunction in vitro viaactivation of the MAPK/Nrf2 pathway and nuclear factor-κB signaling. *Int. J. Nanomed.* **2015**, *20*, 1463–1477. [CrossRef] [PubMed]

17. Krętowski, R.; Stypułkowska, A.; Cechowska-Pasko, M. Efficient apoptosis and necrosis induction by proteasome inhibitor: Bortezomib in the DLD-1 human colon cancer cell line. *Mol. Cell. Biochem.* **2015**, *398*, 165–173. [CrossRef] [PubMed]

18. Tokgun, O.; Demiray, A.; Bulent, K.; Akca, H. Silica nanoparticles can induce apoptosis via dead receptor and caspase 8 pathway on A549 cells. *Adv. Food Sci.* **2015**, *2*, 65–70.

19. Fubini, B.; Hubbard, A. Reactive oxygen species (ROS) and reactive nitrogen species (RNS) generation by silica in inflammation and fibrosis. *Free Radic. Biol. Med.* **2003**, *34*, 1507–1516. [CrossRef]

20. Ahmad, J.; Ahamed, M.; Akhtar, M.J.; Alrokayan, S.A.; Siddiqui, M.A.; Musarrat, J.; Al-Khedhairy, A.A. Apoptosis induction by silica nanoparticles mediated through reactive oxygen species in human liver cell line HepG2. *Toxicol. Appl. Pharmacol.* **2012**, *259*, 160–168. [CrossRef] [PubMed]

21. Ahmed, M. Silica nanoparticles-induced cytotoxicity oxidative stress and apoptosis in cultured A431 and A549 cells. *Hum. Exp. Toxicol.* **2013**, *32*, 186–195. [CrossRef] [PubMed]

22. Joshi, G.N.; Knecht, D.A. Silica phagocytosis causes apoptosis and necrosis by different temporal and molecular pathways in alveolar macrophages. *Apoptosis* **2013**, *18*, 271–285. [CrossRef] [PubMed]

23. Corbalan, J.J.; Medina, C.; Jacoby, A.; Malinski, T.; Radomski, M.W. Amorphous silica nanoparticles trigger nitric oxide/peroxynitrite imbalance in human endothelial cells: Inflammatory and cytotoxic effects. *Int. J. Nanomed.* **2011**, *6*, 2821–2835.

24. Bauer, A.T.; Strozyk, E.A.; Gorzelanny, C.; Malinski, T.; Radomski, M.W. Cytotoxicity of silica nanoparticles through exocytosis of von Willebrand factor and necrotic cell death in primary human endothelial cells. *Biomaterials* **2011**, *32*, 8385–8393. [CrossRef] [PubMed]

25. Zabirnyk, O.; Yezhelyev, M.; Seleverstov, O. Nanoparticles as a novel class of autophagy activators. *Autophagy* **2007**, *3*, 278–281. [CrossRef] [PubMed]

26. Duan, J.; Yu, Y.; Yu, Y.; Li, Y.; Wang, J.; Geng, W.; Jiang, L.; Li, Q.; Zhou, X.; Sun, Z. Silica nanoparticles induce autophagy and endothelial dysfunction via the PI3K/Akt/mTOR signaling pathway. *Int. J. Nanomed.* **2014**, *5*, 5131–5141. [CrossRef] [PubMed]

27. Yu, Y.; Duan, J.; Yu, Y.; Li, Y.; Liu, X.; Zhou, X.; Ho, K.F.; Tian, L.; Sun, Z. Silica nanoparticles induce autophagy and autophagic cell death in HepG2 cells triggered by reactive oxygen species. *J. Hazard. Mater.* **2014**, *15*, 176–186. [CrossRef] [PubMed]

28. Guo, C.; Yang, M.; Jing, L.; Wang, J.; Yu, Y.; Li, Y.; Duan, J.; Zhou, X.; Li, Y.; Sun, Z. Amorphous silica nanoparticles trigger vascular endothelial cell injury through apoptosis and autophagy via reactive oxygen species-mediated MAPK/Bcl-2 and PI3K/Akt/mTOR signaling. *Int. J. Nanomed.* **2016**, *11*, 5257–5276. [CrossRef] [PubMed]

29. Ventresca, E.M.; Lecht, S.; Jakubowski, P.; Chiaverelli, R.A.; Weaver, M.; Del Valle, L.; Ettinger, K.; Gincberg, G.; Priel, A.; Braiman, A.; et al. Association of p75 (NTR) and α9β1 integrin modulates NGF dependent cellular responses. *Cell Signal.* **2015**, *27*, 1225–1236. [CrossRef] [PubMed]

30. Carmichael, J.; DeGraff, W.G.; Gazdar, A.F.; Minna, J.D.; Mitchell, J.B. Evaluation of a tetrazolium-based semiautomated colorimetric assay: Assessment of chemosensitivity testing. *Cancer Res.* **1987**, *47*, 936–942. [PubMed]

31. Kusaczuk, M.; Krętowski, R.; Bartoszewicz, M.; Cechowska-Pasko, M. Phenylbutyrate—A pan-HDAC inhibitor—Suppresses proliferation of glioblastoma LN-229 cell line. *Tumour Biol.* **2016**, *37*, 931–942. [CrossRef] [PubMed]

32. Kusaczuk, M.; Krętowski, R.; Stypułkowska, A.; Cechowska-Pasko, M. Molecular and cellular effects of a novel hydroxamate-based HDAC inhibitor—Belinostat—In glioblastoma cell lines: A preliminary report. *Investig. New Drugs* **2016**, *34*, 552–564. [CrossRef] [PubMed]

33. Alirezaei, M.; Fox, H.S.; Flynn, C.T.; Moore, C.S.; Hebb, A.L.; Frausto, R.F.; Bhan, V.; Kiosses, W.B.; Whitton, J.L.; Robertson, G.S.; et al. Elevated ATG5 expression in autoimmune demyelination and multiple sclerosis. *Autophagy* **2009**, *5*, 152–158. [CrossRef] [PubMed]

34. Pfaffl, M.W. A new mathematical model for relative quantification in real-time RT–PCR. *Nucleic Acids Res.* **2001**, *29*, 2002–2007. [CrossRef]

35. Laemmli, U.K. Cleavage of structural proteins during the assembly of the head of bacteriophage T4. *Nature* **1970**, *227*, 680–685. [CrossRef] [PubMed]

36. Smith, P.K.; Krohn, R.I.; Hermanson, G.T.; Mallia, A.K.; Gartner, F.H.; Provenzano, M.D.; Fujimoto, E.K.; Goeke, N.M.; Olson, B.J.; Klenk, D.C. Measurement of protein using bicinchoninic acid. *Anal. Biochem.* **1985**, *150*, 76–85. [CrossRef]

37. Amaralm, C.; Borges, M.; Melo, S.; da Silva, E.T.; Correia-da-Silva, G.; Teixeira, N. Apoptosis and Autophagy in Breast Cancer Cells following Exemestane Treatment. *PLoS ONE* **2012**, *7*, e42398.

38. De Oliveira, L.F.; Bouchmella, K.; Goncalves Kde, A.; Bettini, J.; Kobarg, J.; Cardoso, M.B. Functionalized Silica Nanoparticles As an Alternative Platform for Targeted Drug-Delivery of Water Insoluble Drugs. *Langmuir* **2016**, *32*, 3217–3225. [CrossRef] [PubMed]

39. Wang, P.; Keller, A.A. Natural and engineered nano and colloidal transport: Role of zeta potential in prediction of particle deposition. *Langmuir* **2009**, *25*, 6856–6862. [CrossRef] [PubMed]

40. Tang, W.; Yuan, Y.; Liu, Ch.; Wu, Y.; Lu, X.; Qian, J. Differential cytotoxicity and particle action of hydroxyapatite nanoparticles in human cancer cells. *Nanomedicine* **2014**, *9*, 397–412. [CrossRef] [PubMed]

41. Tang, L.; Yang, X.; Yin, Q.; Cai, K.; Wang, H.; Chaudhury, I.; Yao, C.; Zhou, Q.; Kwon, M.; Hartman, J.A.; et al. Investigating the optimal size of anticancer nanomedicine. *Proc. Natl. Acad. Sci. USA* **2014**, *28*, 15344–15349. [CrossRef] [PubMed]

42. Mu, Q.; Hondow, N.S.; Krzemiński, L.; Brown, A.P.; Jeuken, L.J.; Routledge, M.N. Mechanism of cellular uptake of genotoxic silica nanoparticles. *Part. Fibre Toxicol.* **2012**, *9*, 29. [CrossRef] [PubMed]

43. Zhou, M.; Xie, L.; Fang, C.-J.; Yang, H.; Wang, Y.-J.; Zhen, X.-Y.; Yan, C.-H.; Wang, Y.; Zhao, M.; Peng, S. Implications for blood-brain-barrier permeability in vitro oxidative stress and neurotoxicity potential induced by mesoporous silica nanoparticles: Effects of surface modification. *RCS Adv.* **2016**, *6*, 2800–2890. [CrossRef]

44. Ye, Y.; Liu, J.; Xu, J.; Sun, L.; Chen, M.; Lan, M. Nano-SiO$_2$ induces apoptosis via activation of p53 and Bax mediated by oxidative stress in human hepatic cell line. *Toxicol. In Vitro* **2010**, *24*, 751–758. [CrossRef] [PubMed]

45. Kim, J.E.; Kim, H.; An, S.S.; Maeng, E.H.; Kim, M.K.; Song, Y.J. In vitro cytotoxicity of SiO$_2$ or ZnO nanoparticles with different sizes and surface charges on U373MG human glioblastoma cells. *Int. J. Nanomed.* **2014**, *15*, 235–241.

46. Napierska, D.; Thomassen, L.C.; Lison, D.; Martens, J.A.; Hoet, P.H. The nanosilica hazard: Another variable entity. *Part. Fibre Toxicol.* **2010**, *7*, 30–39. [CrossRef] [PubMed]

47. Lin, W.; Huang, Y.W.; Zhou, X.D.; Ma, Z. In vitro toxicity of silica nanoparticles in human lung cancer cells. *Toxicol. Appl. Pharmacol.* **2006**, *217*, 252–259. [CrossRef] [PubMed]

48. Chang, J.S.; Chang, K.L.; Hwang, D.F.; Kong, Y.L. In vitro cytotoxicitiy of silica nanoparticles at high concentrations strongly depends on the metabolic activity type of the cell line. *Environ. Sci. Technol.* **2007**, *41*, 2064–2068. [CrossRef] [PubMed]

49. Lu, X.; Qian, J.; Zhou, H.; Gan, Q.; Tang, W.; Lu, J.; Yuan, Y.; Liu, C. In vitro cytotoxicity and induction of apoptosis by silica nanoparticles in human HepG2 hepatoma cells. *Int. J. Nanomed.* **2011**, *6*, 1889–1901.

50. Li, Y.; Monteiro-Riviere, N.A. Mechanisms of cell uptake inflammatory potential and protein corona effects with gold nanoparticles. *Nanomedicine* **2016**, *24*, 3185–3203. [CrossRef] [PubMed]

51. Sikora, A.; Shard, A.G.; Minelli, C. Size and ζ-Potential Measurement of Silica Nanoparticles in Serum Using Tunable Resistive Pulse Sensing. *Langmuir* **2016**, *32*, 2216–2224. [CrossRef] [PubMed]

52. Moore, T.L.; Rodriguez-Lorenzo, L.; Hirsch, V.; Balog, S.; Urban, D.; Jud, C.; Rothen-Rutishauser, B.; Lattuada, M.; Petri-Fink, A. Nanoparticle colloidal stability in cell culture media and impact on cellular interactions. *Chem. Soc. Rev.* **2015**, *44*, 6287–6305. [CrossRef] [PubMed]

53. Wittmaack, K. In search of the most relevant parameter for quantifying lung inflammatory response to nanoparticle exposure: Particle number surface area or what? *Environ. Health Perspect.* **2007**, *115*, 187–194. [CrossRef] [PubMed]

54. Mytych, J.; Wnuk, M. Nanoparticle Technology as a Double-Edged Sword: Cytotoxic Genotoxic and Epigenetic Effects on Living Cells. *J. Biomater. Nanobiotechnol.* **2013**, *4*, 53–63.

55. Bellezza, I.; Scarpelli, P.; Pizzo, S.; Grottelli, S.; Costanzi, E.; Minelli, A. ROS-independent Nrf2 activation in prostate cancer. *Oncotarget* **2017**. Available online: https://www.researchgate.net/publication/317381564 (accessed on 7 June 2017).

56. Oguz, S.; Kanter, M.; Erboga, M.; Toydemir, T.; Sayhan, M.B.; Onur, H. Effects of Urtica dioica on oxidative stress proliferation and apoptosis after partial hepatectomy in rats. *Toxicol. Ind. Health* **2015**, *31*, 475–484. [CrossRef] [PubMed]

57. Sun, L.; Li, Y.; Liu, X.; Jin, M.; Zhang, L.; Du, Z.; Guo, C.; Huang, P.; Sun, Z. Cytotoxicity and mitochondrial damage caused by silica nanoparticles. *Toxicol. In Vitro* **2011**, *25*, 1619–1629. [CrossRef] [PubMed]

58. Ahamed, M.; Alhadlaq, H.A.; Ahmad, J.; Siddiqui, M.A.; Khan, S.T.; Musarrat, J.; Al-Khedhairy, A.A. Comparative cytotoxicity of dolomite nanoparticles in human larynx HEp2 and liver HepG2 cells. *J. Appl. Toxicol.* **2015**, *35*, 640–650. [CrossRef] [PubMed]

59. Gao, J.; Sana, R.; Calder, V.; Calonge, M.; Lee, W.; Wheeler, L.A.; Stern, M.E. Mitochondrial permeability transition pore in inflammatory apoptosis of human conjunctival epithelial cells and T cells: Effect of cyclosporin A. *Investig. Ophthalmol. Vis. Sci.* **2013**, *54*, 4717–4733. [CrossRef] [PubMed]

60. Sun, L.; Wang, H.; Wang, Z.; He, S.; Chen, S.; Liao, D.; Wang, L.; Yan, J.; Liu, W.; Lei, X.; et al. Mixed lineage kinase domain-like protein mediates necrosis signaling downstream of RIP3 kinase. *Cell* **2012**, *148*, 213–227. [CrossRef] [PubMed]

61. Nowak, J.S.; Mehn, D.; Nativo, P.; García, C.P.; Gioria, S.; Ojea-Jiménez, I.; Gilliland, D.; Rossi, F. Silica nanoparticle uptake induces survival mechanism in A549 cells by the activation of autophagy but not apoptosis. *Toxicol. Lett.* **2014**, *224*, 84–92. [CrossRef] [PubMed]

62. Krętowski, R.; Borzym-Kluczyk, M.; Stypułkowska, A.; Cechowska-Pasko, M. Low glucose dependent decrease of apoptosis and induction of autophagy in breast cancer MCF-7 cells. *Mol. Cell. Biochem.* **2016**, *417*, 35–47. [CrossRef] [PubMed]

63. Eisenberg-Lerner, A.; Bialik, S.; Simon, H.U.; Kimchi, A. Life and death partners: Apoptosis autophagy and the cross talk between them. *Cell Death Differ.* **2009**, *16*, 966–975. [CrossRef] [PubMed]

64. Gump, J.M.; Thorburn, A. Autophagy and apoptosis: What is the connection? *Trends Cell Biol.* **2011**, *21*, 387–392. [CrossRef] [PubMed]

65. Shimizu, S.; Yoshida, T.; Tsujioka, M.; Arakawa, S. Autophagic cell death and cancer. *Int. J. Mol. Sci.* **2014**, *2*, 3145–3153. [CrossRef] [PubMed]

© 2017 by the authors. Licensee MDPI, Basel, Switzerland. This article is an open access article distributed under the terms and conditions of the Creative Commons Attribution (CC BY) license (http://creativecommons.org/licenses/by/4.0/).

*nanomaterials*

MDPI

*Article*

# Optimized Photodynamic Therapy with Multifunctional Cobalt Magnetic Nanoparticles

Kyong-Hoon Choi [1], Ki Chang Nam [2], Un-Ho Kim [3], Guangsup Cho [1], Jin-Seung Jung [3,*] and Bong Joo Park [1,*]

[1] Department of Electrical & Biological Physics, Kwangwoon University, Nowon-gu, Seoul 139-701, Korea; solidchem@hanmail.net (K.-H.C.); gscho@kw.ac.kr (G.C.)

[2] Department of Medical Engineering, Dongguk University College of Medicine, Gyeonggi-do 10326, Korea; kichang.nam@gmail.com

[3] Department of Chemistry, Gangneung-Wonju National University, Gangneung 210-702, Korea; rladydanr06@naver.com

* Correspondence: jjscm@gwnu.ac.kr (J.-S.J.); parkbj@kw.ac.kr (B.J.P.); Tel.: +82-33-640-2305 (J.-S.J.); +82-2-940-8629 (B.J.P.)

Received: 2 May 2017; Accepted: 7 June 2017; Published: 10 June 2017

**Abstract:** Photodynamic therapy (PDT) has been adopted as a minimally invasive approach for the localized treatment of superficial tumors, representing an improvement in the care of cancer patients. To improve the efficacy of PDT, it is important to first select an optimized nanocarrier and determine the influence of light parameters on the photosensitizing agent. In particular, much more knowledge concerning the importance of fluence and exposure time is required to gain a better understanding of the photodynamic efficacy. In the present study, we synthesized novel folic acid-(FA) and hematoporphyrin (HP)-conjugated multifunctional magnetic nanoparticles ($CoFe_2O_4$-HPs-FAs), which were characterized as effective anticancer reagents for PDT, and evaluated the influence of incubation time and light exposure time on the photodynamic anticancer activities of $CoFe_2O_4$-HPs-FAs in prostate cancer cells (PC-3 cells). The results indicated that the same fluence at different exposure times resulted in changes in the anticancer activities on PC-3 cells as well as in reactive oxygen species formation. In addition, an increase of the fluence showed an improvement for cell photo-inactivation. Therefore, we have established optimized conditions for new multifunctional magnetic nanoparticles with direct application for improving PDT for cancer patients.

**Keywords:** photodynamic therapy; optimized nano-carrier; multifunctional magnetic nanoparticle; fluence; anticancer activity; prostate cancer cell

---

## 1. Introduction

Over the last few decades, photosensitizer (PS)-mediated photodynamic therapy (PDT) has been introduced as a possible alternative non-invasive localized therapeutic modality for treating cancer as well as cardiovascular, ophthalmic, dermatological, and dental diseases [1–7]. PDT is a two-step procedure that involves the administration of a photosensitizing agent [8], followed by activation of the drug with non-thermal light of a specific wavelength [9]. In particular, this photodynamic process rapidly generates reactive oxygen species (ROS) including peroxides, hydroxyl radicals, superoxide ions, and singlet oxygen, with the latter implicated as the major causative agent of cellular damage in the photodynamic process [10]. However, the results of recent clinical and preclinical studies of PDT indicate that this process still suffers from disadvantages such as the wavelength-dependent tissue penetration depth of the light; inefficient delivery of PS to the target area; loss of PDT efficacy owing to PS aggregation, degradation, or reduction; and toxicity of the PS [11–13].

Several approaches have been proposed to enhance the efficacy of PDT. In some cases, PDT efficacy was found to be significantly improved when nanoparticles were applied as PS carriers, suggesting that the use of nanoparticles can help to overcome the aforementioned limitations [14–16]. Among the various nanoparticles available, such as liposomal vesicles, quantum dots, nanotubes, and gold nanoparticles, the latter have attracted substantial attention because of their chemical inertness, excellent optical properties, and minimal biological toxicity [17,18]. Recently, new synergistic treatment modalities that combine PDT with hyperthermia by using Au nanocomposites have shown the potential to overcome the current limitations of PDT and enhance anticancer efficacy [19–21]. However, the Au nanocomposites must overcome many disadvantages, including higher cost, low conjugation efficiency on the surface of particles, and lack of bio-imaging capability. To improve PDT efficacy, it is also important to understand the photophysical and photochemical properties of as-prepared photosensitizing agents. In particular, the illumination parameters might play an important role in determining PDT efficacy.

Herein, we report the development of new multifunctional magnetic nanoparticles conjugated with hematoporphyrin (HP) and folic acid (FA) ($CoFe_2O_4$-HPs-FAs) for use as potential PDT agents, which were tested by targeting prostate cancer PC-3 cells with FA. The biocompatibility and photodynamic anticancer activity of the $CoFe_2O_4$-HPs-FAs were evaluated in vitro. In addition, we evaluated the effect of variations in the fluence and exposure time on the outcome of the photodynamic anticancer activity of $CoFe_2O_4$-HPs-FAs in PC-3 cells to corroborate the importance of optimizing the irradiation parameters.

## 2. Results and Discussion

### 2.1. Characteristics of Multifunctional $CoFe_2O_4$-HPs-Fas

As illustrated in Scheme 1, novel multifunctional magnetic nanoparticles ($CoFe_2O_4$-HPs-FAs) were prepared by simple surface modification of magnetic nanoparticles with a photosensitizer, HP, and a targeting molecule, FA. First, two carboxyl terminal groups of HP are chemically bonded to metal cations on the surface of the $CoFe_2O_4$ nanoparticles via esterification reaction. Similarly, the FA molecules were introduced to the surface of the $CoFe_2O_4$ nanoparticles to improve the targeting ability.

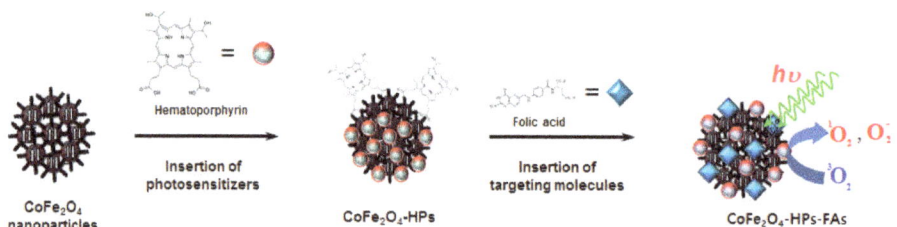

**Scheme 1.** Fabrication procedure for the multifunctional magnetic nanoparticles.

The morphology and particle size of the as-prepared $CoFe_2O_4$ nanoparticles were characterized by transmission electron microscopy (TEM; JEOL, JEM-2100F) and scanning electron microscopy (SEM; Hitachi, SU-70), as shown in Figure 1a,b, respectively. The SEM and TEM images showed that these nanoparticles composed of irregular nanograins are spherical and have a diameter of approximately 70 nm with a rough surface. In addition, the sizes of these nanoparticles were quite uniform. The high-resolution TEM image on the edge of a nanoparticle indicated that the distance between two neighboring planes was 0.269 nm at (220), which is in good agreement with the (220) plane of the spinel $CoFe_2O_4$, as shown in the inset of Figure 1b. Figure 1c shows a histogram of the distribution of the nanoparticles size with a Gaussian fit curve (solid line); the particle size ranged from 45 to 85 nm, and the average particle size ($D_{SEM}$), defined as the size at the peak of the Gaussian-fitting curve, was 69.2 nm. These results indicated that our CoFe$_2$O$_4$ nanoparticles were well dispersed and had a narrow size distribution.

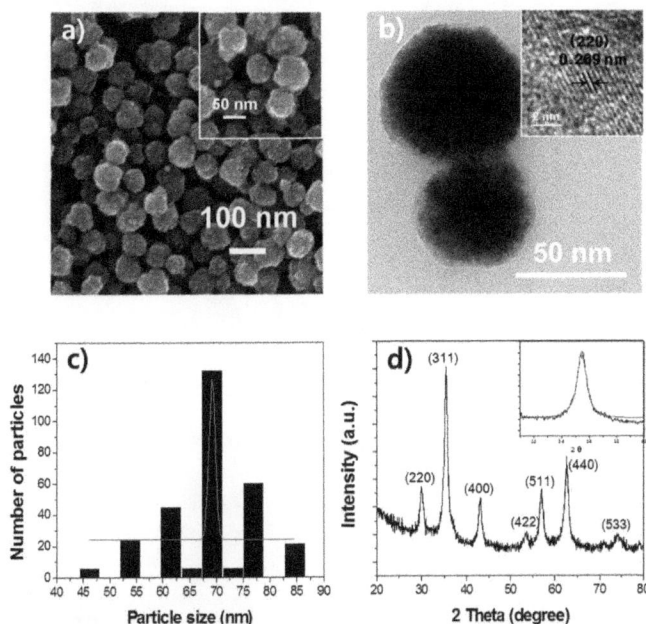

**Figure 1.** Morphology and crystal structure of the CoFe$_2$O$_4$ nanoparticle. (**a**) Field-emission scanning electron microscopy image and (**b**) transmission electron microscopy micrographs of the CoFe$_2$O$_4$ nanoparticle; (**c**) Histogram for the particle size distribution of the CoFe$_2$O$_4$ nanoparticles; (**d**) X-ray diffraction pattern of the CoFe$_2$O$_4$ nanoparticles.

The structure and phase purity of the nanoparticles were confirmed by analysis of the X-ray diffraction (XRD; PANalytical, X'Pert Pro MPD) patterns and the results are presented in Figure 1d. The diffraction peaks matched well with the characteristic peaks of the cubic spinel-type lattice of CoFe$_2$O$_4$, which in turn is well matched to the standard XRD pattern (JCPDS Card No. 22-1086). The peaks observed at 30.1°, 35.5°, 43.1°, 53.6°, 57.1°, 62.7°, and 74.2° can be assigned to the (220), (311), (400), (422), (511), (440), and (533) planes of spinel CoFe$_2$O$_4$, respectively. This result indicates that the obtained high-purity CoFe$_2$O$_4$ nanoparticles have good crystallinity. The average crystallite size of the CoFe$_2$O$_4$ nanograin was estimated to be approximately 9.25 nm via X-ray line broadening using Scherrer's equation.

The CoFe$_2$O$_4$ nanoparticles and CoFe$_2$O$_4$-HPs-FAs showed good magnetic properties. Figure 2a presents the room-temperature hysteresis loop as a function of the applied magnetic field, or the M versus H curve. The magnetization curves of both samples exhibited no hysteresis, and no coercivity was reached, even at the highest magnetic field applied. This indicates that both magnetic particles show superparamagnetic behavior. The CoFe$_2$O$_4$ nanoparticles showed a high-saturation magnetization value of 67.3 emu/g, whereas the high-saturation value of the surface-modified CoFe$_2$O$_4$-HPs-FAs was lower at 39.7 emu/g. The difference in the saturation values is attributed to the diamagnetic contribution of the diamagnetic organic molecules that are chemically bonded to the nanoparticle surface.

From the photoluminescence and photoluminescence excitation spectra shown in Figure 2b, the HP solution showed excitation peaks at 401 (Soret band), 500, 532, and 574 nm (Q band), and the CoFe$_2$O$_4$-HPs-FAs solution showed the same characteristic peaks. No significant shift in the excitation wavelength was observed in comparison to the dissolved CoFe$_2$O$_4$-HPs-FAs, suggesting that the HP molecules, as a PS, remained stable after conjugation to the nanoparticles. At the excitation wavelength of 400 nm, the pure HP produced two strong emission peaks located at 631 nm and 696 nm, respectively, and the CoFe$_2$O$_4$-HPs-FAs exhibited slightly blue-shifted peaks at 628 nm and 694 nm. The blue-shifted emission peaks are attributed to the strong bonding between HP and the magnetic CoFe$_2$O$_4$ nanoparticles.

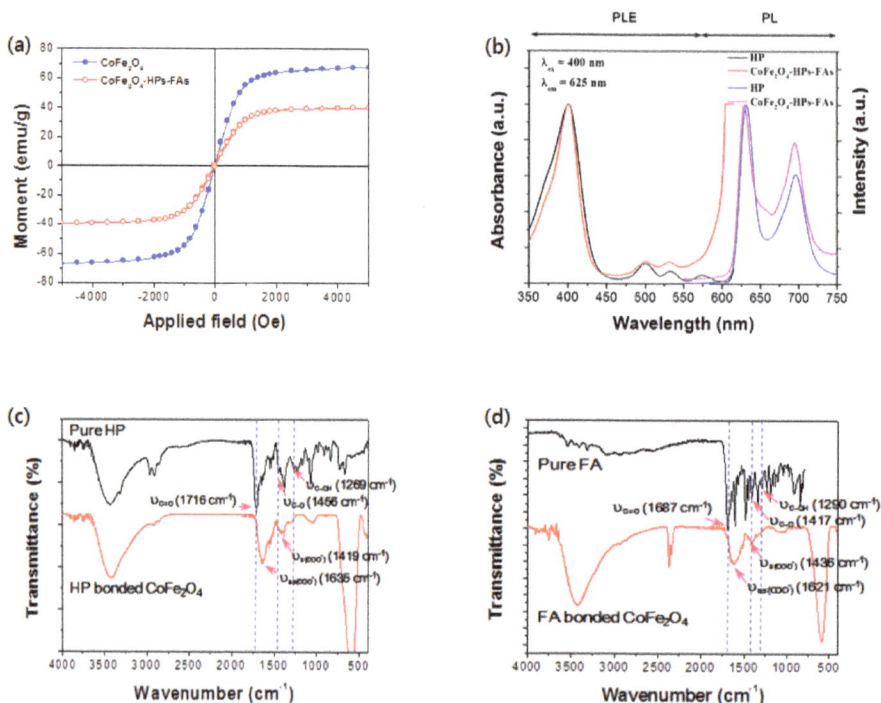

**Figure 2.** Photophysical and magnetic properties of multifunctional magnetic nanoparticles. (a) Room-temperature magnetic hysteresis loops of the CoFe$_2$O$_4$ nanoparticles and the CoFe$_2$O$_4$-HPs-FAs; (b) photoluminescence and photoluminescence excitation spectra of pure HP and CoFe$_2$O$_4$-HPs-FAs in THF; FT-IR spectra of (c) pure HP and HP bound with CoFe$_2$O$_4$ and of (d) pure FA and FA bonded with CoFe$_2$O$_4$.

To confirm the formation of the metal-organic complex, the Fourier Transform InfraRed (FT-IR) spectra of pure HP, HP-coated CoFe$_2$O$_4$, and FA-coated CoFe$_2$O$_4$ nanoparticles were compared. As shown in Figure 2c,d, the absorption peaks were mainly detected in the fingerprint region. Before complex formation, the IR spectra of pure HP and pure FA exhibited a peak in the range of 1687–1716 cm$^{-1}$, indicating the presence of a C=O stretching band of the -COOH groups. In addition, coupled vibrations involving C-O stretching and the O-H deformation ($\upsilon_{C-OH}$) were observed in the range of 1417–1456 cm$^{-1}$ and 1269–1290 cm$^{-1}$, respectively. These results indicate that the pure HP and FA molecules have protonated carboxyl groups (-COOH), as previously described [22]. After the carboxyl acid was converted to the complexes, the IR spectra of HP-coated CoFe$_2$O$_4$ and FA-coated CoFe$_2$O$_4$ nanoparticles showed that the absorption bands of the protonated carboxyl groups significantly changed. Three absorption bands corresponding to the stretching vibrations of the C=O group, (C-O), and $\upsilon_{C-OH}$ of the -COOH group at 1260–1720 cm$^{-1}$ disappeared, whereas the bands assigned to asymmetric vibrations $\upsilon_{as}$(COO), at 1621–1635 cm$^{-1}$, and symmetric vibrations

$v_s(COO)$, at 1419–1436 cm$^{-1}$, appeared. These spectral changes can also be caused by the formation of cation–carboxylate complexes owing to covalent chemical bonding, as described previously [22,23].

The loading capacity with HP molecules of the multifunctional CoFe$_2$O$_4$-HPs-FAs was determined by UV–Vis spectroscopy (Ultraviolet–visible spectroscopy). From the calculated results, when the CoFe$_2$O$_4$ nanoparticle weights varied at 1.56, 3.13, 6.25, 12.5, and 25 µg, the weights of the HP molecules bonded to the surfaces of the CoFe$_2$O$_4$ nanoparticles were 0.2, 0.4, 0.8, 1.60, and 3.22 µg, respectively. Similarly, the concentrations of the FA molecules bonded to the surfaces of the CoFe$_2$O$_4$ nanoparticles were 0.09, 0.17, 0.35, 0.69, and 1.38 µg according to the weights of the CoFe$_2$O$_4$ nanoparticles of 1.56, 3.13, 6.25, 12.5, and 25 µg, respectively.

## 2.2. Singlet Oxygen Generation

In a PDT process, absorption of light by PSs eventually results in the generation of singlet oxygen and other ROS. Singlet oxygen is the major cytotoxic species leading to cell death through the so-called type II mechanism [24,25]. To evaluate the capability of $^1O_2$ generation of CoFe$_2$O$_4$-HPs-FAs, 1,3-diphenylisobenzofuran (DPBF) was employed as a probe molecule. Figure 3 shows the extensive bleaching of DPBF as a function of time (amplitude reduction of spectral features at 424 nm) when incubated with CoFe$_2$O$_4$-HPs-FAs in THF and irradiated with a Xe lamp. Control experiments with only DPBF using the same excitation wavelength showed no bleaching. Therefore, the multifunctional magnetic nanoparticles could be a very important PDT reagent.

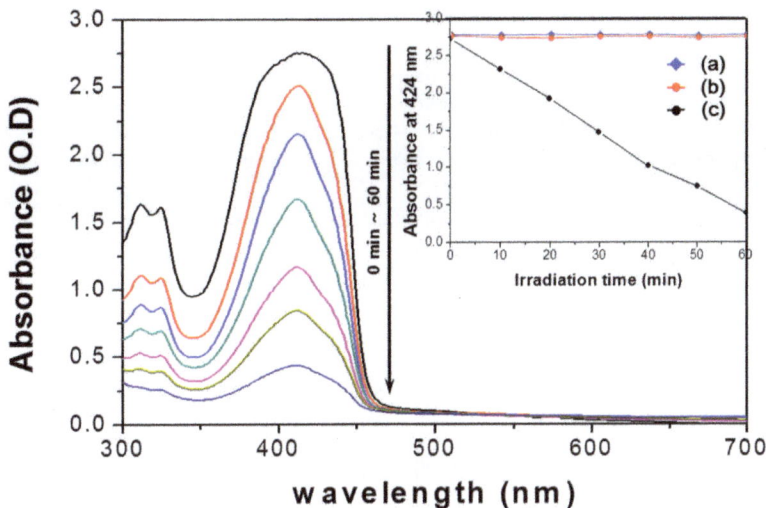

**Figure 3.** UV–Vis spectra of DPBF according to irradiation time in THF solution with the CoFe$_2$O$_4$-HPs-FAs under a Xe lamp. The inset presents the absorption (OD) of DPBF in THF at 424 nm as a function of irradiation time. (**a**) DPBF only plus light; (**b**) DPBF with the CoFe$_2$O$_4$-HPs-FAs without light; (**c**) DPBF with the CoFe$_2$O$_4$-HPs-FAs plus light.

## 2.3. Biocompatibility of Multifunctional CoFe$_2$O$_4$-HPs-Fas

As superparamagnetic CoFe$_2$O$_4$ nanoparticles are good T$_2$-type (negative) contrast agents in MRI, and FA and HP are biocompatible cancer-targeting and therapeutic agents, the anti-cancer effect of CoFe$_2$O$_4$-HPs-FAs was investigated by evaluating the MR signal-enhancing property. With increasing concentrations of CoFe$_2$O$_4$-HPs-FAs in the cells, the MR signal was significantly enhanced (negative in brightness in the T$_2$-weighted image) in vitro (Figure 4a). These results indicate that the nanoparticles can generate high magnetic-field gradients near the surface of the

CoFe$_2$O$_4$-HPs-FAs. Additionally, the relaxivity r$_2$ (1/T$_2$) increases linearly under these conditions (Figure 4b), indicating that the CoFe$_2$O$_4$-HPs-FAs generated MRI contrasts on T$_2$-weighted spin-echo sequences. Transverse relaxivity r$_2$ values were determined from the slope of the linear fit to the data points in 1/T$_2$ vs. the CoFe$_2$O$_4$-HPs-FAs concentration plot. The r$_2$ value obtained for CoFe$_2$O$_4$-HPs-FAs was 177.3 mM$^{-1}$s$^{-1}$. As shown in Figure 4a,b, the T$_2$-weighted phantom images of the CoFe$_2$O$_4$-HPs-FAs exhibited a significant negative dose-dependent contrast enhancement, suggesting that these nanoparticles are promising for theragnostic purposes.

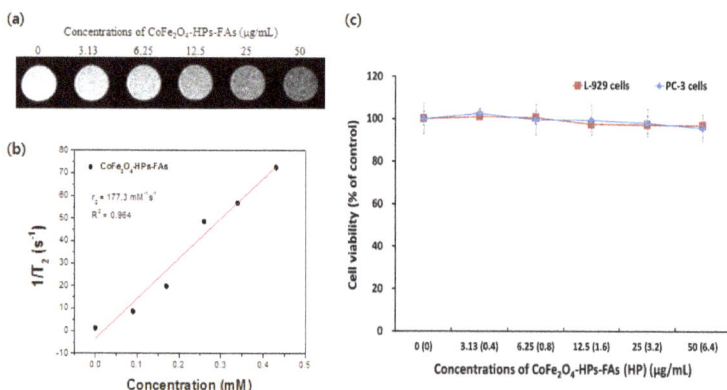

**Figure 4.** T$_2$-weighted MR imaging and biocompatibility of CoFe$_2$O$_4$-HPs-FAs. (a) T$_2$-weighted MR images of prostate cancer cells (PC-3 cells) treated with CoFe$_2$O$_4$-HPs-Fas; (b) Plot of T$_2$ relaxation rate r$_2$ (1/T$_2$) for CoFe$_2$O$_4$-HPs-Fas; (c) Cytotoxicity of CoFe$_2$O$_4$-HPs-FAs (60 nm) in fibroblasts (L-929 cells) and prostate cancer cells (PC-3 cells). Data are expressed as the mean ± standard deviation (*n* = 6).

To evaluate the biocompatibility of the CoFe$_2$O$_4$-HPs-FAs, cytotoxicity tests were carried out with fibroblasts (L-929 cell) and prostate cancer cells (PC-3 cells) using a method recommended by the International Organization for Standardization (ISO 10993-5) [26]. As shown in Figure 4b, the viability of both cell types was not decreased when incubated with CoFe$_2$O$_4$-HPs-FAs as compared to the untreated control cells, and cell viabilities at each concentration of CoFe$_2$O$_4$-HPs-FAs were more than 95%, indicating that the CoFe$_2$O$_4$-HPs-FAs have no cytotoxicity in L-929 and PC-3 cells. Collectively, these results demonstrate that CoFe$_2$O$_4$-HPs-FAs have good biocompatibility and can be used for clinical cancer therapy.

## 2.4. Optimization of the Cellular Uptake and Light Irradiation Time of CoFe$_2$O$_4$-HPs-Fas

Cellular uptake and the intracellular distribution of the CoFe$_2$O$_4$-HPs-FAs are the most important factors for their anticancer efficacy by PDT. Therefore, we carried out cell staining with the Prussian blue staining method and TEM analysis after incubating PC-3 prostate cancer cells with the CoFe$_2$O$_4$-HPs-FAs for 1, 2, and 4 h to confirm the optimal cellular uptake time and intracellular distribution. As shown in Figure 5a, incubation time had a substantial effect on the cellular uptake of the CoFe$_2$O$_4$-HPs-FAs. The number of CoFe$_2$O$_4$-HPs-FAs in the cells was proportional to the incubation time and the accumulated CoFe$_2$O$_4$-HPs-FAs in PC-3 cells appeared to be located in the cytosol. As shown in Figure 5b, the TEM images also clearly demonstrated that most of the CoFe$_2$O$_4$-HPs-FAs were located in the cytoplasm, and the number of CoFe$_2$O$_4$-HPs-FAs in the cytoplasm was also increased depending on the incubation time with cells.

To further evaluate the optimal cellular uptake time of the CoFe$_2$O$_4$-HPs-FAs in prostate cancer cells, the PC-3 cells were incubated with the CoFe$_2$O$_4$-HPs-FAs for 1, 2, and 4 h, and each cell was irradiated with LED light at a dose of 18.36 J/cm$^2$ to confirm the anticancer activity of the

CoFe$_2$O$_4$-HPs-FAs depending on the incubation time. As shown in Figure 5c, the cell viabilities of PC-3 cells were decreased in a dose-dependent manner, regardless of the incubation time of the CoFe$_2$O$_4$-HPs-FAs with PC-3 cells. The cell viability with 1 h incubation was 100, 74, 53.6, 47.6, and 37.1% with increasing CoFe$_2$O$_4$-HPs-FA concentrations, respectively. However, the number of viable cells significantly decreased at 2 and 4 h incubation with increasing doses of the CoFe$_2$O$_4$-HPs-FAs, from 100, 33.6, 9.3, 3.4, and 0.4% for 2 h and from 100, 34.6, 9.6, 8.9, and 5.8% for 4 h compared to control levels. These results suggested that an increased incubation time—i.e., 2 and 4 h—resulted in significantly better photo-killing efficacy of CoFe$_2$O$_4$-HPs-FAs in PC-3 cells compared with a 1-h incubation time. Moreover, the photodynamic anticancer activity at 2 h of incubation was higher than that at 4 h of incubation at high concentrations (12.5 (1.60) and 25 (3.22) μg/mL) of CoFe$_2$O$_4$-HPs-FAs (HPs). Specifically, the photo-killing efficacy of 12.5 (1.60) and 25 (3.22) μg/mL CoFe$_2$O$_4$-HPs-FAs (HPs) ranged from over 96% ($p < 0.005$) to almost 100%. These results confirmed a close correlation between cellular uptake time and anticancer efficacy by PDT, although there was no difference in the photo-killing efficacy between 2 h and 4 h of incubation. Therefore, we selected 2 h as the optimal incubation time for the subsequent photodynamic anticancer activity test of the CoFe$_2$O$_4$-HPs-FAs.

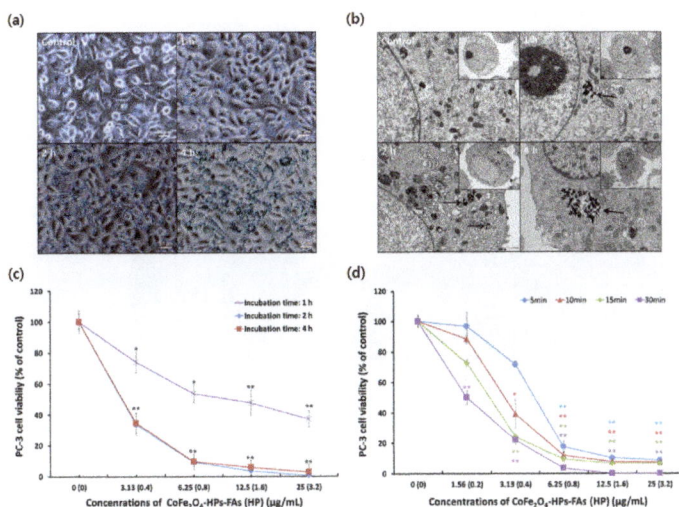

**Figure 5.** Cellular uptake, intracellular localization, and photodynamic anticancer activities of CoFe$_2$O$_4$-HPs-FAs in prostate cancer cells (PC-3 cells). (**a**) Microscopic and (**b**) transmission electron microscopic images of CoFe$_2$O$_4$-HPs-FAs in PC-3 cells to evaluate their cellular uptake and intracellular localization. PC-3 cells treated with 6.25 (0.8) μg/mL CoFe$_2$O$_4$-HPs-FAs (HPs) were incubated for 1, 2, and 4 h in the dark. The TEM images are magnified from a whole cell image (inset). Black arrows indicate the CoFe$_2$O$_4$-HPs-FAs. The scale bars represent 50 μm and 2 μm; (**c**) Photodynamic anticancer activity of CoFe$_2$O$_4$-HPs-FAs according to the incubation time of CoFe$_2$O$_4$-HPs-FAs with prostate cancer cells (PC-3 cells); (**d**) Photodynamic anticancer activity of CoFe$_2$O$_4$-HPs-FAs according to the exposure dose of light emitting diode (LED) light to PC-3 cells. Data are expressed as the mean ± standard deviation ($n = 6$) and were analyzed by Student's $t$-tests. Statistical significance was defined as $p < 0.05$ (* $p < 0.05$, ** $p < 0.005$ vs. control at the same time).

## 2.5. Anticancer Activity of CoFe$_2$O$_4$-HPs-Fas

To confirm the photodynamic anticancer activity according to the exposure dose of LED light exposed to PC-3 cells, the cells were incubated with various concentrations of the CoFe$_2$O$_4$-HPs-FAs and irradiated by LED light at doses of 3.06, 6.12, 9.18, and 18.36 J/cm$^2$ after incubation for 2 h. The photo-killing efficacy was also quantified using the Cell Counting Kit-8 (CCK-8) method, as shown in Figure 5d. The photodynamic anticancer activities by LED irradiation were significantly increased under each dose, even at the lowest dose of 3.06 J/cm$^2$, and cell viabilities dramatically decreased with increased concentrations of the CoFe$_2$O$_4$-HPs-FAs. The photo-killing efficacies of the CoFe$_2$O$_4$-HPs-FAs were markedly increased in a dose-dependent manner. These results demonstrated a close correlation between exposure dose of light and dose of the CoFe$_2$O$_4$-HPs-FAs on the photo-killing efficacy.

Finally, to confirm the anticancer mechanism by PDT in PC-3 cells, we conducted morphological analysis to evaluate the rate of apoptotic cell death induced by LED irradiation after incubation with the CoFe$_2$O$_4$-HPs-FAs for 2 h using an Annexin V-fluorescein isothiocyanate (FITC) apoptosis detection kit and Hoechst 33342 fluorescence dye. Annexin V is an intracellular protein that binds to phosphatidylserine. Phosphatidylserine is normally only found on the intracellular face of the plasma membrane in healthy cells, whereas during the early stage of apoptosis, membrane asymmetry is lost and phosphatidylserine translocates to an external site. Therefore, FITC-labeled Annexin V can be used to specifically target and identify apoptotic cells.

Figure 6a shows the stained images of normal and apoptotic cells at 2 h post-irradiation. In the control (untreated), cells stained by Annexin V-FITC and propidium iodide (PI) were not detected, whereas the cells treated with the CoFe$_2$O$_4$-HPs-FAs were stained green by Annexin V-FITC and red by PI. Cells showing only green fluorescence indicated early-stage apoptotic cells for cell membrane translocation, whereas double-stained cells in green and red indicated late-stage apoptotic cells. The results demonstrated that LED irradiation after treatment of the CoFe$_2$O$_4$-HPs-FAs to cancer cells is mediated by the induction of cell death, and the majority of cell death was mediated via apoptosis.

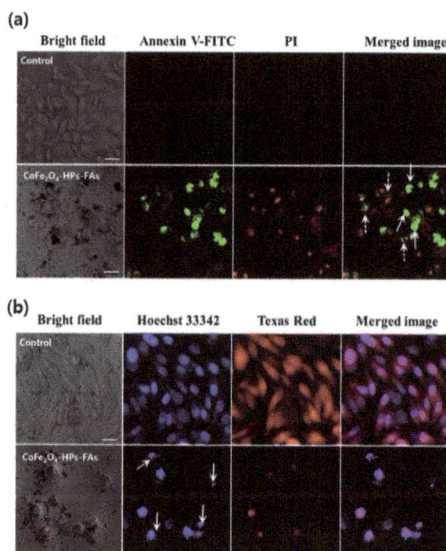

**Figure 6.** Confocal fluorescence images for cell membrane translocation and nuclear fragmentation in apoptotic PC-3 cells. (**a**) Confocal fluorescence images for cell membrane translocation stained with Annexin V-FITC (green) for the cell membrane and propidium iodide (PI, red) for the nucleus in PC-3 cells at 6 h post-irradiation. Cells stained green are apoptotic cells, and solid line and dashed arrows indicate early- and late-stage apoptotic cells, respectively; (**b**) Confocal fluorescence images for nuclear fragmentation (solid line arrows) stained with Hoechst 33342 (blue) for the nucleus and with Texas Red (red) for whole cells in PC-3 cells at 6 h post-irradiation. Apoptotic cell death was induced by LED irradiation at an irradiation dose of 18.36 J/cm$^2$ after treatment with 6.25 (0.8) µg/mL CoFe$_2$O$_4$-HPs-FAs (HPs) for 2 h. The scale bar represents 25 µm.

Moreover, we evaluated the extent of nucleus fragmentation in PC-3 cells with Hoechst 33342 dye. As shown in Figure 6b, the cells irradiated with LED light after treatment of the $CoFe_2O_4$-HPs-FAs for 2 h were quickly condensed and some showed a granular nucleus body, which was not detected in the control cells.

These results are consistent with the results of photodynamic anticancer efficacy shown in Figure 5c,d, further indicating that the photodynamic anticancer effects may be induced via apoptosis.

## 3. Materials and Methods

### 3.1. Synthesis of Multifunctional CoFe₂O₄-HPs-Fas

The $CoFe_2O_4$ magnetic nanoparticles were prepared by applying a similar method as that described in the previous report [27]. In brief, 0.36 g $FeCl_3 \cdot 6H_2O$, 0.11 g $CoCl_2 \cdot 2H_2O$, and 1.5 g NaOAc were dissolved in ethylene glycol/diethylene glycol mixture solvent (5:15, $v/v$), and then this mixture was vigorously stirred for 30 min. The solution was transferred to an 80-mL Teflon-lined autoclave, which was sealed and maintained at 200 °C for 10 h, and was then cooled to room temperature naturally. The black precipitation was collected by magnetic decantation, washed with deionized water and absolute alcohol several times, and then dried in a vacuum oven at 60 °C for 12 h.

To conjugate more PS molecules, the surfaces of the $CoFe_2O_4$ nanoparticles were treated with micro-dielectric barrier discharge plasma for 30 min according to a previously reported method [22].

The photo-functionality and targeting functionality on the $CoFe_2O_4$ nanoparticles were achieved using a wet chemical process similar to the method described in our previous report [28].

### 3.2. Characterization of Multifunctional CoFe₂O₄-HPs-Fas

Field-emission scanning electron microscopy (FE-SEM; SU-70, Hitachi, Tokyo, Japan) and transmission electron microscopy (TEM; JEM-2100F, JEOL, Tokyo, Japan) were applied to determine the size and surface morphology of the multifunctional sub-micron particles. The X-ray diffraction (XRD; X′ Pert Pro MPD, PANalytical, Almelo, The Netherlands) pattern of the product was determined on a PANalytical Pert Pro MPD X-ray diffractometer with a Cu Kα radiation source ($\lambda$ = 0.15405 nm) operated at 40 kV and 150 mA in a 2θ range of 20–80°. A vibrating sample magnetometer (VSM; Lakeshore 7300, Lake Shore Cryotronics, Westerville, OH, USA) was utilized to measure the magnetization versus magnetic field loop at room temperature up to 10 kOe. Photoluminescence and photoluminescence excitation spectra were measured on a spectrophotometer (F-4500, Hitachi, Tokyo, JApan). Infrared (IR) spectra were obtained using a Fourier transform (FT)-IR spectrometer (Spectrum 100, Perkin-Elmer, Waltham, MA, USA). For IR measurements, samples were prepared in an agate mortar and then prepared in the form of pressed wafers (ca. 1% sample in KBr).

### 3.3. Detection of Singlet Oxygen

Degradation of 1,3-diphenylisobenzofuran (DPBF) as a singlet oxygen quencher was applied to determine the release of singlet oxygen into the solution [29]. In the photochemical experiment, an aliquot of 3.0075 mL of tetrahydrofuran (THF) solution containing the $CoFe_2O_4$-HPs-FAs and DPBF ($4.61 \times 10^{-8}$ M) was introduced to a 1-cm quartz cell in the dark. The light source was a Xe lamp (150 W, Abet Technologies, Milford, MA, USA). A 480-nm glass cutoff filter was used to remove the ultraviolet (UV) light, which prevents direct photodegradation of DPBF. Photodegradation of DPBF was monitored by recording the optical density (OD) of the absorption peak at 424 nm. At every 10 min of irradiation, the absorption spectra of the samples were monitored on a UV–Vis spectrophotometer.

### 3.4. Magnetic Resonance Imaging (MRI) Analysis In Vitro

All MRI experiments were performed on a 3.0-Tesla whole-body MRI scanner (Philips Achieva X-series, Amsterdam, The Netherlands) using a dedicated phased array receiver coil for high-resolution MRI as previously described [22]. In brief, PC-3 cells (a prostate cancer cell line) pre-cultured for

24 h were incubated with various concentrations (0, 3.13, 6.25, 12.5, 25, and 50 μg/mL) of the CoFe$_2$O$_4$-HPs-FAs for 2 h in a 24-well culture plate. The cells were then fixed with a 2% glutaraldehyde and paraformaldehyde solution and mixed with a 1.5% agar solution in 1.5-mL micro-centrifuge tubes for MRI.

*3.5. Biocompatibility of Multifunctional CoFe$_2$O$_4$-HPs-Fas*

Cellular toxicity tests on fibroblasts (L-929 cells) and prostate cancer cells (PC-3 cells) were carried out to confirm the biocompatibility of the CoFe$_2$O$_4$-HPs-FAs, as previously described [22,30–32]. In brief, each cell type was plated in a 24-well plate at 2.0 × 10$^5$ cells/mL for L-929 cells and at 1.0 × 10$^5$ cells/mL for PC-3 cells, incubated at 37 °C in 5% CO$_2$ for 24 h, treated with various concentrations (0 (0), 3.13 (0.4), 6.25 (0.8), 12.5 (1.60), 25 (3.22), and 50 (6.44) μg/mL) of CoFe$_2$O$_4$-HPs-FAs (HPs), and then incubated at 37 °C in 5% CO$_2$ for a further 24 h under dark conditions. After incubation, the cells were washed three times with phosphate-buffer saline (PBS) and their viabilities were determined using Cell Counting Kit-8 (CCK-8, Dojindo Laboratories, Kumamoto, Japan). The OD value of each well was measured at 450 nm with a microplate reader (Synergy$^{TM}$ HT, BioTek Instruments, Inc., Winooski, VT, USA) and the cell viabilities are presented as the survival percentage relative to the untreated control.

*3.6. Optimization of the Cellular Uptake and Light Irradiation Time for Photodynamic Anticancer Activity of Multifunctional CoFe$_2$O$_4$-HPs-Fas*

To confirm the optimal cellular uptake time of CoFe$_2$O$_4$-HPs-FAs into the cells, PC-3 cells were plated in a 24-well plate at 1.0 × 10$^5$ cells/mL and incubated at 37 °C in 5% CO$_2$ for 24 h. The cells were further incubated with different concentrations (0 (0), 3.13 (0.4), 6.25 (0.8), 12.5 (1.60), and 25 (3.22) μg/mL) of CoFe$_2$O$_4$-HPs-FAs (HPs) for 1, 2, and 4 h under dark conditions. The cells were washed three times with PBS, the medium was refreshed, and the cells were irradiated by a general green light-emitting diode (LED) at 18.36 J/cm$^2$ as previously reported [22,23,30–32]. The LED had a wavelength range of 480–580 nm, a maximum wavelength of 515 nm, and a maximum dose of 18.36 J/cm$^2$. After irradiation, the cells were incubated for another 24 h and the following day, their viabilities were measured using the CCK-8 kit as described above.

To confirm the cellular uptake and intracellular localization of CoFe$_2$O$_4$-HPs-FAs in PC-3 cell with image analysis, two kinds of methods were adopted: Prussian blue staining for cellular uptake and TEM analysis for intracellular localization, as previously described [22]. In brief, PC-3 cells treated with CoFe$_2$O$_4$-HPs-FAs (6.25 μg/mL) were washed with PBS three times after incubating for 1, 2, and 4 h, and fixed with ice-cold acetone for 10 min to stain the cells with Prussian blue staining reagents. The fixed cells were stained with a 2% potassium ferrocyanide II and 1 M hydrochloric acid mixture (ratio 1:1) for 10 min at 37 °C after washing with PBS, and subsequently counterstained with a 0.1% solution of Nuclear Fast Red in distilled water with 5% aluminum sulfate for 1 min after washing again with PBS. Finally, the stained cells were observed with an inverted microscope (Eclipse Ti, Nikon, Instruments, Inc., New York, NY, USA) after washing with distilled water. To evaluate the intracellular localization of CoFe$_2$O$_4$-HPs-FAs in prostate cancer cells, PC-3 cells treated with CoFe$_2$O$_4$-HPs-FAs (6.25 μg/mL) for 1, 2, and 4 h were washed with PBS three times and fixed with 2.5% glutaraldehyde and 4% formaldehyde for 4 h at 4 °C. Fixed cells were washed with PBS twice and treated with 2% osmium tetroxide for 2 h at 4 °C. The fixed cells were washed again with PBS and dehydrated with increasing concentrations of ethanol and embedded in araldite. The fixed PC-3 cells were then observed with TEM (JEM-1011, JEOL, Tokyo, Japan).

To evaluate the appropriate irradiation energy with green LED for targeting the cancer cells, PC-3 cells were plated in a 24-well plate and incubated for 24 h as described above. After incubation, the cells were incubated with different concentrations (0 (0), 1.56 (0.2), 3.13 (0.4), 6.25 (0.8), 12.5 (1.60), and 25 (3.22) μg/mL) of CoFe$_2$O$_4$-HPs-FAs (HPs) for 2 h and further incubated for 24 h after irradiation at doses of 3.06, 6.12, 9.18, and 18.36 J/cm$^2$. The following day, the cell viabilities were measured using the CCK-8 kit as described above.

*3.7. Morphological Analysis of Apoptotic Cell Death in Prostate Cancer Cells*

To confirm the cell death after LED irradiation, the cell membranes and nuclei were stained with an EzWay™ Annexin V-FITC apoptosis detection kit (K29100, Komabiotech Inc., Seoul, Korea) and Texas Red C2-maleimide dye. The cell images were taken by a laser scanning microscope (LSM 700, Carl Zeiss, Oberkochen, Germany) with fluorescence optics (excitation at 488 nm for FITC, 530 nm for PI, 595 nm for Texas Red C2-maleimide, and 352 nm for Hoechst 33342; emission at 518 nm for FITC, 615 nm for Texas Red C2-maleimide, and 620 nm for PI) as previously described [22,23].

*3.8. Statistical Analysis*

All quantitative data ($n = 6$) are expressed as the means $\pm$ standard deviation, and statistical comparisons were evaluated with a Student's *t*-test. Significant differences were indicated by $p < 0.05$.

## 4. Conclusions

In the present study, we synthesized novel FA- and HP-conjugated multifunctional magnetic nanoparticles (CoFe$_2$O$_4$-HPs-FAs), which were characterized as an effective anticancer reagent for PDT, and demonstrated the dependency of the photodynamic anticancer activities on the incubation time and the exposure dose of LED light in prostate cancer cells (PC-3 cells). These results indicate that the same fluence at different exposure doses results in dissimilar levels of anticancer activities on PC-3 cancer cells as well as in ROS formation. In addition, the increase of the fluence showed an improvement for cell photo-inactivation.

**Acknowledgments:** This work was supported financially by the National Research Foundation of Korea (NRF, #2015M3A9E2066855 and #2015M3A9E2066856) grant funded by the Korean government (MSIP) and by the Research Grant from Kwangwoon University in 2017.

**Author Contributions:** The manuscript was written through contribution of all authors. All authors have given approval to the final version of the manuscript. K.-H.C., K.C.N., and U.-H.K. contributed equally to this work.

**Conflicts of Interest:** The authors declare that there are no conflicts of interest regarding the publication of this paper.

## References

1. Xie, J.; Pan, X.; Wang, M.; Yao, L.; Liang, X.; Ma, J.; Fei, Y.; Wang, P.-N.; Mi, L. Targeting and photodynamic killing of cancer cell by nitrogen-doped titanium dioxide coupled with folic acid. *Nanomaterials* **2016**, *6*, 113. [CrossRef] [PubMed]
2. Dougherty, T.J.; Gomer, C.J.; Henderson, B.W.; Jori, G.; Kessel, D.; Korbelik, M.; Moan, J.; Peng, Q. Photodynamic therapy. *J. Natl. Cancer Inst.* **1998**, *90*, 889–905. [CrossRef] [PubMed]
3. Chatterjee, D.K.; Fong, L.S.; Zhang, Y. Nanoparticles in photodynamic therapy: An emerging paradigm. *Adv. Drug Deliv. Rev.* **2008**, *60*, 1627–1637. [CrossRef] [PubMed]
4. Avula, U.M.; Kim, G.; Lee, Y.E.; Morady, F.; Kopelman, R.; Kalifa, J. Cell-specific nanoplatform-enabled photodynamic therapy for cardiac cells. *Heart Rhythm* **2012**, *9*, 1504–1509. [CrossRef] [PubMed]
5. Sibata, C.H.; Colussi, V.C.; Oleinick, N.L.; Kinsella, T.J. Photodynamic therapy: A new concept in medical treatment. *Braz. J. Med. Biol. Res.* **2000**, *33*, 869–880. [CrossRef] [PubMed]
6. Konopka, K.; Goslinski, T. Photodynamic therapy in dentistry. *J. Dent. Res.* **2007**, *86*, 694–707. [CrossRef] [PubMed]
7. Gursoy, H.; Ozcakir-Tomruk, C.; Tanalp, J.; Yilmaz, S. Photodynamic therapy in dentistry: A literature review. *Clin. Oral Investig.* **2013**, *17*, 1113–1125. [CrossRef] [PubMed]

8.  Mroz, P.; Bhaumik, J.; Dogutan, D.K.; Aly, Z.; Kamal, Z.; Khalid, L.; Kee, H.L.; Bocian, D.F.; Holten, D.; Lindsey, J.S.; et al. Imidazole metalloporphyrins as photosensitizers for photodynamic therapy: Role of molecular charge, central metal and hydroxyl radical production. *Cancer Lett.* **2009**, *282*, 63–76. [CrossRef] [PubMed]

9.  Vrouenraets, M.B.; Visser, G.W.; Snow, G.B.; van Dongen, G.A. Basic principles, applications in oncology and improved selectivity of photodynamic therapy. *Anticancer Res.* **2003**, *23*, 505–522. [PubMed]

10. Konan, Y.N.; Gurny, R.; Allémann, E. State of the art in the delivery of photosensitizers for photodynamic therapy. *J. Photochem. Photobiol. B Biol.* **2002**, *66*, 89–106. [CrossRef]

11. Vivero-Escoto, J.L.; Elnagheeb, M. Mesoporous silica nanoparticles loaded with cisplatin and phthalocyanine for combination chemotherapy and photodynamic therapy in vitro. *Nanomaterials* **2015**, *5*, 2302–2316. [CrossRef] [PubMed]

12. Hah, H.J.; Kim, G.; Lee, Y.E.; Orringer, D.A.; Sagher, O.; Philbert, M.A.; Kopelman, R. Methylene blue-conjugated hydrogel nanoparticles and tumor-cell targeted photodynamic therapy. *Macromol. Biosci.* **2011**, *11*, 90–99. [CrossRef] [PubMed]

13. Lin, J.; Wang, S.; Huang, P.; Wang, Z.; Chen, S.; Niu, G.; Li, W.; He, J.; Cui, D.; Lu, G.; et al. Photosensitizer-loaded gold vesicles with strong plasmonic coupling effect for imaging-guided photothermal/photodynamic therapy. *ACS Nano* **2013**, *7*, 5320–5329. [CrossRef] [PubMed]

14. Simon, T.; Boca-Farcau, S.; Gabudean, A.M.; Baldeck, P.; Astilean, S. LED-activated Methylene blue-loaded pluronic-nanogold hybrids for in vitro photodynamic therapy. *J. Biophotonics* **2013**, *6*, 950–959. [CrossRef] [PubMed]

15. Yan, F.; Zhang, Y.; Kim, K.S.; Yuan, H.K.; Vo-Dinh, T. Cellular uptake and photodynamic activity of protein nanocages containing Methylene blue photosensitizing drug. *Photochem. Photobiol.* **2010**, *86*, 662–666. [CrossRef] [PubMed]

16. Wilson, B.C.; Patterson, M.S. The physics, biophysics and technology of photodynamic therapy. *Phys. Med. Biol.* **2008**, *53*, R61–R109. [CrossRef] [PubMed]

17. Huang, X.; Tian, X.J.; Yang, W.L.; Ehrenberg, B.; Chen, J.Y. The conjugates of gold nanorods and chlorin E6 for enhancing the fluorescence detection and photodynamic therapy of cancers. *Phys. Chem. Chem. Phys.* **2013**, *15*, 15727–15733. [CrossRef] [PubMed]

18. Chen, C.L.; Kuo, L.R.; Chang, C.L.; Hwu, Y.K.; Huang, C.K.; Lee, S.Y.; Chen, K.; Sin, S.J.; Huang, J.D.; Chen, Y.Y. In situ real-time investigation of cancer cell photothermolysis mediated by excited gold nanorod surface plasmons. *Biomaterials* **2010**, *31*, 4104–4112. [CrossRef] [PubMed]

19. Zhao, Z.; Shi, S.; Huang, Y.; Tang, S.; Chen, X. Simultaneous photodynamic and photothermal therapy using photosensitizer functionalized Pd nanosheets by single continuous wave laser. *ACS Appl. Mater. Interfaces* **2014**, *6*, 8878–8885. [CrossRef] [PubMed]

20. Vankayala, R.; Lin, C.C.; Kalluru, P.; Chiang, C.S.; Hwang, K.C. Gold nanoshells-mediated bimodal photodynamic and photothermal cancer treatment using ultra-low doses of near infra-red light. *Biomaterials* **2014**, *35*, 5527–5538. [CrossRef] [PubMed]

21. Song, X.; Liang, C.; Gong, H.; Chen, Q.; Wang, C.; Liu, Z. Photosensitizer-conjugated albumin-polypyrrole nanoparticles for imaging-guided in vivo photodynamic/photothermal therapy. *Small* **2015**, *11*, 3932–3941. [CrossRef] [PubMed]

22. Park, B.J.; Choi, K.H.; Nam, K.C.; Ali, A.; Min, J.E.; Son, H.; Uhm, H.S.; Kim, H.J.; Jung, J.S.; Choi, E.H. Photodynamic anticancer activities of multifunctional cobalt ferrite nanoparticles in various cancer cells. *J. Biomed. Nanotechnol.* **2015**, *11*, 226–235. [CrossRef] [PubMed]

23. Choi, K.H.; Nam, K.C.; Malkinski, L.; Choi, E.H.; Jung, J.S.; Park, B.J. Size-dependent photodynamic anticancer activity of biocompatible multifunctional magnetic submicron particles in prostate cancer cells. *Molecules* **2016**, *21*, 1187. [CrossRef] [PubMed]

24. Castano, A.P.; Mroz, P.; Hamblin, M.R. Photodynamic therapy and anti-tumour immunity. *Nat. Rev. Cancer* **2006**, *6*, 535–545. [CrossRef] [PubMed]

25. Collins, H.A.; Khurana, M.; Moriyama, E.H. Blood-vessel closure using photosensitizers engineered for two-photon excitation. *Nat. Photonics* **2008**, *2*, 420–424. [CrossRef]

26. International Organization for Standardization (ISO). *International Standard ISO 10993-5:2009, Biological Evaluation of Medical Devices—Part 5: Tests for In Vitro Cytotoxicity*; International Organization for Standardization: Geneva, Switzerland.

27. Choi, K.H.; Choi, E.W.; Min, J.E.; Son, H.; Uhm, H.S.; Choi, E.H.; Park, B.J.; Jung, J.S. Comparison study on photodynamic anticancer activity of multifunctional magnetic particles by formation of cations. *IEEE Trans. Magn.* **2014**, *50*, 5200704. [CrossRef]

28. Choi, K.H.; Lee, H.J.; Park, B.J.; Wang, K.K.; Shin, E.P.; Park, J.C.; Kim, Y.K.; Oh, M.K.; Kim, Y.R. Photosensitizer and vancomycin-conjugated novel multifunctional magnetic particles as photoinactivation agents for selective killing of pathogenic bacteria. *Chem. Commun.* **2012**, *48*, 4591–4593. [CrossRef] [PubMed]

29. Choi, K.H.; Wang, K.K.; Shin, E.P.; Oh, S.L.; Jung, J.S.; Kim, H.K.; Kim, Y.R. Water-soluble magnetic nanoparticles functionalized with photosensitizer for photocatalytic application. *J. Phys. Chem. C* **2011**, *115*, 3212–3219. [CrossRef]

30. Choi, K.H.; Nam, K.C.; Kim, H.J.; Min, J.; Uhm, H.S.; Choi, E.H.; Park, B.J. Synthesis and characterization of photo-functional magnetic nanoparticles ($Fe_3O_4$@HP) for applications in photodynamic cancer therapy. *J. Korean Phys. Soc.* **2014**, *65*, 1658–1662. [CrossRef]

31. Park, B.J.; Choi, K.H.; Nam, K.C.; Min, J.; Lee, K.D.; Uhm, H.S.; Choi, E.H.; Kim, H.J.; Jung, J.S. Photodynamic anticancer activity of $CoFe_2O_4$ nanoparticles conjugated with hematoporphyrin. *J. Nanosci. Nanotechnol.* **2015**, *15*, 7900–7906. [CrossRef] [PubMed]

32. Nam, K.C.; Choi, K.H.; Lee, K.D.; Kim, J.H.; Jung, J.S.; Park, B.J. Particle size dependent photodynamic anticancer activity of hematophorphyrin-conjugated $Fe_3O_4$ particles. *J. Nanomater.* **2016**, *2016*, 1278393. [CrossRef]

© 2017 by the authors. Licensee MDPI, Basel, Switzerland. This article is an open access article distributed under the terms and conditions of the Creative Commons Attribution (CC BY) license (http://creativecommons.org/licenses/by/4.0/).

*nanomaterials*

MDPI

*Article*

# Eu, Gd-Codoped Yttria Nanoprobes for Optical and T$_1$-Weighted Magnetic Resonance Imaging

Timur Sh Atabaev [1,*], Jong Ho Lee [2], Yong Cheol Shin [3], Dong-Wook Han [3,*], Ki Seok Choo [4], Ung Bae Jeon [4], Jae Yeon Hwang [4], Jeong A. Yeom [4], Hyung-Kook Kim [5,*] and Yoon-Hwae Hwang [5,*]

[1] Department of Physics and Astronomy, Seoul National University, Seoul 08826, Korea
[2] Center for Biomaterials, Biomedical Research Institute, Korea Institute of Science and Technology, Seoul 02792, Korea; pignunssob@naver.com
[3] Department of Cogno-Mechatronics Engineering, Pusan National University, Busan 46241, Korea; choel15@naver.com
[4] Department of Radiology, Pusan National University Yangsan Hospital, Yangsan 50612, Korea; kschoo0618@naver.com (K.S.C.); junwb73@hanmail.net (U.B.J.); yhwang79@gmail.com (J.Y.H.); sigmajeonga@hanmail.net (J.A.Y.)
[5] Department of Nano Energy Engineering, Pusan National University, Miryang 50463, Korea
* Correspondence: timuratabaev@yahoo.com (T.S.A.); nanohan@pusan.ac.kr (D.-W.H.); hkkim@pusan.ac.kr (H.-K.K.); yhwang@pusan.ac.kr (Y.-H.H.); Tel.: +82-10-3326-0512 (T.S.A.)

Academic Editor: Thomas Nann
Received: 13 December 2016; Accepted: 3 February 2017; Published: 10 February 2017

**Abstract:** Nanoprobes with multimodal functionality have attracted significant interest recently because of their potential applications in nanomedicine. This paper reports the successful development of lanthanide-doped Y$_2$O$_3$ nanoprobes for potential applications in optical and magnetic resonance (MR) imaging. The morphology, structural, and optical properties of these nanoprobes were characterized by transmission electron microscope (TEM), field emission scanning electron microscope (FESEM), X-ray diffraction (XRD), energy-dispersive X-ray (EDX), and photoluminescence (PL). The cytotoxicity test showed that the prepared lanthanide-doped Y$_2$O$_3$ nanoprobes have good biocompatibility. The obvious contrast enhancement in the T$_1$-weighted MR images suggested that these nanoprobes can be used as a positive contrast agent in MRI. In addition, the clear fluorescence images of the L-929 cells incubated with the nanoprobes highlight their potential for optical imaging. Overall, these results suggest that prepared lanthanide-doped Y$_2$O$_3$ nanoprobes can be used for simultaneous optical and MR imaging.

**Keywords:** yttria; nanoprobes; optical imaging; magnetic resonance imaging; cytotoxicity; T$_1$-weighted contrast agent

## 1. Introduction

In recent years, nanoprobes with multimodal functionality have been investigated extensively for a range of biomedical applications, such as optical imaging (OI), magnetic resonance imaging (MRI), bioseparation, controlled drug delivery, etc. [1–4]. Among them, gadolinia nanoprobes are used widely to improve the T$_1$-weighted contrast between different tissues [5–8]. These gadolinia nanoprobes can efficiently induce the longitudinal relaxation of water protons and brighten the imaging place. On the other hand, the main limitation for industrial applications of gadolinia nanoprobes is the high production cost of gadolinium due mainly to its rarity on Earth. To resolve this limitation, gadolinium ions can be doped into another earth-abundant and low-cost host matrix. Surface-localized gadolinium ions in the low-cost host matrix can also change the longitudinal relaxation of the nearby water protons for MRI contrast enhancement. The similarity in chemical properties make the matrix of yttrium oxide $Y_2O_3$ (yttria) a promising host candidate for lanthanide doping [9,10]. For example, recent studies have suggested that a homogeneous yttria-lanthanides solid mixture can be obtained using the homogeneous urea precipitation method [11,12]. In addition, yttria-based nanostructures can potentially be used for optical biomaging purposes [13,14].

To the best of the authors' knowledge, $Eu^{3+}$ and $Gd^{3+}$ codoped yttrium oxide nanospheres were not utilized as potential nanoprobes for simultaneous optical and magnetic resonance imaging. Therefore, the main aim of this study was to explore the possibilities of lanthanide-doped $Y_2O_3$ nanoprobes for bimodal imaging. The optical properties and MR relaxivity rate of the prepared lanthanide-doped $Y_2O_3$ nanoprobes were examined as a function of the $Gd^{3+}$ concentration. The results are expected to make a strong contribution to other nanoprobes' development.

## 2. Results and Discussion

For a comparative study of the optical and magnetic properties, samples with different $Gd^{3+}$ codoping concentrations (0, 3, 7 and 10 mol %) were prepared. Transmission electron microscope (TEM), field emission scanning electron microscope (FESEM) and energy-dispersive X-ray (EDX) were used to examine the morphology and chemical composition of the samples. Figure 1a–d shows that all nanoprobes prepared had a spherical morphology within the range, 61–69 nm. For example, the mean diameters of the 10 mol % $Gd^{3+}$ codoped nanoprobes, as measured by dynamic light scattering (DLS) (Figure S1, Supplementary Materials) were in accordance with the estimated sizes from FESEM analysis (Figure S2, Supplementary Materials). Obviously, codoping with $Gd^{3+}$ did not alter the morphology, even at relatively high $Gd^{3+}$ codoping concentrations as shown by FESEM measurements (Figure S2, Supplementary Materials). Bare $Y_2O_3$:$Eu^{3+}$ and 3 mol % $Gd^{3+}$ codoped $Y_2O_3$:$Eu^{3+}$ nanoparticles were used to determine the elemental composition. EDX (Figure S3, Supplementary Materials) clearly indicated the presence of specific dopants (Eu in $Y_2O_3$:$Eu^{3+}$) and (Eu, Gd in $Y_2O_3$:$Eu^{3+}$, $Gd^{3+}$) in the synthesized nanoparticles. Zeta potential was further measured at pH 7.5 to check the colloidal stability of the prepared samples. Measurement results show that zeta potential for all nanoprobes were in the range of 7–9 mV. Thus, prepared colloidal solutions of nanoprobes were stable for several hours only. However, homogeneous suspensions were formed again when solutions were ultrasonicated for several seconds. It is worth to mention that special surface coating can increase their colloidal stability in biological environment [6,8]. However, one should keep in mind that surface coating may affect the $T_1$-relaxivity of these nanoprobes.

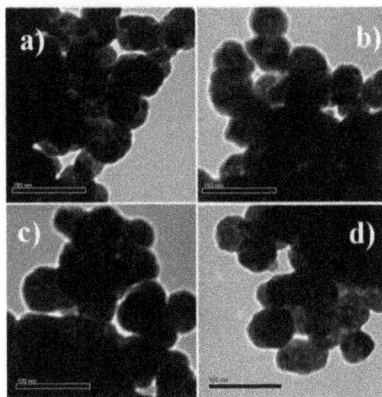

**Figure 1.** Transmission electron microscope (TEM) images of (**a**) bare $Y_2O_3$:$Eu^{3+}$; (**b**) 3 mol % $Gd^{3+}$ codoped $Y_2O_3$:$Eu^{3+}$; (**c**) 7 mol % $Gd^{3+}$ codoped $Y_2O_3$:$Eu^{3+}$; and (**d**) 10 mol % $Gd^{3+}$ codoped $Y_2O_3$:$Eu^{3+}$.

Figure S4 (Supplementary Materials) shows typical XRD patterns of the prepared samples. The X-ray diffraction (XRD) peaks for bare $Y_2O_3$:$Eu^{3+}$ were assigned to the standard cubic structure of $Y_2O_3$ (JCPDS No. 86-1107, space group *Ia3* (206)). A careful examination of the Gd-codoped samples showed that the position of all the XRD peaks shifted slightly towards lower angles. On the other hand, the samples still retained the cubic structure of $Y_2O_3$. The observed similarity with the $Y_2O_3$ structure suggests that the Gd-codoped samples are solid $Y_2O_3$-based solutions rather than mechanical mixtures of $Y_2O_3$, $Eu_2O_3$, and $Gd_2O_3$.

The prepared samples were tested further by photoluminescence (PL) at room temperature. Although the main reason for Gd-codoping was to achieve the paramagnetic functionality of bare $Y_2O_3$:$Eu^{3+}$, Gd-codoping had some interesting effects on PL emission. All samples were measured under identical conditions so that the emission ratio could be compared. Figure 2 shows the PL emission spectrum revealed several main groups of emission lines, which were assigned to the $^5D_1 \rightarrow {}^7F_1$ and $^5D_0 \rightarrow {}^7F_j$ (where $j$ = 0, 1, 2, 3) transitions within the $Eu^{3+}$ [12–14]. Partial replacement of $Y^{3+}$ with $Gd^{3+}$ in the host matrix allows easier charge transfer from $Gd^{3+}$ to $Eu^{3+}$ [15]. Therefore, Gd-codoping does not alter the peak position, but greatly improves the PL emission intensity. For example, the relative intensity of the strongest $^5D_0 \rightarrow {}^7F_2$ (at 612 nm) peak increased monotonically with increasing Gd-concentration, which is in good agreement with the previously reported study [12].

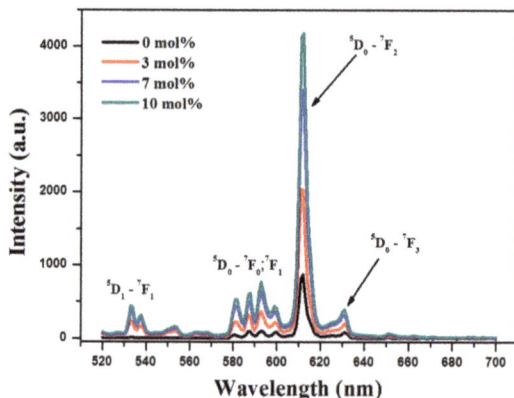

**Figure 2.** Photoluminescence (PL) emission spectra of prepared samples.

A 1.5 T clinical MRI scanner was used to demonstrate the applicability of the Gd-codoped samples for $T_1$-weighted MR imaging. The slope of the linear fit of $1/T_1$ vs. the nanoparticle concentration yielded a longitudinal relaxivity ($R_1$) of the samples. Figure 3 shows that the $R_1$ values of the 3%, 7%, and 10% codoped samples were approximately $2.59 \pm 0.03$, $2.64 \pm 0.04$, and $2.67 \pm 0.08$ $s^{-1} \cdot mM^{-1}$ respectively. As expected, the $R_1$ value increased with increasing Gd-concentration in the samples. Figure 3 (inset) shows that the $T_1$-relaxation time of the water protons was reduced significantly, and the $T_1$-weighted images became brighter with increasing concentration of 10 mol % $Gd^{3+}$ codoped nanoparticles. The resulting $R_1$ values were comparable to commercially available Gd chelates, such as gadopentetic acid Gd-DTPA ($\sim$3 $s^{-1} \cdot mM^{-1}$) [16], which means that these nanoparticles can be used as bimodal contrast agents for MR and optical imaging.

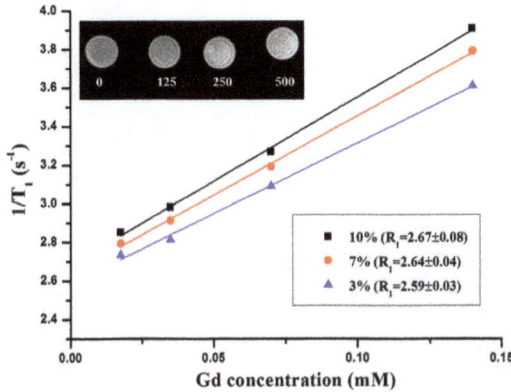

**Figure 3.** Longitudinal relaxivity rate $R_1$ vs. various concentrations of Gd-codoped nanoparticles measured at room-temperature. Inset is $T_1$-weighted images of the 10 mol % $Gd^{3+}$ codoped nanoparticles at various concentrations (ppm).

The cytotoxicity was measured to test the biosafety of the 10 mol % $Gd^{3+}$ codoped nanoparticles. Figure 4 shows the cytotoxicity profiles of the nanoparticles in L-929 fibroblastic cells, which were determined using a WST-8 assay. The L-929 fibroblastic cells showed a noticeable concentration-dependent decrease in their relative cell viability. The prepared nanoparticles caused no significant decrease in cell viability at concentrations less than 60 ppm. This value is much higher than reported 8–10 ppm for $Gd_2O_3$-based nanoparticles [6,8]. Therefore, considering the in vitro cytotoxicity only, 10 mol % $Gd^{3+}$ codoped nanoparticles can be used safely for bio-imaging at doses lower than 60 ppm.

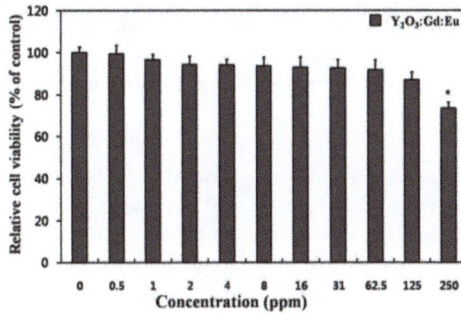

**Figure 4.** Relative cell viability of L-929 cells exposed to increasing concentrations (0–250 ppm) of the 10 mol % $Gd^{3+}$ codoped nanoparticles. An asterisk (*) denotes a significant difference compared with the control, $p < 0.05$.

To reveal the optical imaging potential of the 10 mol % $Gd^{3+}$ codoped nanoparticles, a cultured monolayer of L-929 cells was incubated in the culture medium with a nanoparticle suspension at 10 ppm. Figure 5 shows that the L-929 cells grow with normal fibroblast-like morphologies after labeling with nanoparticles. The bright red fluorescence from the nanoparticles was observed mainly in the cytoplasm rather than inside the nuclei, suggesting that the nanoparticles could make cell imaging possible through efficient internalization into the cells with a uniform distribution in the cytoplasm. Therefore, the Gd-codoped nanoparticles can be utilized easily for clinical MRI applications and optical cell tracking.

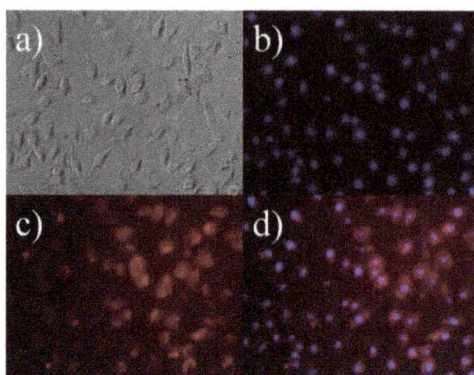

**Figure 5.** Fluorescence micrograps (200×) of L-929 cells treated with 10 ppm of 10 mol % Gd$^{3+}$ codoped nanoparticles, followed by cell nuclei counterstaining with 10 μmol/L of 4′6-diamidino-2-phenylindole (DAPI). (**a**) Phase contrast image of the cells co-labelled with nanoparticles and DAPI; (**b**,**c**) Fluorescence images of the cells collected from DAPI (blue) and nanoprobes (red) respectively; (**d**) Merged image of (**b**,**c**).

## 3. Materials and Methods

### 3.1. Nanoprobes Preparation

Analytical graded Y$_2$O$_3$ (99.99%), Eu$_2$O$_3$ (99.99%), Gd$_2$O$_3$ (99.99%), HNO$_3$ (70.0%), and urea (99.0%–100.5%) were purchased from Sigma-Aldrich (St. Louis, MO, USA) and used as received. Spherical Y$_2$O$_3$ nanoprobes co-doped with Eu$^{3+}$ and Gd$^{3+}$ were fabricated using a urea homogeneous precipitation method using the reported protocols [8–12]. Briefly, a sealed beaker with a freshly prepared aqueous solution of rare-earth nitrates (0.0005 mol in 40 mL of H$_2$O) was placed into an electrical furnace and heated to 90 °C for 1.5 h. The dried synthesized precipitates were then calcined in air at 800 °C for 1 h to produce the oxide NPs. In all cases, the Eu$^{3+}$ doping concentration was kept constant at 1 mol %, whereas the Gd$^{3+}$ concentration was varied from 0 to 10 mol %.

### 3.2. Characterization

The structure of the prepared powders was examined by XRD Bruker D8 Discover (Billerica, MA, USA) using Cu-Kα radiation (λ = 0.15405 nm) at a 2θ scan range 20°–60° 2θ. The morphology of the particles was characterized by field emission scanning electron microscopy FESEM Carl Zeiss Supra 25 (Oberkochen, Germany) equipped with EDX. Size distribution and zeta potentials of the obtained nanoprobes were measured using a Nano ZS Zetasizer (Malvern, UK). The photoluminescence PL measurements were performed using a Hitachi F-7000 (Tokyo, Japan) spectrophotometer equipped with a 150 W Xenon lamp as the excitation source. The T$_1$-weighted images were obtained using a 1.5 T MRI scanner Siemens (Munich, Germany) using the T$_1$-weighted spin-echo method (TR/TE = 500 ms/15 ms, field of view (FOV) = 100 mm × 100 mm, slice thickness = 2 mm, matrix = 256 × 204, number of excitations (NEX) = 2). All measurements were performed at a room temperature of 22 ± 1 °C.

### 3.3. Cell Culture and Cytotoxicity Assay

A murine fibroblast cell line (L-929 cells from subcutaneous connective tissue) was obtained from the American Type Culture Collection (ATCC CCL-1™, Rockville, MD, USA). The cells were routinely maintained in Dulbecco's modified Eagle's medium (Sigma-Aldrich, St. Louis, MO, USA), supplemented with 10% fetal bovine serum (Sigma-Aldrich, St. Louis, MO, USA) and 1%

antibiotic antimycotic solution (including 10,000 units penicillin, 10 mg streptomycin and 25 mg amphotericin B per mL, Sigma-Aldrich, St. Louis, MO, USA) at 37 °C in 95% humidity and 5% $CO_2$. The number of viable cells was indirectly quantified using highly water-soluble tetrazolium salt [WST-8,2-(2-methoxy-4-nitrophenyl)-3-(4-nitrophenyl)-5-(2,4-disulfophenyl)-2*H*-tetrazolium, monosodium salt] (Dojindo Lab., Kumamoto, Japan), reduced to a water-soluble formazan dye by mitochondrial dehydrogenases. The cell viability was found to be directly proportional to the metabolic reaction products obtained in WST-8. Briefly, the WST-8 assay was conducted as follows. L-929 cells were treated with increasing concentration (0–250 ppm) of nanoprobes and then incubated with WST-8 for the last 4 h of the culture periods (24 h) at 37 °C in the dark. Parallel sets of wells containing freshly cultured nontreated cells were regarded as negative controls. The absorbance was determined to be 450 nm using an ELISA reader (SpectraMax® 340, Molecular Device Co., Sunnyvale, CA, USA). The relative cell viability was determined as the percentage ratio of the optical densities in the medium (containing the nanoprobes at each concentration) to that of the fresh control medium.

### 3.4. Fluorescence Microscopy

To examine the cellular uptake and distribution of nanoprobes within the L-929 cells and subsequent cell imaging, the cells were treated with 10 ppm of nanoprobes for 4 h. After treatment, the cells were fixed with 3.5% paraformaldehyde (Sigma-Aldrich) in 0.1 M phosphate buffer (pH = 7) for 10 min at room temperature and immediately observed under a fluorescence microscope (IX81-F72, Olympus Optical, Osaka, Japan).

### 3.5. Statistical Analysis

All variables were tested in three independent cultures for cytotoxicity assay, which was repeated twice ($n = 6$). Quantitative data are expressed as the mean $\pm$ standard deviation (SD). Data were tested for homogeneity of variances using the test of Levene, prior to statistical analysis. Statistical comparisons were carried out by a one-way analysis of variance (ANOVA), followed by a Bonferroni test for multiple comparisons. A value of $p < 0.05$ was considered statistically significant.

## 4. Conclusions

In summary, $Eu^{3+}$ and $Gd^{3+}$ codoped $Y_2O_3$ nanoparticles were prepared for potential MRI and optical imaging applications. We showed that Gd-codoping into the $Y_2O_3$ host matrix resulted in PL emission enhancement. MRI relaxivity studies of the Gd-codoped samples suggested that the prepared nanoparticles can also be used for $T_1$-weighted contrast enhancement. In addition, the cytotoxicity results showed that these nanoparticles are safe for bio-imaging at doses lower than 60 ppm. Therefore, the bimodal functionality and low toxicity makes these nanoparticles suitable for nanomedical applications.

**Supplementary Materials:** The following are available online at http://www.mdpi.com/2079-4991/7/2/35/s1.

**Acknowledgments:** This work was supported by a National Research Foundation of Korea (NRF) grant funded by the Korea government (MSIP) (No. 2016R1A2B4007611 and 2014R1A2A1A11051146). This work was also supported by Basic Science Research Program through the NRF of Korea funded by the Ministry of Education (No. 2016R1D1A1B03931076).

**Author Contributions:** Timur Sh Atabaev, Hyung-Kook Kim and Yoon-Hwae Hwang conceived and designed the experiments; Timur Sh Atabaev performed the preparation and chacterization of nanoprobes; Jong Ho Lee and Yong Cheol Shin performed cytotoxicity and cellular imaging experimens; Ki Seok Choo, Ung Bae Jeon, Jae Yeon Hwang, and Jeong A. Yeom performed the $T_1$-weighted MRI experiments, Timur Sh Atabaev, Dong-Wook Han, Hyung-Kook Kim, and Yoon-Hwae Hwang analyzed the data; Timur Sh Atabaev wrote the paper.

**Conflicts of Interest:** The authors declare no conflict of interest.

## References

1. Shen, J.; Sun, L.D.; Zhang, Y.W.; Yan, C.H. Superparamagnetic and upconversion emitting $Fe_3O_4$/$NaYF_4$:Yb,Er hetero-nanoparticles via a crosslinker anchoring strategy. *Chem. Commun.* **2010**, *46*, 5731–5733. [CrossRef] [PubMed]
2. Atabaev, T.S.; Lee, J.H.; Han, D.W.; Kim, H.K.; Hwang, Y.H. Fabrication of carbon coated gadolinia particles for dual-mode magnetic resonance and fluorescence imaging. *J. Adv. Ceram.* **2015**, *4*, 118–122. [CrossRef]
3. Atabaev, T.S.; Kim, H.K.; Hwang, Y.H. Fabrication of bifunctional core-shell $Fe_3O_4$ particles coated with ultrathin phosphor layer. *Nanoscale Res. Lett.* **2013**, *8*, 357. [CrossRef] [PubMed]
4. Atabaev, T.S.; Urmanova, G.; Hong, N.H. Highly mesoporous silica nanoparticles for potential drug delivery applications. *NanoLife* **2014**, *4*, 1441003. [CrossRef]
5. Faucher, L.; Gossuin, Y.; Hocq, A.; Fortin, M.A. Impact of agglomeration on the relaxometric properties of paramagnetic ultra-small gadolinium oxide nanoparticles. *Nanotechnology* **2011**, *22*, 295103. [CrossRef] [PubMed]
6. Atabaev, T.S.; Lee, J.H.; Han, D.W.; Kim, H.K.; Hwang, Y.H. Ultrafine PEG-capped gadolinia nanoparticles: Cytotoxicity and potential biomedical applications for MRI and luminescent imaging. *RSC Adv.* **2014**, *4*, 34343–34349. [CrossRef]
7. Chen, F.; Chen, M.; Yang, C.; Liu, J.; Luo, N.; Yang, G.; Chen, D.; Li, L. Terbium-doped gadolinium oxide nanoparticles prepared by laser ablation in liquid for use as a fluorescence and magnetic resonance imaging dual-modal contrast agent. *Phys. Chem. Chem. Phys.* **2015**, *17*, 1189–1196. [CrossRef] [PubMed]
8. Atabaev, T.S.; Lee, J.H.; Han, D.W.; Choo, K.S.; Jeon, U.B.; Hwang, J.Y.; Yeom, J.A.; Kang, C.; Kim, H.K.; Hwang, Y.H. Multicolor nanoprobes based on silica-coated gadolinium oxide nanoparticles with highly reduced toxicity. *RSC Adv.* **2016**, *6*, 19758–19762. [CrossRef]
9. Atabaev, T.S.; Vu, H.H.T.; Kim, H.K.; Hwang, Y.H. The optical properties of $Eu^{3+}$ and $Tm^{3+}$ codoped $Y_2O_3$ submicron particles. *J. Alloys Compd.* **2012**, *525*, 8–13. [CrossRef]
10. Atabaev, T.S.; Vu, H.H.T.; Kim, H.K.; Hwang, Y.H. Synthesis and optical properties of $Dy^{3+}$-doped $Y_2O_3$ nanoparticles. *J. Korean Phys. Soc.* **2012**, *60*, 244–248. [CrossRef]
11. Li, J.G.; Li, X.; Sun, X.; Ikegami, T.; Ishigaki, T. Uniform colloidal spheres for $(Y_{1-x}Gd_x)_2O_3$ ($x$ = 0–1): Formation mechanism, compositional impacts, and physicochemical properties of the oxides. *Chem. Mater.* **2008**, *20*, 2274–2281. [CrossRef]
12. Ajmal, M.; Atabaev, T.S. Facile fabrication and luminescent properties enhancement of bimodal $Y_2O_3$:$Eu^{3+}$ particles by simultaneous $Gd^{3+}$ codoping. *Opt. Mater.* **2013**, *35*, 1288–1292. [CrossRef]
13. Atabaev, T.S.; Jin, O.S.; Lee, J.H.; Han, D.W.; Vu, H.H.T.; Hwang, Y.H.; Kim, H.K. Facile synthesis of bifunctional silica-coated core-shell $Y_2O_3$:$Eu^{3+}$,$Co^{2+}$ composite particles for biomedical applications. *RSC Adv.* **2012**, *2*, 9495–9501. [CrossRef]
14. Atabaev, T.S.; Lee, J.H.; Han, D.W.; Hwang, Y.H.; Kim, H.K. Cytotoxicity and cell imaging potentials of submicron color-tunable yttria particles. *J. Biomed. Mater. Res. A* **2012**, *100*, 2287–2294. [CrossRef] [PubMed]
15. Liu, G.X.; Hong, G.Y.; Sun, D.X. Synthesis and characterization of $SiO_2$/$Gd_2O_3$:Eu core-shell luminescent materials. *J. Colloid Interface Sci.* **2004**, *278*, 133–138. [CrossRef] [PubMed]
16. Hou, Y.; Qiao, R.; Fang, F.; Wang, X.; Dong, C.; Liu, K.; Liu, C.; Liu, Z.; Lei, H.; Wang, F.; et al. $NaGdF_4$ nanoparticle-based molecular probes for magnetic resonance imaging of intraperitoneal tumor xenografts in vivo. *ACS Nano* **2013**, *7*, 330–338. [CrossRef] [PubMed]

© 2017 by the authors. Licensee MDPI, Basel, Switzerland. This article is an open access article distributed under the terms and conditions of the Creative Commons Attribution (CC BY) license (http://creativecommons.org/licenses/by/4.0/).

![nanomaterials logo] *nanomaterials*

MDPI

*Communication*

# Toxicity and $T_2$-Weighted Magnetic Resonance Imaging Potentials of Holmium Oxide Nanoparticles

Timur Sh. Atabaev [1,*], Yong Cheol Shin [2], Su-Jin Song [2], Dong-Wook Han [2,*] and Nguyen Hoa Hong [1,*]

1   Department of Physics and Astronomy, Seoul National University, Seoul 08826, Korea
2   Department of Cogno-Mechatronics Engineering, Pusan National University, Busan 46241, Korea;
    choel15@naver.com (Y.C.S.); songsj86@gmail.com (S.-J.S.)
*   Correspondence: timuratabaev@yahoo.com (T.S.A.); nanohan@pusan.ac.kr (D.-W.H.);
    nguyenhong@snu.ac.kr (N.H.H.); Tel.: +82-10-3326-0512 (T.S.A.)

Received: 29 June 2017; Accepted: 2 August 2017; Published: 7 August 2017

**Abstract:** In recent years, paramagnetic nanoparticles (NPs) have been widely used for magnetic resonance imaging (MRI). This paper reports the fabrication and toxicity evaluation of polyethylene glycol (PEG)-functionalized holmium oxide ($Ho_2O_3$) NPs for potential $T_2$-weighted MRI applications. Various characterization techniques were used to examine the morphology, structure and chemical properties of the prepared PEG–$Ho_2O_3$ NPs. MRI relaxivity measurements revealed that PEG–$Ho_2O_3$ NPs could generate a strong negative contrast in $T_2$-weighted MRI. The pilot cytotoxicity experiments showed that the prepared PEG–$Ho_2O_3$ NPs are biocompatible at concentrations less than 16 μg/mL. Overall, the prepared PEG–$Ho_2O_3$ NPs have potential applications for $T_2$-weighted MRI imaging.

**Keywords:** holmium oxide; paramagnetic nanoparticles; $T_2$-weighted magnetic resonance imaging; toxicity

---

## 1. Introduction

Magnetic and optical metal oxide nanoparticles (NPs) have attracted considerable attention over the past three decades for biomedical imaging and diagnosis [1]. In particular, iron oxide NPs [2,3], manganese oxide NPs [4,5], and gadolinium oxide NPs [6,7] have been investigated for potential magnetic resonance imaging (MRI). However, superparamagnetic iron oxide NPs show saturation magnetization at approximately 1.5 T, which limits their MRI applicability in high magnetic fields [2]. From this point of view, paramagnetic rare-earth metal oxide NPs with higher magnetic moments and higher density of magnetic ions per surface unit are more promising for MRI applications. For example, $Gd_2O_3$ NPs were reported to show higher longitudinal relaxivity ($r_1$) compared to commercially available Gd-based chelates [8,9]. $Gd_2O_3$ NPs brightens the imaging place (positive contrast), because it changes the spin-lattice relaxation of water protons. On the other hand, the main limitation associated with the broad use of $Gd_2O_3$ NPs is in their high toxicity. In this regard, surface modification or Gd-doping into a less toxic material can be used [10,11], but these alterations can also deteriorate the relaxivity rates of the $Gd_2O_3$ NPs.

Other rare-earth ions, such as $Dy^{3+}$ and $Ho^{3+}$, have larger magnetic moments (~10.5 μB) than $Gd^{3+}$ (~8.1 μB). On the other hand, both $Dy_2O_3$ and $Ho_2O_3$ NPs are more suitable for $T_2$-weighted MRI (negative contrast) due to the fast spin relaxation of their $4f$ electrons. For example, a number of studies demonstrated the suitability of $Dy_2O_3$ NPs for $T_2$-weighted MRI [12,13]. In particular, the reported transverse $r_2$ relaxivities of $Dy_2O_3$ NPs were much higher than that of commercially available iron oxide NPs [12,13]. On the other hand, there are almost no reports of the potential toxicity and applications of $Ho_2O_3$ NPs as a MRI contrast nanoprobe [14]. Therefore, this study examined the PEG-grafted $Ho_2O_3$ NPs to explore their toxicity and applicability as a new potential $T_2$-weighted

MRI contrast agent. A murine fibroblast L-929 cell line was used as a pilot in-vitro model to check the cytotoxicity of the PEG–Ho$_2$O$_3$ NPs. This study suggests that the prepared PEG–Ho$_2$O$_3$ NPs can be potentially used as a new $T_2$-weighted MRI contrast agent at concentrations less than 16 µg/mL.

## 2. Materials and Methods

### 2.1. Synthesis Process

Analytical grade Ho$_2$O$_3$ (99.9%), HNO$_3$ (70%), polyethylene glycol (PEG, average M$_n$ = 4000) and urea (99.0–100.5%) were purchased from Sigma-Aldrich (St. Louis, MO, USA) and used as received. Ho$_2$O$_3$ NPs were prepared using the reported protocols [15,16]. In brief, holmium oxide powder was converted to a holmium nitrate salt with the help of nitric acid. Later on, a sealed beaker with a freshly prepared aqueous solution of holmium nitrate (0.5 mmol in 40 mL of H$_2$O) was placed into a forced convection drying oven (J-300M, Jisico Co., Ltd., Seoul, South Korea) and heated to 90 °C for 1.5 h. The collected precipitates were then calcined in air at 600 °C for 1 h to produce the Ho$_2$O$_3$ NPs. PEG-functionalization of Ho$_2$O$_3$ NPs was performed according to a reported protocol [8]. The obtained colloidal solution was then dialyzed in deionized ultrapure water for 24 h to eliminate the unreacted products.

### 2.2. Characterization

The structure of the prepared powders was examined by X-ray diffraction (XRD, Bruker D8 Discover, Billerica, MA, USA) using Cu-Kα radiation (λ = 0.15405 nm) at a 2θ scan range 20–60°. The morphology of the particles was characterized by transmission electron microscopy (TEM, JEM-2100, JEOL Ltd., Tokyo, Japan). Energy dispersive X-ray spectroscopy (EDX, JEOL Ltd., Tokyo, Japan) was used to perform an elemental analysis. Hydrodynamic sizes and zeta potentials of the obtained nanoprobes were measured using a Nano ZS Zetasizer (Malvern Instruments Ltd., Malvern, UK). Fourier transform infrared infrared spectroscopy (FTIR, Jasco FT/IR6300, Tokyo, Japan) was used to examine the structural properties of prepared samples. The magnetization measurements were performed using a magnetic properties measurement system (MPMS-5XL/Quantum Design Inc., San Diego, CA, USA). The $T_2$-weighted images were obtained using a 1.5 T small animal MRI scanner (Siemens Healthinners, Enlargen, Germany). The measurement parameters used were as follows: the repetition time (TR) = 2009 ms, the time to echo (TE) = 9 ms, the field of view (FOV) = 160 × 160 mm, slice thickness = 5 mm, matrix = 256 × 256, number of excitations (NEX) = 1. All characterization measurements were performed at a room temperature of 22 ± 1 °C. The conditions for cell culture, cytotoxicity assay, fluorescence assay and statistical analysis were reported in our previous report [11].

## 3. Results and Discussion

Paramagnetic NPs for multimodal imaging have attracted considerable interest in recent years for potential nanomedical applications. Ultrasmall holmium oxide Ho$_2$O$_3$ NPs were proposed recently for potential MRI imaging applications [14]. However, the biocompatibility of Ho$_2$O$_3$ NPs is still a big issue to be addressed. Furthermore, it is very important to develop eco-friendly and low-cost synthesis method for fabricating the highly monodispersed Ho$_2$O$_3$ NPs at large scales. To address these issues, we designed a simple two-step approach to synthesize the highly monodispersed PEG–Ho$_2$O$_3$ NPs. The successful synthesis of PEG–Ho$_2$O$_3$ NPs is confirmed with several analysis techniques. XRD was used to examine the structural properties of the as-prepared Ho$_2$O$_3$ NPs. Figure 1a shows an XRD pattern of the as-prepared Ho$_2$O$_3$ NPs. The XRD peaks were assigned to the standard cubic (*Ia*$_3$) Ho$_2$O$_3$ structure (JCPDS no. 43-1018) [17]. No additional impurity peaks were detected; thus, the obtained nanoprobes can be considered a pure cubic Ho$_2$O$_3$ phase. Energy dispersive X-ray spectroscopy (Figure 1b) revealed the presence of Ho and O elements only, indicating the formation of a pure Ho$_2$O$_3$ structure after a calcination process.

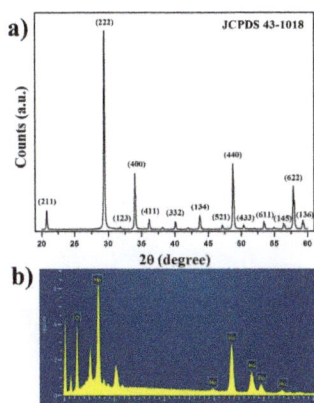

**Figure 1.** (a) X-ray diffraction (XRD) and (b) EDX analysis of as-prepared $Ho_2O_3$ nanoparticles (NPs).

Figure 2 presents the morphology and size distribution of the as-prepared PEG–$Ho_2O_3$ NPs. According to transmission electron microscopy, the prepared nanoprobes had an almost spherical morphology within the range 67–81 nm. On the other hand, the measured hydrodynamic sizes of PEG–$Ho_2O_3$ NPs were in the range of 80–90 nm (polydispersity index PDI = 1.67). The difference between observed and measured sizes can be explained by hydration coverage and existence of a thin PEG layer on the $Ho_2O_3$ NPs surface [18]. FTIR analysis (Figure 3) was used to examine the successful PEG-functionalization on the surface of $Ho_2O_3$ NPs. The PEG–$Ho_2O_3$ NPs showed the angular deformation of water molecules (~1660 $cm^{-1}$) and the stretching vibrations of the OH group (~3600 $cm^{-1}$). In addition, FTIR analysis showed the scissoring (~1470 $cm^{-1}$) and waging (~1340 $cm^{-1}$) modes of the $CH_2$ group of the PEG chain. A most prominent peak at ~1100 $cm^{-1}$ was also assigned to the PEG chain C–O–C vibration [8]. Thus, FTIR analysis confirmed the presence of water and PEG molecules on the surface of $Ho_2O_3$ NPs. A thin PEG layer on the $Ho_2O_3$ NPs surface can enhance the steric repulsion and prolong the blood circulation time [8]. In addition, one can achieve higher biocompatibility of prepared NPs through the PEG surface functionalization. The zeta potential was further measured at the physiological pH of 7.4 to ensure the colloidal stability of the PEG–$Ho_2O_3$ NPs. The measured zeta potential for PEG–$Ho_2O_3$ NPs was approximately (−16.7 mV). Therefore, the colloidal solution of PEG–$Ho_2O_3$ NPs can be stable for a relatively long time.

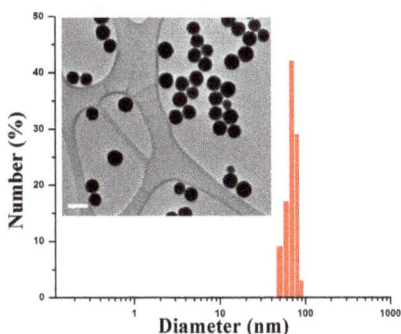

**Figure 2.** Measured hydrodynamic sizes of polyethylene glycol functionalized holmium oxide (PEG–$Ho_2O_3$) NPs. Inset is a transmission electron microscopy (TEM) image of PEG–$Ho_2O_3$ NPs (bar scale = 100 nm).

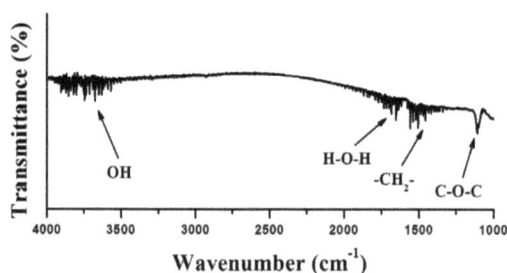

**Figure 3.** Fourier transform infrared spectroscopy (FTIR) analysis of PEG–Ho$_2$O$_3$ NPs.

The magnetic properties of prepared PEG–Ho$_2$O$_3$ NPs were investigated further using an MPMS. Figure 4a shows the M(H) curve for the prepared PEG–Ho$_2$O$_3$ NPs at room temperature ($T$ = 300 K). The observed linear relationship between the magnetization and applied field shows typical paramagnetic behavior of PEG–Ho$_2$O$_3$ NPs at room temperature. Figure 4b shows the measured inverse $1/T_2$ relaxation times vs. Ho$^{3+}$ concentration. The transverse $r_2$ relaxivity rate was estimated from a linear fit of $1/T_2$ vs. Ho$^{3+}$ concentration. The slope of the linear fit revealed a transverse relaxation rate ($r_2$) of 23.47 mM$^{-1}$·s$^{-1}$. One can also easily observe that the $r_2$ map images become darker with increasing Ho$^{3+}$ concentration (Figure 4b, Inset). The obtained $r_2$ value of PEG–Ho$_2$O$_3$ NPs is much higher than the reported transverse relaxation rate ($r_2$ = 17.95 mM$^{-1}$·s$^{-1}$) for Mn-doped iron oxide NPs [19]. It should be also noted that magnetic moment of Ho$_2$O$_3$ is not saturated at room temperature compared to widely employed iron oxide NPs [14]. As a result, the magnetic moment of PEG–Ho$_2$O$_3$ NPs will further increase with an increase in the applied magnetic fields. Therefore, the prepared PEG–Ho$_2$O$_3$ NPs can be applied as a $T_2$-weighted MRI agent, particularly at high magnetic fields, because their contrast enhancements will increase with an increase magnetic field [12,13].

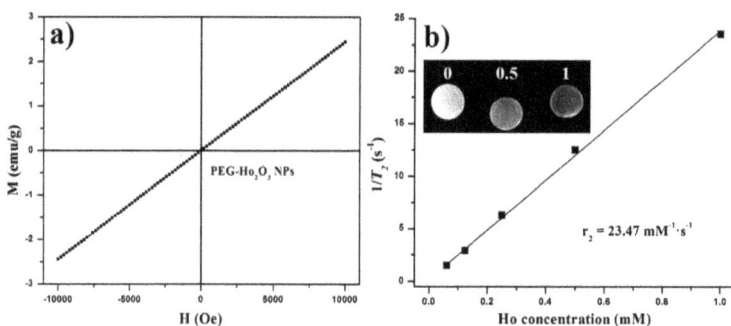

**Figure 4.** (a) Measured M(H) curve at 300 K; (b) $r_2$ relaxation rate ($1/T_2$) vs. Ho$^{3+}$ concentration (mM). Inset is $r_2$ map images of PEG–Ho$_2$O$_3$ NPs aqueous solution at different concentrations.

The toxicity of the prepared NPs is another important factor that should be taken into consideration for potential nanomedical applications. Figure 5 presents the cytotoxicity profiles of the PEG–Ho$_2$O$_3$ NPs in L-929 fibroblastic cells using a WST-8 assay [10,11]. Metal ions can generate reactive oxygen species in the cell interior ("Trojan horse" mechanism), which leads to oxidative stress to living cells [20]. Therefore, the cytotoxicity results showed an obvious dose-dependent decrease in their relative cell viability. Obviously, PEG–Ho$_2$O$_3$ NPs caused no significant decrease in cell viability up to 16 μg/mL. Considering the in-vitro cytotoxicity only, the PEG–Ho$_2$O$_3$ NPs can be used at concentrations less than 16 μg/mL. However, the cytotoxicity against the cells exposed to PEG–Ho$_2$O$_3$ NPs must be tested by other viability end-point measurements.

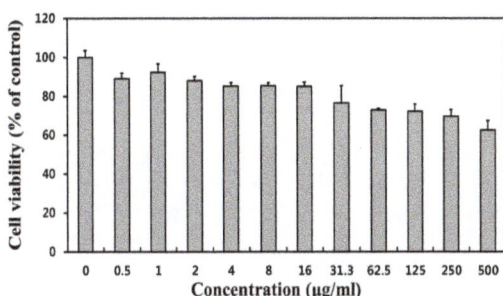

**Figure 5.** Cytotoxicity profiles of PEG–Ho$_2$O$_3$ NPs in L-929 fibroblastic cells.

Fluorescence microscopy (IX81-F72, Olympus Optical, Osaka, Japan) was used to visualize the cellular uptake and distribution of PEG–Ho$_2$O$_3$ NPs within the cultured L-929 cells. Figure 6a shows the phase contrast image of L-929 cells after incubation with PEG–Ho$_2$O$_3$ NPs suspension (10 μg/mL). The phase contrast image showed that the L-929 cells labeled with PEG–Ho$_2$O$_3$ NPs spread well with normal fibroblast-like morphologies. Although a detailed study for the cellular uptake was not performed, it is believed that the PEG–Ho$_2$O$_3$ NPs permeated into the cell membrane by non-specific endocytosis rather than pinocytosis [10,20]. Figure 6b shows that the prepared PEG–Ho$_2$O$_3$ NPs can also emit green light due to the intra 4$f$-transitions in holmium ions [21]. Therefore, prepared PEG–Ho$_2$O$_3$ NPs can be simultaneously utilized as a bimodal nanoprobe for MRI and optical imaging.

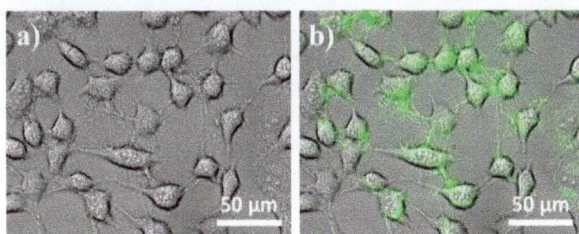

**Figure 6.** Phase contrast (**a**) and fluorescence (**b**) images of L-929 fibroblastic cells incubated with 10 μg/mL of PEG–Ho$_2$O$_3$ NPs.

## 4. Conclusions

In summary, PEG–Ho$_2$O$_3$ NPs were prepared and their applicability as new $T_2$-weighted MRI contrast nanoprobes was assessed. Cytotoxicity measurements showed that the prepared PEG–Ho$_2$O$_3$ NPs were nontoxic at concentrations less than 16 μg/mL. MRI relaxivity studies revealed high transverse relaxivity ($r_2$ = 23.47 mM$^{-1}$·s$^{-1}$), suggesting that the prepared PEG–Ho$_2$O$_3$ NPs can be used as an efficient $T_2$-weighted nanoprobe. In addition, green fluorescence was also detected from the PEG–Ho$_2$O$_3$ NPs due to intra 4$f$-transitions in holmium ions. Therefore, the prepared PEG–Ho$_2$O$_3$ NPs could be used as a dual-imaging nanoprobe.

**Acknowledgments:** This work was supported by the BK21 PLUS program at the Department of Physics and Astronomy, Seoul National University. This research was financially supported by Basic Science Research Program through the National Research Foundation of Korea (NRF) funded by the Ministry of Education

(No. 2016R1D1A1B03931076). Part of this study has been performed using facilities at IBS Center for Correlated Electron Systems, Seoul National University.

**Author Contributions:** Timur Sh. Atabaev and Nguyen Hoa Hong conceived and designed the experiments; Timur Sh. Atabaev performed the synthesis and characterization of the samples; Yong Cheol Shin and Su-Jin Song performed the cytotoxicity and cellular imaging experiments; Timur Sh. Atabaev, Dong-Wook Han and Nguyen Hoa Hong analyzed the data; Timur Sh. Atabaev wrote the paper.

**Conflicts of Interest:** The authors declare no conflict of interest.

## References

1. Atabaev, T.S.; Kim, H.K.; Hwang, Y.H. Fabrication of bifuctional core-shell Fe$_3$O$_4$ particles coated with ultrathin phosphor layer. *Nanoscale Res. Lett.* **2013**, *8*, 357. [CrossRef] [PubMed]
2. Gautam, A.; van Veggel, F.C.J.M. Synthesis of nanoparticles, their biocompatibility, and toxicity behavior for biomedical applications. *J. Mater. Chem. B* **2013**, *1*, 5186–5200. [CrossRef]
3. Sun, C.; Sze, R.; Zhang, M. Folic acid-PEG conjugated superparamagnetic nanoparticles for targeted cellular uptake and detection by MRI. *J. Biomed. Mater. Res. A* **2006**, *78A*, 550–557. [CrossRef] [PubMed]
4. Ding, X.; Liu, J.; Li, J.; Wang, F.; Wang, Y.; Song, S.; Zhang, H. Polydopamine coated manganese oxide nanoparticles with ultrahigh relaxivity as nanotheranostic agents for magnetic resonance imaging guided synergetic chemo-/photothermal therapy. *Chem. Sci.* **2016**, *7*, 6695–6700. [CrossRef] [PubMed]
5. Shin, J.; Anisur, R.M.; Ko, M.K.; Im, G.H.; Lee, J.H.; Lee, I.S. Hollow manganese oxide nanoparticles as multifunctional agents for magnetic resonance imaging and drug delivery. *Angew. Chem. Int. Ed. Engl.* **2009**, *48*, 321–324. [CrossRef] [PubMed]
6. Atabaev, T.S.; Lee, J.H.; Han, D.W.; Kim, H.K.; Hwang, Y.H. Fabrication of carbon coated gadolinia particles for dual-mode magnetic resonance and fluorescence imaging. *J. Adv. Ceram.* **2015**, *4*, 118–122. [CrossRef]
7. Faucher, L.; Gossuin, Y.; Hocq, A.; Fortin, M.A. Impact of agglomeration on the relaxometric properties of paramagnetic ultra-small gadolinium oxide nanoparticles. *Nanotechnology* **2011**, *22*, 295103. [CrossRef] [PubMed]
8. Atabaev, T.S.; Lee, J.H.; Han, D.W.; Kim, H.K.; Hwang, Y.H. Ultrafine PEG-capped gadolinia nanoparticles: Cytotoxity and potential biomedical applications for MRI and luminescent imaging. *RSC Adv.* **2014**, *4*, 34343–34349. [CrossRef]
9. Chen, F.; Chen, M.; Yang, C.; Liu, J.; Luo, N.; Yang, G.; Chen, D.; Li, L. Terbium-doped gadolinium oxide nanoparticles prepared by laser ablation in liquid for use as a fluorescence and magnetic resonance imaging dual-modal contrast agent. *Phys. Chem. Chem. Phys.* **2015**, *17*, 1189–1196. [CrossRef] [PubMed]
10. Atabaev, T.S.; Lee, J.H.; Han, D.W.; Choo, K.S.; Jeon, U.B.; Hwang, J.Y.; Yeom, J.A.; Kang, C.; Kim, H.K.; Hwang, Y.H. Multicolor nanoprobes based on silica-coated gadolinium oxide nanoparticles with highly reduced toxicity. *RSC Adv.* **2016**, *6*, 19758–19762. [CrossRef]
11. Atabaev, T.S.; Lee, J.H.; Shin, Y.C.; Han, D.W.; Choo, K.S.; Jeon, U.B.; Hwang, J.Y.; Yeom, J.A.; Kim, H.K.; Hwang, Y.H. Eu, Gd-codoped yttria nanoprobes for optical and T$_1$-weighted magnetic resonance imaging. *Nanomaterials* **2017**, *7*, 35. [CrossRef] [PubMed]
12. Norek, M.; Kampert, E.; Zeitler, U.; Peters, J.A. Tuning of the size of Dy$_2$O$_3$ nanoparticles for optimal performance as an MRI contrast agent. *J. Am. Chem. Soc.* **2008**, *130*, 5335–5340. [CrossRef] [PubMed]
13. Norek, M.; Pereira, G.A.; Geraldes, C.F.G.C.; Denkova, A.; Zhou, W.; Peters, J.A. NMR transversal relaxivity of suspensions of lanthanide oxide nanoparticles. *J. Phys. Chem. C* **2007**, *111*, 10240–10246. [CrossRef]
14. Kattel, K.; Kim, C.R.; Xu, W.; Kim, T.J.; Park, J.W.; Chang, Y.; Lee, G.H. Synthesis, magnetic properties, map images, and water proton relaxivities of D-glucuronic acid coated Ln$_2$O$_3$ nanoparticles (Ln = Ho and Er). *J. Nanosci. Nanotechnol.* **2015**, *15*, 7311–7316. [CrossRef] [PubMed]
15. Atabaev, T.S.; Vu, H.H.T.; Kim, H.K.; Hwang, Y.H. The optical properties of Eu$^{3+}$ and Tm$^{3+}$ co-doped Y$_2$O$_3$ submicron particles. *J. Alloys Compd.* **2012**, *525*, 8–13. [CrossRef]
16. Atabaev, T.S.; Vu, H.H.T.; Kim, H.K.; Hwang, Y.H. Ratiometric pH sensor based on fluorescent core-shell nanoparticles. *J. Nanosci. Nanotechnol.* **2017**, *17*, 8313–8316.
17. Pandey, S.D.; Samanta, K.; Singh, J.; Sharma, N.D.; Bandyopadhyay, A.K. Raman scattering of rare earth sesquioxide Ho$_2$O$_3$: A pressure and temperature dependent study. *J. Appl. Phys.* **2014**, *116*, 133504. [CrossRef]

18. Faucher, L.; Tremblay, M.; Lagueux, J.; Gossuin, Y.; Fortin, M.A. Rapid synthesis of PEGylated ultrasmall gadolinium oxide nanoparticles for cell labeling and tracking with MRI. *ACS Appl. Mater. Interfaces* **2012**, *4*, 4506–4515. [CrossRef] [PubMed]

19. Zhang, M.; Cao, Y.; Wang, L.; Ma, Y.; Tu, X.; Zhang, Z. Manganese doped iron oxide theranostic nanoparticles for combined $T_1$ magnetic resonance imaging and photothermal therapy. *ACS Appl. Mater. Interfaces* **2015**, *7*, 4650–4658. [CrossRef] [PubMed]

20. Atabaev, T.S.; Lee, J.H.; Han, D.W.; Hwang, Y.H.; Kim, H.K. Cytotoxicity and cell imaging potentials of submicron color-tunable yttria particles. *J. Biomed. Mater. Res. A* **2012**, *100A*, 2287–2294. [CrossRef] [PubMed]

21. Atabaev, T.S.; Vu, H.H.T.; Kim, Y.D.; Lee, J.H.; Kim, H.K.; Hwang, Y.H. Synthesis and luminescence properties of $Ho^{3+}$ doped $Y_2O_3$ submicron particles. *J. Phys. Chem. Solids* **2012**, *73*, 176–181. [CrossRef]

© 2017 by the authors. Licensee MDPI, Basel, Switzerland. This article is an open access article distributed under the terms and conditions of the Creative Commons Attribution (CC BY) license (http://creativecommons.org/licenses/by/4.0/).

MDPI AG

St. Alban-Anlage 66

4052 Basel, Switzerland

Tel. +41 61 683 77 34

Fax +41 61 302 89 18

http://www.mdpi.com

*Nanomaterials* Editorial Office

E-mail: nanomaterials@mdpi.com

http://www.mdpi.com/journal/nanomaterials

www.ingramcontent.com/pod-product-compliance
Lightning Source LLC
Chambersburg PA
CBHW051856210326
41597CB00033B/5923